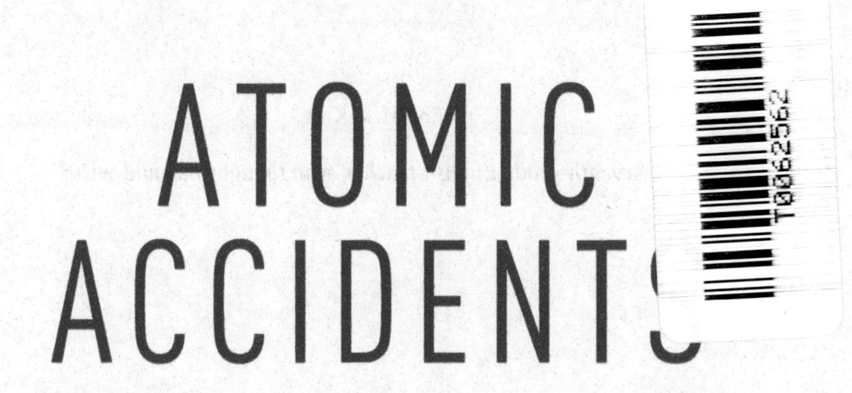

A HISTORY OF NUCLEAR MELTDOWNS AND DISASTERS: FROM THE OZARK MOUNTAINS TO FUKUSHIMA

JAMES MAHAFFEY

PEGASUS BOOKS
NEW YORK LONDON

For Mrs. King

My seventh-grade English teacher, who thought I should write

ATOMIC ACCIDENTS

Pegasus Books, Ltd.
148 West 37th Street, 13th Floor
New York, NY 10018

First Pegasus Books paperback edition February 2015
First Pegasus Books hardcover edition January 2014

Interior design by Maria Fernandez

Library of Congress Cataloging-in-Publication Data is available.

ISBN: 978-1-60598-680-7

10 9 8 7

Printed in the United States of America
Distributed by Simon & Schuster
www.pegasusbooks.com

CONTENTS

AUTHOR'S NOTE

A TRIUMPH OF SOVIET TECHNOLOGY

NUCLEAR ENGINEERS LOVE AND ADMIRE hydroelectric power. It's perfectly clean. It makes no smoke, no carbon monoxide, no radioactive waste, no toxic exhaust or lingering byproduct of any kind. Unlike nuclear power, it's very simple. Dam up a river. Let the water from the top of the resulting lake fall through a pipe, gain energy, and spin a turbine. Connected directly to the turbine shaft, without any gears or transmission, is a multi-pole electrical generator. Three wires emerge from the side of the generator. Connect those to the transformer yard, and you're sending electricity to paying customers 24 hours a day, or as long as there is water in the lake.

It is shockingly quiet on the turbine deck. The spinning machinery, usually covered with ceramic tiles made in subdued pastel colors or mounted flush with the floor, makes a high-pitched but subdued whine. The walls tremble, but only slightly, sometimes just beyond perception. The deck is spotlessly clean, and above, looking down on the machinery, is the glassed-in control room. Banks of instruments read out the status of the turbines, which is invariably good. Being a hydro-plant operator has got to be a boring job. You sit there as the generators go round and round, decade after decade, and the silt slowly builds up at the base of your dam. Excitement is when water starts to trickle down where it isn't supposed to on the face of the dam, or a big snow-melt upstream has to be diverted over the spillway, but aside from that it's fairly dull. If all the electrical power could be generated with hydro dams, then the world would be a cleaner, calmer place.

There are many fine hydroelectric dams in the United States, each demonstrating the American ability to bend nature to our needs with well-thought-out engineering practice and the skill necessary to build large things. In 1961 the United States and the Soviet Union were engaged in an all-out war. It was a cold war, in which the purpose was not to see how many of the other side we could kill but to prove who had the superior experimental economic system. It was an American form of capitalism against a Russian form of communism, and the battles raged on many fronts. The Soviets had already cleaned our plow in the race for manned space flight, putting Vostok 1 with Yuri Gagarin into orbit on April 14. We were falling way behind.

Thinking to show the U.S., which had built the magnificent Hoover Dam in Nevada, how to really build a hydroelectric plant, the Soviet Union decided to construct the world's largest reservoir to produce an impressive 6.4 billion watts of power using a line of ten turbo-generators. Work began to build an enormous dam, 3,497 feet wide and 807 feet high, across the Yenisei River near Sayanogorsk in Khakassia. The Sayano-Shushenskaya Hydro Power Plant took 17 years to build. This single dam generated one tenth of the power used in Siberia, and 70 percent of it was used to smelt aluminum. In 2006 production peaked, and the dam produced 26.8 trillion watt-hours of electricity. It was the sixth largest hydroelectric plant in the world, and it was built to withstand a Richter 8.0 earthquake. It was listed in the Guinness Book of World Records as the sturdiest dam in existence.

The water intake pipes on a dam are called penstocks, which is a carryover from the days of wooden water wheels. The sieve at the intake that keeps floating stuff like canoes or swimming bears out of the penstocks is the trash rack. The rotor, or the part of the turbine that spins, is called the runner, and it is hit with water from all sides. The force of the water and therefore the power that turns the generator is controlled precisely by hydraulically actuated wicket gates encircling the runner.

The turbine hall was magnificent, finished in white with gray and sky blue accents, with a curved picture window a couple of stories tall forming the front wall and looking out over the gently flowing Yenisei river, reformed from the turbine exhausts into an idyllic, park-like setting. An extremely large traveling crane, capable of easily picking up a 156-ton turbine runner, ran the length of the hall on rails in the ceiling.

On the morning of August 17, 2009, Unit 2 was the master turbine, setting the pace for the others to follow and running at a precise 142.86 revolutions per minute. Each turbine had its own penstock down from the top of the dam and its wicket gates constantly followed the directions from Unit 2, but there were only nine turbines running that morning. Unit 6 was down for maintenance, and an unusually large number of workers were down on the deck worrying over it.

At 8:13 local time something in the water, probably a loose log, made it through the trash rack at the top of the dam, fell 636 feet through the Unit 2 penstock, and lodged in the runner. Not good. In a fraction of a second, the log had spun around and slapped shut all the wicket gates, and the turbine jerked to a low speed. The turbines were extremely smooth and stable in two conditions: running at full speed and standing still. In any other condition while connected to the electrical grid, they would vibrate. Unit 2, the master turbine, suddenly lost speed as its water was cut off. It was no longer generating power, and was instead pulling power out of the other generators, acting as an electric motor. Unit 2 started hopping up and down, wrestling with its sudden change of identity and status. The bolts holding the top on Unit 2 blew off, and the 900-ton turbo-generator jumped out of the floor. Suddenly having no speed control from Unit 2, Units 7 and 9 quickly reached runaway speed and started flinging parts through what was left of the picture window. The Unit 2 penstock collapsed and destroyed everything around it.[1]

Having no penstock to contain it, Lake Sayano-Shushenskoe started emptying into the powerhouse. Units 3, 4, and 5 were still generating power, but they could not do it correctly while under water. The main transformer exploded, sending 40 metric tons of cooling oil, mixed with highly toxic polychlorinated biphenyls, down the Yenisei. The overhead crane wrenched loose from its rails and crashed through the floor, followed by the ceiling caving in on the only units left running. In a matter of seconds, a sweetly running hydroelectric plant was reduced to a twisted mass of water-soaked wreckage. Pieces were scattered hundreds of feet away, and the once-beautiful powerhouse looked as if it had been crushed under a giant's foot and then ground up just to make a point. After weeks of searching, the remains of 74 workers were found in the ruins. One was never found, making the toll 75.

The potential power of a simple mass of water is amazing, particularly when it is leveraged by a 636-foot drop. In a few moments, a placid lake of life-giving water had killed 75 people, destroyed a large power plant, and contaminated the drinking water for everyone downstream with a virulent carcinogen called PCB. In 1986 it would take the Chernobyl-4 RBMK nuclear reactor several weeks to end the lives of 54 men. Both power plants had been built about the same time, and both were the pride of Soviet technological advancement.

The problem of water inappropriately forced on a large power plant would come up again, this time in Japan in 2011. We now call this incident "Fukushima."

---

1 This is one of three possible explanations for the Sayano-Shushenskaya Dam disaster, and it is my favorite. The official report on the accident, released on October 3, 2009, concludes that six bolts holding the cover on the generator were missing (only 49 were found), and excessive vibration caused the cover to fly off. Maybe. You say you didn't hear about this on the news? At the time, a popular singer named Michael Jackson had died from the improper use of an anesthetic, and the world was too spellbound to consider any other disasters.

INTRODUCTION

BILL CRUSH AND THE HAZARDS OF STEAM UNDER PRESSURE

My first remembrance to which I can assign a date was in 1954. I was three months shy of my fourth birthday, and the event has stuck clearly in my mind for all these years.

Back then, there was a regional railroad in north Georgia named the Gainesville Midland. It was a small operation, always strapped for cash, and it was probably the last railroad in Georgia to run steam locomotives. The flagship of the line was a decapod, a heavy freight engine having ten driver wheels, number GM207, named "the Russian." It was so named because it was built by the Baldwin Locomotive Works in Eddystone, Pennsylvania, in 1916, under contract with Czar Nicolas II, Emperor and Autocrat of All the Russias. It was ready to ship in 1917, but Nicolas was under severe stress at the time, and payment was not forthcoming. Finding the Russian government completely collapsed, Baldwin sold its entire inventory of oddly specified 2-10-0 decapods at auction in an attempt to recover manufacturing costs. The Gainesville Midland wound up with three of them, and Baldwin was happy to readjust the gauge for the light tracks in Georgia.

The Russian looked incomprehensibly huge to me. How could something so big, so massive, move at all? It blotted out the sun when it passed, throwing black soot high into the crisp autumn air and causing the ground to move under my feet like a Japanese earthquake. The boiler sat so high, I could see daylight through the spokes in the drive wheels as it thundered by at top speed, making 35 miles per hour pulling a mixed string of five cars. On the downhill, you could outrun it on a bicycle. It ran back and forth, between Athens and Gainesville, roughly alongside the Winder Highway.

One Sunday afternoon we were at my grandparents' house in Hoschton, Georgia. The town used to be on the Gainesville Midland line, but the tracks had been torn up in 1947 when the route was cancelled. The train station was still there, empty of purpose like other buildings in the hamlet that had seen more prosperous times. It was a slow day.

It had been raining constantly for the past week, and everything in Georgia was soaking wet, including the fellow who came to the door with urgent news. "Colonel!" he cried. "The Midland done wrecked!"

Granddaddy dropped his *New York Times* and rose to his feet. "Wrecked?"

"Yes, sir! It's the Russian. She's off the rails, up yonder, nearly t' Gainesville." He pointed vaguely west.

This was no time to be sitting around listening to the house settle. We piled into the Studebaker and hot-wheeled it up the road to a wide spot that no longer exists, called Candler, just south of Gainesville. You could feel the spectacle growing as we approached. Cars were parked or abandoned off the road. First a few, then clumps, then seemingly every car in the world. People were walking, jogging, and sprinting, all in one direction, pointing and shouting. We pulled off and started walking.

After trudging about a hundred miles we reached a sharp turn, where the tracks veered off to the left, and there it was, lying on its side, wheels in the air, like a dead dinosaur. The heavy Russian had taken the turn too fast, and the red clay under the tracks, saturated with water, just slid out from under it. I could swear the thing was still breathing. Periodically you could hear steam burbling somewhere deep inside its enormous body. People were just standing there in awe of the spectacle, uncountable hundreds, quietly staring and whispering to each other. Someone said that the engineer had to be cut out of the wreckage with an acetylene torch. I stood on my tiptoes and tried to see the twisted wreck of the cab. It was too far away, down a hill.[2]

I learned something that day, and it had nothing to do with going too fast around a curve: there's a great deal of entertainment value in a train wreck. Even the aftermath of a crash, with the engine upside down and cars scattered all over the place, is surprisingly theatrical—a tragedy in hot steel, plowed mud, and scattered coal. There was sport in just analyzing the disaster, thinking what could have happened, back-tracing the last moments of the engine's life, and imagining it digging the long trench as its energy dissipated into the ground. If it were roped off, you could sell tickets.

As is almost always the case, I was not the first to think of this. In 1896 a passenger agent for the Missouri, Kansas & Texas Railway (Katy), William "Bill" Crush, came up with a brilliant publicity stunt that would drum up passenger business. Being a natural-born salesman, he was able to convince his boss that they should stage a head-on collision between two locomotives. With a little advertising, it would attract thousands of people! There would be no charge to see the crash, but they could sell train tickets to bring people to the event. At two dollars per roundtrip ticket, they would not only gain publicity for their railroad, they would clear a profit as well.

In the 19th century, rail travel was the premier form of ground transportation, and just about everybody spent time in a railcar, gazing out the window as the rural terrain sped by or sleeping in the sitting position. Steam trains were large, heavy, fearsome beasts, breathing

2 I'm not sure what happened to GM207. I've found GM206, GM208, and GM209, all resting comfortably in display settings. GM208 is in Winder, GM209 is in Gainesville, and GM206 is somewhere in North Carolina at a railway museum. All are Russian-pattern 2-10-0 locomotives, but only GM206, built by Alco-Brooks, is said to have been built for Russian export in 1918. I swear the wrecked engine was called "The Russian," but the story is hazy, and I don't even know why such a tiny railroad needed so many engines.

fire and looking dangerous. Some people were excited by the technical advances that had made this mass transportation possible, and some were terrified of it. There were too many newspaper stories every day about train wrecks. It seemed that engines were always blowing up for no obvious reason, crashing into each other, tilting off the rails, or plunging off a trestle into a gorge. There were citizens who could not be forced onto a train at gunpoint. Engineers blamed impossible schedules and poorly maintained tracks. Conductors blamed engineers. Railyard workers blamed brakemen, and railroad owners blamed the newspapers for lurid prose.

In 1891, a particularly bad year, 7,029 Americans lost their lives in railroad accidents. There were only about 64.4 million Americans at the time, so that makes the fatality rate 1.4 times that of automobile travel in 2011. The idea of staging a train wreck in 1896 was a superb piece of psychology. Instead of assuring passengers that all trains were safe and nobody could get hurt, show them the worst that could possibly happen. Let them feel the heat blast, the steam escape, and the ground-trembling thud. Allow them to get as close as they dared, and, what was most essential, let them see it coming. There would be no buried dread of the random, completely unexpected accident. The fear of the unknown would be replaced by the excitement of expectation.

A bare patch of ground outside the city limits of Waco, Texas, was staked out, and a set of temporary tracks was laid. Two obsolete 4-4-0 American pattern locomotives, looking like Civil War relics, were purchased and dolled up. One was painted green with red trim, and the other was painted red with green trim. Boxcars were added, with advertising for the Oriental Hotel in Dallas and the Ringling Brothers Circus painted on the sides. Tents were erected. A temporary restaurant was built, as well as a jail, and a 2,100-foot-long platform was banged together to give people a place to stand and watch the show. Eight tank cars filled with water were brought in to prevent spectator dehydration.

The event was scheduled for September 15, and by then the crowd had grown to over 40,000 souls. As an afterthought, Bill Crush was asked, "Is this safe? Them old boilers ain't gonna explode, are they?"

Since the invention of high-pressure steam earlier that century, boiler explosions had become the number one fear of everyone participating in the steam-power revolution. Boiler explosions had been killing anyone standing near an over-pressurized locomotive since 1831. Steam carried a lot of pent-up energy. It wasn't just the immediate fire under the boiler that was the problem, it was the heat energy built up and stored in the steel vessel that was so dangerous. A steam explosion could happen at any time, out of the blue, without a hint of warning. A boiler would disintegrate, sending hot, knife-like pieces ripping mercilessly through a crowd. It was not the sort of publicity that a railroad ever needed.

"Naw," said Bill, patting the still-sticky paint. "These old engines are tough. It's just going to make a big noise and crush it like a tomato can. No blow-up. I'm sure." Of all the employees in the Katy, Bill Crush probably knew the least about steam and mechanical stress.

The afternoon was getting hot, and the crowd was growing restless. Two hundred men were hired to control the mob, but it was beginning to get out of hand. The two engineers were ready at the throttles, the boilers were redlined, and the steam relief valves had sprung open and were blowing mist. Crush rode out in front of the crowd on a borrowed white horse, raised his hat high, let it hang for a moment, then dropped it. The crowd went wild, and the engineers jerked their throttles full open. C. E. Stanton in the green engine and Charles Cain in the red one coolly waited for 12 puffs from the cylinders and bailed out, with the lightly loaded engines gaining speed. People pushed and shoved for an unobstructed view.

On they came, blowing dark clouds of smoke and setting off emergency signal torpedoes placed all along the track. Bang. Bang. Bang bang bang. Faster and faster, reaching a combined collision speed of 100 miles per hour. The official event photographer, J. C. Deane, tripped his high-speed shutter just as the two cowcatchers met. The two old engines, weighing about 35 tons each, suddenly occupied the same spot on the track. There was a terrific sound of crashing, bending metal as the two locomotives melted together, lifted their front trucks off the track, and seemed to hang for an instant. The wooden cars behind splintered and crushed as the two trains telescoped together.

Then, something bad happened. At least one of the boilers exploded with a heavy roar, sending a rain of jagged metal into the crowd. The first casualty was Deane, the photographer, stationed closest to the crash point. A piece of hot locomotive hit him in the face, cleaned out an eye-socket, and left a bolt and washer embedded in his forehead. He spun around to face the audience and went limp. Louis Bergstrom, also on the photography team, was cold-cocked by a flying plank. Ernest Darnall, a boy with a rare viewing opportunity sitting high in a tree, caught a heavy iron hook trailing a length of chain right between the eyes, splitting his skull down the middle. DeWitt Barnes, in a dignified standing position between his wife and another woman, was killed instantly by an unidentified fragment. People in the front row were scalded, screaming, and dripping blood. In all, three people were killed on the spot and six were very seriously injured. A Civil War veteran was visibly shaken, saying that it reminded him of seeing a line of men dropped by a Yankee rifle volley.

Instant tragedy, however, did not dampen the crowd's enthusiasm. They rushed the scene by the thousands in an incoming wave, poring over the wreckage to pick up or wrench loose the largest pieces they could carry. Many palms were singed as people pounced on bolts, rivets, bits of boiler tubes, and all manner of unidentifiable relics. To appease grieving families, Bill Crush was immediately and visibly fired from his job at the Katy. He was quietly re-hired the next day. From that day forward, the Katy Railroad flourished, and the many who had decided not to go to the event regretted the decision for the rest of their lives, as the stories of "The Crash at Crush" were told over and over in song, ragtime march, musical play, and *Sports Illustrated*.

Bill Crush wasn't even the first to think of this. Incredibly, there were four independently staged engine head-butts in September 1896. None was as spectacular as Crush's 100-mile-per-hour boiler bust, but the clustering indicates an unfulfilled need in the human psyche, peaking in 1896. Just outside Denver on September 30, two old narrow-gauge 2-6-0 Union Pacific and Denver & Gulf engines were smushed together for a crowd as a fund-raiser for the Democratic Party.[3] The crash made a lot of smoke and noise, but the engines were so feeble, the railroad was able to rebuild them and put them back into service.

3 For those who may wonder, "2-6-0" is the standard way of specifying a steam locomotive configuration. This particular engine has two wheels on the pilot truck, six steam-driven wheels, and no trailing truck behind the drivers.

On September 18 at the county fair in Sioux City, Iowa, two ancient Mason Bogey engines were smashed together to a cheering mob. In Des Moines at the State Fair on September 9, just six days before Crush's spectacle, "Head-On Joe" Connolly arranged the collision of two really old 4-6-0 engines bought as junk from the Des Moines Northern & Western Railroad. The teeming masses numbered 70,000, and the gate receipts exceeded $10,000. That was a lot of money in 1896. Connolly was more adept at staging a crash than was Crush, and he knew to avoid a steam explosion. He had nothing to worry about. The elderly, arthritic engines were leaking steam at every joint. One was able to make 10 miles per hour, and the other 20. They hit at almost the right spot in front of the stands, there were the obligatory smoke and noise, and parts cartwheeled through the air, but the crowd was slightly disappointed. Still, they swarmed over the heap of steaming wreckage and carried off everything that was loose. Connolly returned home with $3,538.

Head-On Joe went on to make a career of locomotive crashing, eventually boasting that he had staged 73 wrecks, without killing a single spectator. He put together shows from Massachusetts to California, mostly at state fairs but anywhere people would gather and pay to see two trains smash together. The city with the most staged crashes was San Antonio, Texas, with four. New York City, Milwaukee, and Des Moines had three each. His biggest audience was at the Brighton Beach Racetrack, New York, on July 4, 1911, where 162,000 people paid at the gate to see two old 4-4-0 engines kill each other. There were imitators, of course, but Head-On Joe had it down to a science. He knew that he had to have at least 1,800 feet of track, or the engines could not make enough speed for a proper spectacle. A track length of 4,000 feet was optimal, as the engines could accelerate to a combined speed of 45 miles per hour. That was fast enough to tear up the machinery and make the tender ride up over the cab without a boiler explosion. It took a mile of track to make 65 miles per hour combined, but that was too fast. Boiler explosions were fine, but you had to have the onlookers so far away, they couldn't see anything. They wanted to be close enough to feel the collision, to hear the iron screaming in agony, and smell the hot metal, without being maimed. The locomotives had to be inexpensive and junky, without being undersized or wheezy. To wreck two nice-looking passenger engines seemed extravagant and in bad taste. To bury two old freight haulers in a moment of glory seemed merciful. Sometimes the engines looked hesitant as they tried to accelerate toward oblivion. Sometimes they looked angry, like pit bulls, not really knowing why they had to kill the other engine, but up to the task and really getting into it. It was art, in a machine-age sort of way.

At 73 years old, Head-On Joe's last staged train wreck was back in Des Moines, on August 27, 1932, at the State Fair. A matched pair of 4-6-0s, just retired from the Chicago, Milwaukee, St. Paul & Pacific, faced off on the field. Both were freshly painted, and they were named "Roosevelt" and "Hoover." Roosevelt was aimed east, toward Washington, D.C. A respectable mob of 45,000 came to see them on their last trip. After a short but suspense-filled run, the engines met, with the drama intensified by a box of dynamite tied to the pilot on each participant and fire-starters in the trailing passenger coaches. Hoover's boiler exploded, rudely injuring two spectators with hurled shrapnel. There would be no lawsuits. They were, after all, standing near where they knew there was going to be a train wreck. What did they expect? Connolly collected his $4,000 and quietly faded away to his home town in Colo, Iowa. When he died in 1948, a brass locomotive bell was found on the family estate, possibly the only souvenir he had kept from the destruction of 146 train engines.

The last staged train wreck in the United States was probably the one near Magnolia, Illinois, on June 30, 1935. Two 2-6-0s from the Mineral Point & Northern, the 50 and 51, were supposed to meet on a bridge going a combined 50 miles per hour, but they missed the point, impacting instead in an open field at a fraction of the required speed. Coal flew vertically out of the 51's tender and a puff of smoke rose, but the damage was so slight and the spectacle was so pitiful, it didn't make the morning paper. The age of the staged train wrecks ended with a whimper. A creative plan to replace them with airplanes crashing into each other in mid-air did not materialize.

The need to see train engines crash together may have played out in the 1930s, but the specter of exploding locomotives would affect engineering for generations. Even today, in the 21st century, most of the safety design effort in a nuclear power plant is devoted to preventing a steam catastrophe. A nuclear plant is, after all, just another steam engine, heating water to a temperature beyond the boiling point and using the resulting vapor to rotate a shaft. The main difference between a nuclear generating station and its equivalent 100 years ago is that disintegrating uranium has replaced burning coal as the source of heat.

Numerous substitutes for steam as the prime mover in a power plant have been tried, but nothing has proven more reliable, efficient, or economical than boiling water. The task of converting heat into electrical current is not straightforward, but using steam as the transfer medium means that a large-output plant can be compact, and the working fluid is neither toxic nor flammable. Sitting on a small plot of land next to a river, a four-boiler steam plant can light up everything for a hundred miles, and if it is nuclear-powered then there is not even a pile of coal cinders and a mile-long line of rail cars waiting to be unloaded. Still, there is the fear of a steam explosion, something that impressed itself on both the public and the technical acolytes long ago.

In the early years of nuclear power development, in the technology scramble after World War II, early experiments and some small disasters pointed out the dangers of a runaway nuclear reaction. In practice, it was possible to increase the power output of a nuclear reactor not as a gradual heat transfer, like boiling water on the stove to make tea, but as a step function, or an abrupt increase in the blink of an eye. If you were standing near such an occurrence, you died, and it had the potential of flashing water directly and promptly into steam. The possibility of a runaway reaction and a resulting steam explosion was seen as the most critical safety concern in nuclear power development. If only this worst possible accident could be designed out of nuclear reactor plants, then everything else would be taken care of. All we had to do was keep the steam from exploding, and nuclear power would be stable enough to unleash on a safety-conscious public.

And so it was. With testing, accident simulations, well-thought-out engineering effort, and unusually robust building standards, the possibility of an explosive steam release was forcibly eliminated from nuclear power plants. In 56 years of commercial nuclear power generation in the United States, there has never been a steam explosion, and not one life has been lost.[4]

No dreaded boilers coming apart, ripping holes in buildings and sending shrapnel into the crowd to worry about, but everything else in the history of nuclear accidents has happened for what seem to be the most insignificant, unpredictable reasons, much to the consternation of engineers everywhere. Entire reactor plants, billions of dollars of investment, have been wrecked because a valve stuck open or an operator turned a switch handle the wrong way. Some water gets into a diesel engine cooling pump, and six reactors are wiped out. Imagine the frustration of having built an industry having the thickest concrete, the best steel, meticulously inspected welds, with every conceivable problem or failure having a written procedure to cover it, and then watch as three levels of backup fail one at a time and the core melts. Obviously, the machinery was more sensitive to simple error than anyone could have thought, and thicker concrete is needed.

All the issues to be addressed concerning accident avoidance are not technical. Some are deeply philosophical. It is painful to notice, but some of the worst nuclear accidents were caused by reactor operator errors in which an automatic safety system was overridden by a thinking human being. Should we turn over the operation of nuclear power plants to machines? Would this eliminate the strongest aspect of human control, which is the ability to synthesize solutions to problems that were never anticipated? The machine thinks in rigid, prescribed patterns, but in dealing with a cascade of problems with alarms going off all over the place, has this proven to be the better mode of thought? Should operators be taught to think like machines, or should they be encouraged to be creative? Study the history of nuclear disasters, and you will have this subject to ponder.

---

4 I have to be careful here not to fall into the usual pro-nuclear trap of overstating a concept. Several people have been killed in nuclear industry accidents, and many of those incidents will be discussed here. The worst nuclear accident in American history was a steam explosion, but it was a military reactor. So far, every death that can be positively linked to nuclear activity has been of military personnel, government workers in the atomic bomb industry, or a civilian working in fuel reprocessing. Nobody has died because he or she was working in a commercial nuclear power plant in the United States. The Soviet Union is another matter.

There is also the elephant in the room: ionizing radiation. Nuclear engineers are acutely aware of this elephant and have designed it out of the way. Concrete thickness helps a lot to keep radiation away from all workers at the plant and certainly out of the public. The human fear of radiation is special and pervasive. As you will see, it originates in the initial shock of discovery, when we were introduced to the unsettling concept of death by an invisible, undetectable phenomenon. We have never quite gotten over it, and, in fact, all the fear of a steam explosion is not connected to the problem of hurtling chunks of metal or the burning sensation, but directly to the problem of radiation dispersal into the public. Steam, when it escapes in an unplanned incident at the reactor plant, takes with it pieces of the hot nuclear fuel. It floats in the air and blows with the wind, transporting with it the dissolved, highly radioactive results of nuclear fission. This undesirable process is at the root of accident avoidance in the nuclear power industry. Employee safety is, of course, very important, but public safety is even more so. To keep the industry alive, thriving, and growing, it is imperative that the general population not feel threatened by it.

Feeling threatened is not the same as being threatened, but the difference gets lost. The danger from low levels of radiation is quite low, as expressed as morbidity statistics or probabilities, but there is an unfortunate lack of connection to probability in the average person. Low probabilities are a particular problem of perception. If they were not, then nobody would play the lottery and the gambling industry would collapse. The impression of radiation, and even the science, can get lost in the numbers. In reading these chronicles of nuclear incidents big and small, I hope that you can develop a sense for the origins and the realities of our collective dread of radioactivity. Will this universal feeling prevent the full acceptance of nuclear power? Will we develop a radioactivity vaccine, or will we gradually evolve into a race that can withstand it? Perhaps.

There is also the problem of the long-term radiation hazard. People do not mind a deadly threat so much if it leaves quickly, like an oil refinery going up in a fireball or a train-load of chlorine gas tankers crashed on the other side of town. For some reason, a cache of thousands of rusting, leaking poisonous nerve-gas cylinders in Aniston, Alabama, does not scare anyone, but the suggestion of fission products stored a mile underground at Yucca Mountain, Nevada, causes great concern.

In this book we will delve into the history of engineering failures, the problems of pushing into the unknown, and bad luck in nuclear research, weapons, and the power industry. When you see it all in one place, neatly arranged, patterns seem to appear. The hidden, underlying problems may come into focus. Have we been concentrating all effort in the wrong place? Can nuclear power be saved from itself, or will there always be another problem to be solved? Will nuclear fission and its long-term waste destroy civilization, or will it make civilization possible?

Some of these disasters you have heard about over and over. Some you have never heard of. In all of them, there are lessons to be learned, and sometimes the lessons require multiple examples before the reality sinks in. In my quest to examine these incidents, I was dismayed to find that what I thought I knew, what I had learned in the classroom, read in textbooks, and heard from survivors could be inaccurate. A certain mythology had taken over in both the public and the professional perceptions of what really happened. To set the record straight, or at least straighter than it was, I had to find and study buried and forgotten original reports and first-hand accounts. With declassification at the federal level, ever-increasing digitization of old documents, and improvements in archiving and searching, it is now easier to see what really happened.[5]

So here, Gentle Reader, is your book of train wrecks, disguised as something in keeping with our 21st century anxieties. In this age, in which we strive for better sources of electrical and motive energy, there exists a deep fear of nuclear power, which makes accounts of its worst

---

5 A good example of this enhanced document availability is in my search for the original report, "The Accident to the NRX Reactor on December 12, 1952, DR-32," by W. B. Lewis. This was a very important accident. It was the world's first core meltdown, and it happened at the Chalk River facility in Canada. I had heard about it many times, but I wanted the raw document. It seemed that every nuclear data repository I could think of, even in Canada, had an abstract of the paper, but not the paper itself. After a lot of digging I found it. The Russians had it, possibly recovered from the old KGB archives. This turned out to be a gold mine of information, including such things as accounts of the "Castor and Pollux" vertical assembly machines used in the development of the French atomic bombs.

moments of destruction that much more important. The purpose of this book is not to convince you that nuclear power is unsafe beyond reason, or that it will lead to the destruction of civilization. On the contrary, I hope to demonstrate that nuclear power is even safer than transportation by steam and may be one of the key things that will allow life on Earth to keep progressing; but please form your own conclusions. The purpose is to make you aware of the myriad ways that mankind can screw up a fine idea while trying to implement it. Don't be alarmed. This is the raw, sometimes disturbing side of engineering, about which much of humanity has been kept unaware. You cannot be harmed by just reading about it.

That story of the latest nuclear catastrophe, the destruction of the Fukushima Daiichi plant in Japan, will be held until near the end. We are going to start slowly, with the first known incident of radiation poisoning. It happened before the discovery of radiation, before the term was coined, back when we were blissfully ignorant of the invisible forces of the atomic nucleus.

CHAPTER 1

WE DISCOVER FIRE

"In Ozma's boudoir hangs a picture in a radium frame. This picture appears to be of a pleasant countryside, but when anyone wishes for the picture to show a particular person or place, the scene will display what is wished for."

—from a description of a plot device in L. Frank Baum's *Land of Oz*, thought to be placed somewhere on the Ozark Plateau.

IT WAS HUNTING SEASON IN the Ozark Mountains in November 1879. Sport hunters Bill Henry, John Dempsey, and Bill Boyceyer of Barry County, Missouri, were out to shoot a wildcat. They had left their hunting party behind, chasing a cat through the dense woods with their enthusiastic hunting dog. The dog, with his seemingly boundless dog-energy, ran full tilt down a gulley, then straight up the side of a steep hill, chasing the cat through previously untrampled territory. The cat looked desperate. Leaping around on the side of the mountain, he disappeared into a black hole, and the hound did not hesitate to dive in after him.

The three men, somewhat winded from the pursuit, knew they had him now. They cocked their pieces, aimed high at the orifice, and waited for the cat to come blasting out. The wait became uncomfortable. Fifteen minutes, and not only was there no cat, but the dog hadn't come back. They half-cocked their firearms and started to climb, but just then they heard the dog barking, somewhere on top of the hill. They whistled him down. He had obviously gone clean through the mountain and come out the other side.

Henry, Dempsey, and Boyceyer immediately found this hole in the side of the mountain more interesting than the wildcat. They had been around here before, but had never noticed the hole. It was oddly placed, and it would be easy to miss. It required investigation.

Cautiously, the three entered the opening. Shortly inside they saw along the wall what appeared to be a vein of pure, silvery metal, and dollar signs came up in their eyes. Could it be? Could they have stumbled into an undisturbed silver mine? It was growing dark, and they decided to retire to the hunting camp and do some planning. Nobody was to say anything to anybody about the hole, and they would return tomorrow for a more thorough exploration. The next morning they returned to the site, dogless this time but with a boy to help carry things. They lit pitch-pine torches and crawled into the opening, single file, with Henry leading. The cavern opened up, and everything in it looked strange and unfamiliar. At about two hundred feet in, the tunnel was partially blocked by what looked like a large tree trunk of solid silver. It was the strangest metal they had ever seen, with the bluish sheen of a peacock's tail. In the yellow glare of the torches it seemed faceted, like a cut diamond. In the tight, unfamiliar surroundings, imaginations ran wild. Henry selected a free rock on the floor and used it to bang on the mineral column. A few unusually heavy pieces chipped off, and they put them in a small tin box for transport.

Still feeling the tingle of adventure, they squeezed one at a time past the silvery obstruction and pressed on. At an estimated five hundred feet from the entrance they entered an arched room, and their perceptions started to veer into hallucinogenic territory. The walls of the room shone like polished silver, the floor was a light blue, and the ceiling was supported by three transparent crystal columns. Hearts raced as the oxygen level dropped. The men each knew that they had found their eternal fortune, and in their minds, gently slipping away, they were already spending it. They pressed past the columns, and the torches started to sputter and die. The walls were starting to get very close, and a blind panic gripped all three hunters simultaneously. They scrambled, crawled, and grabbed their ways to the cave portal as quickly as possible, with Henry dragging the box of samples.

Boyceyer was first out into the fresh air and sunlight. He took a deep breath, and his legs stopped working. He keeled over in a heap at the entrance, and shortly thereafter Henry tripped over him and passed out cold. Dempsey emerged in a strangely talkative mood, babbling and making no sense at all. The boy, left sitting out under a tree, had quickly seen and heard enough. He leaped to his feet and ran in the opposite direction, down the mountain in free fall, bursting into the campsite winded and trying to explain what had happened up there, pointing. Eventually calming him down and extracting a coherent message, the men quickly assembled a rescue team and hurried to the site.

It is now clear that the hunters were suffering the classic symptoms of oxygen deprivation. When the rescuers arrived, Boyceyer and Dempsey were coming around, but Henry was enfeebled, dazed, and unable to hike out. The men decided to cut the hunting expedition short and take him home. On the way his condition deteriorated. Fearing the specter of a new form of plague, they took him to a hospital in Carthage, Missouri. The doctors had no idea what was ailing him. His symptoms were puzzling. Sores resembling burns broke out all over his body, and his legs seemed paralyzed. Bill Henry remained hospitalized for several weeks, and he had time to plan for extracting his fortune from the hole in the mountain.

When he had recovered enough to leave under his own power, he staggered back to the cave to stake out a claim and work his silver mine, but the person who actually owned the land on which the mountain stood did not share his optimism, and no mining agreement could be reached between the two men. The guy wouldn't even come out and see the cave with its sparkling silver, just sitting there ready to be hauled away. Perhaps he knew more than he would admit about that mountain. He wanted no part of a mining venture, and he advised Bill Henry to find something else to do.

Exasperated and angry beyond words, Henry returned to the site and avalanched as much material as he could move into the portal, making a hole that had been hard to find impossible to see. He would come back later, once he had figured out some further strategy.

There is no record of Henry having returned again, and he disappeared into the murk of history. The cave location faded away, and the story became one of the colorful, spooky legends to be told around campfires after dark up in the Ozarks. That's the story, but it was not written down until 34 years after the incident, and facts could have drifted. There are questions. The initial problem was obviously oxygen deprivation, but what had taken the place of normal air in this cave? It could have been methane, the scourge of coal mining, but the cave was not lined with coal and there was not a hint of tool marks anywhere. And what had caused the burn-like lesions all over Bill Henry? Was he alone allergic to some mineral on the walls? What was the bright, iridescent stuff lining the cave? That is not what silver, or even gold, looks like in its native state. Later explorations of the cave would provide unexpected answers to these questions.

Meanwhile, in the formal physics lecture theaters and laboratories in Europe in 1879, the danger of being in a certain cave in Missouri and what it had to do with anything were unknown. Scientists across the Continent and in the United Kingdom, working at well-established universities, were busy studying the interesting properties of electricity in evacuated glass tubing. A thrillingly dangerous piece of equipment called a Ruhmkorff coil produced high-voltage electricity for these experiments. They were essentially inventing and refining what would become the neon sign. Research was progressing at an appropriate pace, gradually unraveling the mysteries of atomic structure.

Working independent of any academic pretension in the United States was a highly intelligent, well-educated immigrant from Croatia, Nikola Tesla. He came ashore in June 1884 with a letter of introduction to Thomas Edison, famous American inventor of the record player and the light bulb. He was given an engineering job at $18 a week improving Edison's awkward and ultimately unusable DC electrical power system, but he quit a year later under intractable disagreements concerning engineering practice, salary, general company philosophy, and his boss's personal hygiene. He immediately started his own power company, lost control of it, and wound up as a day laborer for the Edison Company laying electrical conduit. Not seeing a need for sleep, he spent nights working on high-voltage apparatus and an alternating-current induction motor.

In Europe they were working with induction coils that could produce a ripping 30,000 volts, stinging the eyes with ozone wafting out of the spark gap and with a little buzzer on the end making the spark semi-continuous. In New York, Tesla was lighting up the lab with 4,000,000 volts and artificial lightning bolts vibrating at radio frequencies. Naturally drawn to the same rut of innovation as his Old World colleagues, he connected an evacuated glass tube to his high-voltage source in April 1887. It had only one electrode. He connected it to his lightning machine and turned it on, just to see what would happen. Electrons on the highly over-driven electrode slammed themselves against the glass face of the tube, trying desperately to get out and find ground somewhere. The glass could not help but fluoresce under the stress, making a weak but interesting light. Tesla had invented something important, but he would not know exactly what it was until years later. He applied for a patent for his single-electrode tube, calling it an "incandescent light bulb" as a finger-poke in Edison's eye.

In 1891 Tesla's fortunes improved considerably when George Westinghouse, Edison's competitor for the electrical power market, became interested in his alternating current concepts. He moved into a new laboratory on Fifth Avenue South, and he had room to spread out and really put his high-voltage equipment to use. One night, he connected up his single-electrode tube built back in 1887. He turned off all the lights so he could see arcs and electron leakage. To his surprise, something invisible was coming out the end of his tube and causing the fresh white paint on the laboratory wall to glow. Curious, he put his hand in the way. His hand did stop the emanations, but only partly. The bones in his hand were dense enough to stop it from hitting the wall, but not the softer parts, and he could see his skeletal structure projected on the paint. Tesla, fooling around in his lab after hours, had invented radiology. In the next days he substituted photographic plates for the wall, and made skeletal photos of a bird, a rabbit, his knee, and a shoe with his foot in it, clearly showing the nails in the sole.

Unfortunately, Tesla was pulled toward greater projects, and he failed to pursue the obvious application of this discovery.

Four years later, on December 28, 1895, the discovery of the unusual radiation was formally announced, not by Tesla, but by Wilhelm Röntgen, working at the University of Munich. Röntgen was also studying fluorescence, using his trusty Rhumkorff apparatus and a two-electrode tube custom-built by his friend and colleague, Phillipp von Lénárd.[6] Like Tesla, he was startled to notice that some sort of invisible emanations from the tube pass through flesh, but are stopped by bones or dense material objects. In his paper in the *Proceedings of the Physical Medical Society*, Röntgen gave the phenomenon a temporary name: x-rays. Amused at reading the paper, Tesla sent Röntgen copies of his old photo plates. "Interesting," replied Röntgen. "How did you make these?" Not trusting his own setup to be kind, Röntgen covered his apparatus with sheets of lead, with a clear hole in the front to direct the energy only forward.

Tesla, on the other hand, put his head in the beam from his invention and turned it up to full power, just to see what it would do. Röntgen had jumped him on the obvious medical usage, but there had to be some other application that could be exploited for profit. After a short while directly under the tube, he felt a strange sensation of warmth in the top of his head, shooting pains, and a shock-effect in his eyes. Seeing the value of publication shown by Röntgen's disclosure, he wrote three articles for the *Electrical Review* in 1896 describing what it felt like to stick your head in an x-ray beam.

The effects were odd. "For instance," he first wrote, "I find there is a tendency to sleep and I find that time seems to pass quickly." He

6 Röntgen could have mail-ordered a mass-produced vacuum tube, called a Pulyui Lamp, from Poland and saved himself some time, if communications and advertising technology had been what they are today. Ivan Pulyui, a college professor at the University of Vienna from the Ukraine, is sometimes credited with having sold the first x-ray tubes, before the x-ray was discovered. The claim is semi-true. His Pulyui Lamp was available perhaps as early as 1882, but it was sold as a light bulb, and Pulyui did not realize that it was streaming x-rays along with a blue glow until he read Röntgen's paper in 1895. Pulyui immediately saw the medical diagnostic use of x-rays, and his lamps became quite useful.

speculated that he had discovered an electrical sleep aid, much safer than narcotics. In his next article for 1896, after having spent a lot of time being x-rayed, he observed "painful irritation of the skin, inflammation, and the appearance of blisters . . . , and in some spots there were open wounds." In his final article of 1896, published on December 1, he advised staying away from x-rays, ". . . so it may not happen to somebody else. There are real dangers of Röntgen radiation."

These writings were the first mention in technical literature of the hazards of over-exposure to the mysterious, invisible rays. For the first time in history, something that human senses were not evolved to perceive was shown to cause tissue damage. The implication was a bit terrifying. It was something that could be pointed at you, and you would not know to get out of the way. Some of the effects were even delayed, and at a low rate of exposure, which was completely undetectable, one could be endangered and not even know it. The effect was cumulative. Tesla's equipment was powerful. He was fortunate not to have set his hair on fire, but his health was never quite the same.

At the Sorbonne in Paris in 1898, Marie Curie, with some help from her husband, Pierre, discovered a new element, named "radium," in trace quantities mixed into uranium ore. It had invisible, energetic influences on photographic plates, just as her thesis advisor, Henri Becquerel, had found in uranium salt two years before. She named the effect "radiation." It was similar in character to Röntgen's x-rays, only these came streaming freely out of a certain mineral, without any necessary electricity. A clue to the relation was its curious property of encouraging the formation of sores on flesh that was exposed to it.

The Curies were among the finest scientists the world had known, and their dedication to task, observational ability, and logic were second to none, but their carelessness with radioactive substances was practically suicidal. Marie loved to carry a vial of a radium salt in a pocket of her lab smock, because it glowed such a pretty blue color, and she could take it out and show visitors. Pierre enjoyed lighting up a party at night using glass tubes, coated inside with zinc sulfide and filled with a radium solution, showing off their discovery to amazed guests. He got it all over his hands, and on swollen digits the skin peeled off. Surely, the cause and effect were obvious.[7]

In 1904 Thomas A. Edison, the "Wizard of Menlo Park," had been experimenting with x-rays for several years. Edison thought of using x-rays to make a fluorescent lamp, and he proceeded to test a multitude of materials to find which one would glow the brightest under x-rays. His faithful assistant was a young, eager fellow, Charles M. Dally, who had worked for him for the past 14 years.

Dally was born in Woodbridge, New Jersey, in 1865, and he had served in the United States Navy for six years as a gunner's mate. After discharge from the Navy he signed on at the Edison Lamp Works in Harrison, New Jersey, as a glass blower, and in 1890 he moved to the Edison Laboratory in West Orange to work directly for Mr. Edison. He was put to work evaluating the new lamp technology. Day after day, he held up screens of fluorescent material in front of an operating x-ray tube, staring directly at it to determine the quality of the light it produced. Nobody gave thought to any danger, but after a while Edison noticed that he could no longer focus his eye that he used briefly to test a new fluoroscope, and "the x-ray had affected poisonously my assistant, Mr. Dally."

In the beginning Dally's hair began to fall out and his face began to wrinkle. His eyelashes and eyebrows disappeared, and he developed a lesion on the back of his left hand. Dally usually held the fluorescent screen in his right hand in front of the x-ray tube, and tested it by waving his left hand in the beam. There was no acute pain, only a soreness and numbness. Dally kept testing the fluorescent screens. His solution to the physical deterioration was to swap hands, using his right to wave in front of the beam.

---

7 Or were they? Much has been repeated about Pierre Curie's radiation burns on his hands and fingers and later on his body. Consider, instead, the fact that for every ton of uranium ore the Curies processed in their crude laboratory, they had to use five tons of concentrated sulfuric and hydrochloric acid. Over a few years, they ran through eight tons of ore. Although the Curies were brilliant, creative scientists, their laboratory hygiene was not really up to the standards of the time, and Pierre was the less careful of the two. He didn't even wear gloves, and he got acid all over himself. Sulfuric acid desiccates living tissue, killing it, and hydrochloric acid digests it.

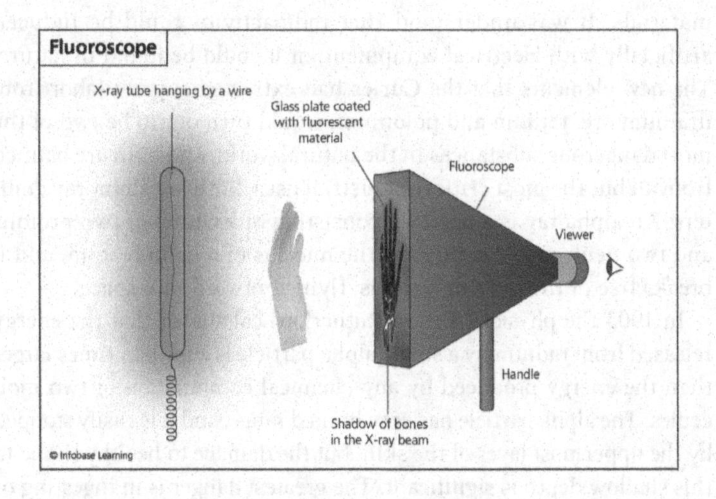

Thomas Edison's radical idea for a new type of light bulb was to use x-rays hitting a fluorescent screen. Clarence Dally tested many types of fluorescent paint for Edison by waving his left hand between an x-ray tube and a fluoroscope screen while viewing the effect through an eyepiece. The cumulative effect of hundreds of hours of x-ray exposure was fatal.

The lesion on his left hand would not heal, and conventional medical practice was at a loss to explain why. The pain became intolerable, and attempts to graft new skin onto the spreading sore were unsuccessful. The vascular system in the hand collapsed, and a cancer was detected at the base of the little finger. The physicians had no choice but to amputate the left hand at the wrist. Dally kept working on the x-ray project, holding the apparatus with his right and waving the stump in front of the screen.

In the meantime a deep ulceration developed on his right hand, and four fingers had to be removed. Eventually, both arms had to be amputated, one at the shoulder and the other above the elbow. All efforts to stop the progression of the disease eventually failed and Dally, after eight years of suffering, died in October of 1904. Edison was shaken, and he dropped all work on the fluorescent lamp. "I am afraid of radium and polonium too," he commented, "and I don't want to monkey with them."

At the time there were no rules, regulations, laws, procedures, or helpful suggestions for the handling and storage of radioactive materials. It was understood that radioactivity could be induced artificially with electrical equipment, or it could be found in nature. The new elements that the Curies had extracted at great labor from uranium ore, radium and polonium, would turn out to be two of the most dangerous substances in the natural world, and both are banned from all but the most critical industrial uses. Both are alpha-ray emitters. An alpha ray is a particle, consisting of a clump of two protons and two neutrons. It is literally the nucleus of a helium atom, and it breaks free of the radium nucleus, flying outward into space.

In 1903 the physicist Ernest Rutherford calculated that the energy released from radium by a single alpha particle is a million times larger than the energy produced by any chemical combination of two molecules. The alpha particle has very limited range, and it is easily stopped by the uppermost layer of the skin, but the damage to healthy tissue to this shallow depth is significant. The greatest danger is in ingesting or breathing radium dust, as the destructive energy of each alpha particle released is fully deposited in body tissues. Atop that danger, there is the continuing breakdown of the decay products, the debris left after an alpha particle has jumped off the radium or polonium nucleus. These damaged nuclei emit an entire range of different radiations from further decays. By the time of Rutherford's calculation, Pierre Curie was suffering unbearable pain from burns all over his body. He would lie in bed all night, unable to sleep, moaning. As a professor at the Sorbonne, the distinguished University of Paris, he asked for a reduced teaching load, complaining of having only "a very feeble capacity for work" due to his work refining radium out of uranium ore.

On April 19, 1906, after a luncheon of the Association of Professors of the Science Faculties, he walked to his publisher's office to go over some proofs of his latest scientific paper. It was raining hard, and the street traffic was heavy. He found his publisher locked and closed down, due to a strike. Curie then turned and stepped into the rue Dauphine to cross, slipped on a wet cobblestone, and sprawled into the street. His head went under the wheel of a 6-ton, horse-drawn wagon loaded down with military uniforms. Curie died instantly.

It took 11 years, but eventually news of the discovery of radium penetrated the Ozark Territory, and in 1909 James L. Leib, a prospector and self-schooled geologist, saw a logical connection between the published properties of radium and the legend of the mysterious cave dating back to 1879. The spot price of radium at the time was, gram for gram, about one hundred times the value of diamonds, or $70,000 per gram. It was the most valuable material in the world, as it had found use in cancer therapy. It was true that radium would kill living tissue, but its working range was very slight. A carefully placed radium needle would wipe out a cancer tumor immediately adjacent to it without harming anything else. There was much demand.

With effort, Leib found the remaining member of the hunting party, Old Bill Boyceyer, still alive in Chance, Oklahoma. Old Bill was glad to tell the story yet again and give what he could remember as directions to the hole, with a caution: Don't go in!

Leib found the cave, right where Boyceyer remembered, and he entered with unusual caution. He went in only far enough to pick up some bits of weird-looking, bluish rocks. Leib corresponded directly with Madame Curie, obtaining instructions for exposing photographic plates to the ore and confirming radioactivity. With the help of a photographer in Bentonville, Missouri, he succeeded. The few rocks he had brought back from the hole burned dark images into the plates, right through the dark-slides and black paper wrapping. Steel nails and a key left atop the plates showed up clearly as shadows, blocking the radiation. The radiographs were displayed at county fairs and apple shows all over the Ozarks, with Leib trying to drum up interest in opening up a radium mine. There were fortunes to be made, far greater than could be extracted from a mere gold mine.

In the spring of 1912 an enterprising man of vision from Chicago named John P. Nagel bought the land out from under Leib and commenced developing it as a mineral excavation site. Nagel proudly owned a mining operation that employed several men, housed in a dormitory built from local materials, and photos show him standing over a production table heaped with big chunks of ore. Within a few years the easy pickings in the mine played out, it was abandoned, and the mystery hole in the Ozarks once again slipped into obscurity.

It is clear that Leib and Nagel saw a connection between the inexplicable burns on Bill Henry after his cave adventure and later tabloid descriptions of burns on lab technicians from handling radium. They reasoned that the hunting trio had stumbled upon a radium mine.

This account of the first documented radiation injury requires clarification.[8] For one thing, there is no such thing as a radium mine. All the radium-266 that may have been in existence when the Earth was assembled from interstellar debris quickly disappeared, in astronomical terms, as its half-life is only 1,600 years. However, there is always a very small supply of radium in the Earth's crust, because it is a decay product of uranium, which has been on this planet from the beginning. The radium also undergoes radioactive decay into radon gas, and an equilibrium exists between production and loss. Radium is therefore available in uranium deposits in trace amounts. Many tons of uranium must be processed to extract a few milligrams of radium-266. Note that none of the many minerals known to contain uranium are shiny, metallic, or particularly interesting looking. Uranium metal does not exist in nature, but if it did, it would quickly turn dark gray and soak up every oxygen molecule that passed its way.

Mining uranium in the confines of tunnels is, of course, dangerous without safety measures, but the danger is slow to affect the human body. Breathing the radioactive dust and gas in a mine for decades can cause lung cancer, but it can take 20 years for it to metastasize. Just standing in a uranium mine, leaning against the wall, or taking a nap in a dark corner will not cause anything. No person before or since has developed radiation burns on skin from being in contact even with pure uranium. It certainly gives off alpha, beta, and gamma radiation plus a chain of radioactive decay products, but the process is so slow, it cannot immediately affect living tissue. How then is this

8 Technically, the initial radiation injury happened the first time someone was sunburned, an incident which is lost to prehistory. The case of Bill Henry in the cave is, however, the first that I have found where a person may have suffered an acute radiation injury of local nuclear origin. Chronic radiation sickness has been documented since the early 15th century, when copper miners in the Scheesberg area of St. Jaochimsthal, Bohemia, began complaining of a mysterious respiratory illness. Miners were dying at an alarming rate of the *bergsucht*, or mountain sickness. It was easily explained as secret machinations of the evil mountain dwarves, who were enraged that people were tearing up their domain with picks and shovels. An alternate explanation is that radon gas from uranium ore mixed in with the copper, diluted in their breathing air, was alpha decaying in their lungs and leaving behind radioactive polonium-218, bismuth-214, and lead-210. Breathe that long enough, and you will develop lung disease.

incident explainable? In 1879 there wasn't even enough knowledge to make up such a story.

Henry, Dempsey, and Boyceyer had ventured into an undisturbed series of caverns lined with uranium ore of exceptional purity.[9] There was no cross-ventilation of the rooms, and radon-222 gas, with a half-life of 3.842 days, had been free to collect, undisturbed, as it seeped out of the walls, floor, and ceiling. It is a heavy, noble gas, not interacting chemically with anything, but emitting powerful alpha particles and associated gamma radiation. The back chambers of the cave may have collected radioactive gas for millions of years, as it displaced the cover gas of atmospheric nitrogen and some oxygen, again reaching an optimum equilibrium state between production and loss by radioactive decay. There was no mention of anything alive in the cave, and the apparently clean floors indicated that no bats had ever lived in there.

Radon-222 is the product of the decay of radium-226, and is, indirectly, a product of the slow decay of uranium-238, the predominant isotope in uranium ore. The rough walls of the cave gave a tremendous surface area of radioactive ore, and the loss of radon by rapid decay was slightly less than the production of radon by radium decaying at or near the inner surface of the cave. Radon production occurring significantly below the surface of the ore would not contribute anything, as the gas would decay into something else before it had a chance to diffuse to the surface. Without the abnormally high production rate due to the large surface area, the radon leaking into the cave would have dissipated faster than it was made, and trivial amounts would have built up. When the hunters advanced deeper into the cavern, they were breathing it instead of normal air. The lack of oxygen made them hallucinate, pass out, and talk crazy.

9 Or had they? The minerals described in the garbled accounts really don't seem like uranium-bearing ore. It actually could be a description of some sort of thorium-phosphate crystal or a pure form of monazite mineral. Thorium has a slow alpha-beta-gamma disintegration much like uranium, resulting in radium-228, which is a fairly active beta-emitter. Also in the decay chain are radium-224 and radon-220, both alpha emitters. The tailings from the old Pea Ridge iron mine in Washington County, Missouri, are now seen as a source of thorium, which could be a potential reactor fuel. However, in 2009 John Gustovson, a geologist, discovered what could be one of the "biggest deposits of undiscovered uranium in the U.S." in southern Missouri.

Uranium or thorium, regardless of how pure or how close to the skin or the length of the exposure, cannot produce the burns described on Henry. These natural materials are simply insufficiently radioactive, and the radium traces must be laboriously extracted and concentrated to start doing harm. The concentrated radon, however, in this highly unusual situation, could have done it. Why it seemed to affect Henry most severely is probably because he was the most aggressive explorer of the three, squeezing through every narrow passage, and perhaps the clothing he wore contributed to the effect. It was probably heavier or had more layers than what the other two explorers wore. The radon gas, not reacting chemically with anything, was free to diffuse into his clothing, subjecting him to alpha and gamma radiation as it decayed, but this would not explain his burns. It is also possible that Henry was the one of the three explorers who was unusually sensitive to radiation.

The decay products of radon-222 are a complex chain of 11 radioactive isotopes, from polonium-218 down to thallium-206, before it ends at stable, non-radioactive lead-206. Half-lives range from 0.1463 milliseconds to 22.3 years. All the radon decay products in the 11-member chain are solids, even at the atomic level, and they would definitely stick to his clothing and his skin, with each product extremely radioactive. As Henry squeezed through the cave, scrubbing the wall and standing in concentrated radon gas, his clothing was loaded up with radon decay products in the form of fine dust. Over the next few days, being portaged to the hospital in Carthage, he could have been hit with eight beta rays coming from each radon atom. His lungs started to clear as soon as he got into fresh air, but his clothing was heavily contaminated.

Alpha radiation consists of a large clump of nuclear particles, or nucleons, and it represents a sudden, radical crumbling of an atomic nucleus, just happening out of the blue. The resulting alpha particle is a helium-4 nucleus, complete, and when hurled at anything solid it can cause damage on a sub-atomic level.

The beta "ray" or "particle" (either term is correct) is actually an electron or its evil twin, the positron, banished from a nucleus and hurtling outward at high speed. It is the result of the sudden, unpredictable change of a neutron into a proton or a proton into a neutron down inside an atomic nucleus. This decay event also completely changes the atom's identity, its chemical properties, and its place in the hallowed Periodic Table of the Elements. Meanwhile, the traveling beta particle, while much lighter than the alpha particle, is still an "ionizing" radiation. If it is a particularly energetic beta example (they come in all strengths), it can hit an atom that's looking the other way with enough force to blow its upper electrons out of orbit, break up molecular bonds, and bounce things around, causing the matter in its way to heat up. On skin this effect turns up as a burn, or a reddening of the surface, just like you get from an aggressive tanning booth.

The gamma ray, yet another form of nuclear radiation, is an electromagnetic wave similar to ultraviolet light or x-rays, only it is far more energetic. A gamma ray of sufficient energy can penetrate your car door, go clean through your body, and out the other side, leaving an ionized trail of molecular corruption in its path. It is the product of a rearrangement or settling of the structure of an atomic nucleus, and it naturally occurs often when a nucleus is traumatized by having just emitted an alpha or a beta particle. Gamma rays can be deadly to living cells, but, unlike the clumsy alpha particle, they can enter and leave without losing all their energy in your flesh. It's the difference between being hit with a full-metal-jacketed .223 or a 12-gauge dumdum. Both hurt.

Improbable as it seems, Bill Henry apparently suffered beta burns from exposure to concentrated radon-222 and radon decay products on the cave floor. He recovered from this acute dose and suffered no lasting effects, as is typical of brief radiation encounters. His exposure was only on the surface and not ingested. With current knowledge and understanding of radiation exposure symptoms, his socks would have been hazmat, held with tongs.[10]

---

10 And, what happened to the dog? How did the dog, close to the ground and gulping radon through the nose, survive? How did he come out the other side of the mountain in a cavern with no cross vent? It may be that neither the cat nor the dog actually went through the cave. I think the cat faked out the dog, looking as if he were plunging into the cave, but skipping left and, under the cover of dense foliage, hot-footed around and down the mountain, never to be seen again. The dog did go in the cave, but he stopped at the strange, overpowering smell of radon. The alpha particles from radon directly activate olfactory nerve endings, without any chemical reaction, and it makes an odd "metallic" odor. Dog backed out, and the hunters, again looking through thick brush, didn't see him leave. Dog tried to clear his nose and reacquired cat-smell. He resumed the pursuit, but by this time the cat was long gone. He wound up on top of the mountain, casting about. A report that they "sent the dog back in, and he again came out on top" is questionable. You don't "send" a dog into a black hole stinking of radon.

Learning can be a slow process. In the first quarter of the 20th century, we at least developed an inkling of the danger of radiation, that unique peril that bedevils all things nuclear, particularly as medical applications were developed. Eventually the practice of testing an x-ray machine by putting an arm in the beam and watching it turn red became taboo, as technicians began failing to show up for work. As radiologists began to suffer from leukemia, bone cancer, and cataracts, the procedure for taking an x-ray picture evolved into assuring the patient in no uncertain terms that this procedure was absolutely harmless, then slipping behind a lead-lined shield before pressing the START button. Still, at the time there were no government-level safety standards in place, and radiation intensity or dosage measurements had not been established.

Radium therapy was widely hailed for definite curative effects in treating cancer, the dreaded disease that killed so many people, and this was the public's introduction to radiation by nuclear decay. Further applications of this miracle metal by enthusiastic entrepreneurs would soon lead to tragic consequences, and the two most publicized disasters would change everything. The public, scientific, legislative, and industrial perceptions of radioactivity were about to be forever carved into stone in a distinctively negative way, and it would affect our basic sense of fear to this day.

William John Aloysius Bailey, one of nine children raised by a widow in a bad section of Boston, was born on May 25, 1884. He grew up poor but bright and ambitious, beginning school at Quincy Grammar and graduating near the top of his class from Boston Public Latin, famous as a launching point for ragamuffins into the Ivy League. He did poorly on his Harvard entrance exam, but he appeared sharp of mind and had a certain intense determination, and he was accepted as a freshman in the fall of 1903. Unfortunately, the cost of being a Harvard man was more than he could bear, and he had to drop out after two years. Not to be held back on a technicality, he would always boast of a Harvard degree and to have earned a fictitious doctorate from the University of Vienna, which if asked would claim to have never heard of him.

Out of school, Bailey hit the street running. He set up an import-export business in New York City, with the master plan to be appointed as the unofficial U.S. trade ambassador to China. This didn't happen.

He bounced around a while in Europe, acquiring a worldly patina, and he wound up in Russia drilling for oil at the beginning of World War I in 1914. This proved unprofitable and life-threatening, so he made it back home, where he worked on several mechanical inventions in his workshop. Barely half a year later, on May 8, 1915, he was arrested in New York on charges of running a mail-order con out of his apartment. He had been accepting mail deposits of $600 each for automobiles to be picked up somewhere in Pittsburgh. No cars showed up, and Bailey had to spend 30 days in jail. His mistake had been trying a small number of grand thefts. Reasoning that punishment would be less likely for making a great number of petty thefts, he turned to patent medicines, researching to find what the public thought they needed most.

Brilliant at this end of commerce, he came up with Las-I-Go For Superb Manhood, guaranteed to treat the symptoms of male impotence. He was finally brought to justice for this outrageous product in May 1918 and fined $200. The interesting part of this turn of events was the active ingredient in Las-I-Go: strychnine.

Known since ancient times as a deadly poison, strychnine is a colorless crystalline alkaloid found in the seeds and bark of plants of the genus Strychnos, family Loganiaceae. It is a powerful neurotoxin, useful if you want to kill small animals and birds.[11] For a human, the lethal dose is about a tenth of a gram.[12] It affects the motor nerves in the spinal cord. Transmission of a nerve impulse requires several chemical actions, one of which is an inhibitor chemical called glycine binding to an assigned port on a nerve structure. The presence of the glycine inhibitor sets the trigger point of a nerve impulse. Strychnine overcomes the glycine and binds to its port, depriving the nerve of its set-point; and without this control, the muscle at the end of the nerve will contract at the slightest impulse. This leads to painful muscle seizures and, with a sufficient number of nerves affected, death.

11 Listen to "Poisoning Pigeons in the Park," by Tom Lehrer, recorded live in 1959 for the album "An Evening Wasted with Tom Lehrer."

12 Technically, the LD-50, or lethal dose with 50 percent probability, can be estimated roughly from a surprisingly small number of recorded strychnine poisonings. It ranges from 15 to 120 milligrams, administered orally.

However, in very low doses strychnine can act as a nerve stimulant, and I can see how Bailey, and most likely others, saw it as a clever treatment for erectile disorder. Known for both its poisonous and medicinal uses in ancient China and India, strychnine made the news in the Olympic Games of 1904, held in St. Louis, Missouri. The winner of the 24.85-mile marathon race was an American, Fred Lorz of the Mohawk Athletic Club of New York, but he was quickly disqualified after loud protests from spectators.[13] It seems that Lorz, complaining of being very tired after having run nine miles, was given a lift in his manager's car, which completed 11 more miles of the race before it broke down. Lorz, somewhat refreshed, dismounted the stalled machine, turned to salute goodbye, and ran the remaining five miles to break the tape at 3:13:00.

Behind Lorz by 00:15:53 was Thomas Hicks, another American runner, but an English import who worked in a brass foundry in Boston. At about 10 miles from the finish, Hicks was exhausted, and he begged his trainers to let him stop running and lie down on the soft gravel for a while. "Not on your life," he was told, and his trainers gave him a sub-lethal dose of strychnine, about a milligram, plus a shot of brandy. Feeling slightly vigorous, Hicks was able to complete a few more miles, but he collapsed and had to have another shot of strychnine. He had to be carried across the finish line by two trainers, and it took four doctors to get his heart going again so he could stagger to the podium and receive his gold medal for the marathon. Another dose of the stimulant probably would have killed him.

Arsenic, also a well-known pesticide and a favorite poison in old murder mysteries, has also been used in sublethal doses as a medicine, treating everything from syphilis to cancer. An arsenic compound is still used to treat promyelocytic leukemia, and the isotope arsenic-74 is used as a radioactive tracer to find tumors. In fact, after World War I, radium sublethal dose treatment had become the glamour field of medicine. The reality that swallowing 0.2 milligrams could kill you simply meant that it was one of the most powerful and exciting of the deadly poisons that surely would cure diseases in trace quantities.

---

13 Fred Lorz (1884-1914) was banned from competition for life by the judgmental Amateur Athletic Union, but he was reinstated after a sincere apology and some palm lubrication. He went on to win the Boston Marathon, cleanly, in 1905, hitting the tape at 2:28:25.

Marie and Pierre Curie set out to find the effects that minute quantities of radium would have on living cells, animals, and ultimately humans. Sensing a Nobel Prize opportunity, British researchers also launched several studies, referring to it as "mild radium therapy" to distinguish it from the more radical radium needle treatment used to kill cancer tumors.

The principle of sublethal radiation treatment can be traced to the homeopathic theories of the 19th century and even to the legendary healing powers of the great European hot springs, dating back at least to Roman times. Just bathing in certain water that bubbled out of the ground seemed to be curative, and there was always the plan to bottle some of it so you could take some of the magic home with you. The enduring mystery of the springs, however, was that bottled water seemed to lose its curative potency after a few days sealed in a bottle. Why? In 1903, with recent discoveries of radioactive elements and their decay rates, it was found that the active ingredient in European springs was radon gas with a half-life of only 3.824 days, introduced into the water underground from the decay of radium traces in the rocks. Might this alpha-particle radiation be the triggering agent that accounts for the puzzling operation of the endocrine system? Could a small radiation flux be not only beneficial, but necessary to sustain life? In 1921 Frederick Soddy received the Nobel Prize in chemistry for his work in radioisotope research, and in 1923 Frederick Banting and John MacLeod won the Nobel Prize in physiology for discovering that the hormone insulin controls the body's transduction of energy. The seeming connection between these two hot topics, the discovery of nuclear energy release and the conversion of sugar into energy, was noticed by scientists and entrepreneurs.[14]

---

14 And the beat goes on. The concept that small doses of radiation and other toxins are not damaging to human health and are actually beneficial remains alive. It is now called "hormesis," and serious research started decades ago continues to this day. Studies have included everything from alcoholism in roundworms to radiation victims in Japan. As a scientific pursuit, hormesis researchers are plagued by an unusually high level of controversy and rancorous debate, and any published study is subject to being summarily torn to pieces in review. At the opposite end of the belief spectrum is the "linear non-threshold" or LNT opinion, holding that any amount of radiation or toxin, no matter how small, is damaging, and there is no threshold to get under. The United States Food and Drug Administration, Environmental Protection Agency, and Nuclear Regulatory Commission are steadfastly of the LNT persuasion, and harbor no acceptance of a beneficial radiation level. Both positions, LNT and hormesis, are probably correct under definable circumstances, but the jury is still out.

A presentation at the 13th International Congress of Physiologists by the German researcher George Wendt only intensified the theoretical atmosphere. Wendt had found that moribund, vitamin-starved rats would be temporarily rejuvenated by exposure to the alpha radiation coming from radium. The old homeopathic principle seemed valid: a poisonous substance in large quantities would destroy life, but in trace amounts it was beneficial, even necessary. By the end of the war in Europe, radioactive liniments, candles, and potions of every kind were available to a buying public. In the U.S. in 1921, interest surged after Marie Curie, twice the winner of a Nobel Prize, made an exhausting whistle-stop tour of the country. If pinned down with the right question, she would acknowledge the medicinal properties of radiation as a catalyst for essential body functions.

William Bailey, always on the lookout for a new way to redistribute wealth, jumped into the fray. He formed a company named Associated Radium Chemists, Inc. in New York City and sold a line of radioactive medicines. There was "Dax" for coughs, "Clax" for influenza, which had recently wiped out 3% of the world's population, and "Arium" for that run-down feeling. Unfortunately, none of these concoctions actually worked, and Bailey's operation was shut down by the Department of Agriculture for fraudulent advertising. Never deterred, he soon started two new corporations: the Thorone Company, making a radioactive treatment for "all glandular, metabolism and faulty chemistry conditions" (impotence), and the American Endocrine Laboratory, producing a device called the Radioendocrinator. This contraption was designed to place a gold-plated radiation source near where it was needed. Around the neck for an inadequate thyroid gland, tied in back for the adrenal glands, and, for men who may be specially concerned, a unique jock-strap held it comfortably under the scrotum. Suggested retail price was $1,000, but the market quickly saturated.

Bailey moved to East Orange, New Jersey, ground zero for making interesting chemicals, in 1925 and began manufacture of his most successful product, Radithor, a triple-distilled water enriched with radium salts. This radioactive elixir was guaranteed to practically raise the dead, curing 150 diseases from high blood pressure to dyspepsia. It was advertised as "Perpetual Sunshine." It came in a tiny, half-ounce clear glass bottle, with a cork in it and a paper wrapper around the stopper. Many radium medicines were being sold at the time, and most were absolute frauds, either having a slight bit of rapidly decaying radon or nothing at all dissolved in the water. Bailey was true to his word. Each dose of Radithor contained one microcurie each of radium-226 and radium-228.[15] It was genuinely poisonous.[16] Bailey bought his material from the nearby American Radium Laboratory, marked it up by about 500 percent, and resold it under the banner "A Cure for the Living Dead."

The public need and the advertising slogans were good, but the most brilliant of Bailey's promotional setups was his rebate plan. He promised physicians a 17-percent kickback for every bottle of Radithor prescribed. The American Medical Association angrily labeled it "fee-splitting quackery," but it helped sell over 400,000 bottles of the stuff in five years. A case of 24 bottles retailed for $30. Dr. William Bailey became comfortably rich.

Into the middle of this campaign fell Eben McBurney Byers, socialite, man about town, free-wheeling bachelor, Yale man, accomplished athlete, and wealthy chairman of the Girard Iron Company, which he inherited from his father. Powerful, handsome, vigorous in all pursuits, and broad of chest, Byers had competed in the U.S. Amateur Golf Tournament every year since 1900 and won it in 1906, two strokes over George Lyon. In his Pittsburgh home was a room dedicated only to skeet-shooting trophies, and he loved keeping racing horses in his stables in New York and England. He also maintained homes in Southampton, Rhode Island, and Aiken, South Carolina.

---

15 Not to sow confusion here, but there are two completely different species of radium at work in Radithor. The radium-226 is a product of uranium-238 decay. It occurs as traces in uranium ore, and it has a half-life of 1,600 years. Radium-228 is a product of thorium-232 decay. It occurs in monazite sand, or thorium ore, and it has a half-life of 5.75 years. Unlike radium-226, this isotope of radium emits beta rays instead of alpha particles, and at the time it was a byproduct of the gas mantle industry. The syllable "thor" in Radithor stands for mesothorium, an obsolete term meaning radium-228, and the "Radi" means radium-226.

16 That doesn't sound like a lot, but it is. It means that the radium-226 and radium-228 in that bottle were decaying at a rate of over 4 million times per second. Even after sitting on a dusty shelf for over 80 years, an old, dried-up bottle of Radithor will swamp a Geiger counter. Each of those 4 million radiation bursts per second finds something to hit and destroy in the body of the consumer. Once the radium is gone or decayed away, the resulting products keep radiating, adding to the injury. In the case of radium-228, the radiation dose rate actually increases by over 10% after the person stops ingesting Radithor.

In the fall of 1927 he was aboard a chartered Pullman, returning from the Yale-Harvard football game, when he fell from his upper berth to the floor, injuring an arm. There was an ache in the bone that wouldn't go away, despite the attentions of his trainers and personal physicians. When it started to affect his golf game and possibly his libido, Byers became very concerned, and he cast about for a doctor who could fix it, finding Dr. Charles Clinton Moyar right in his home town. Dr. Moyar, finding the source of the pain difficult to pin down, prescribed Radithor.

In December 1927 Byers began drinking three bottles a day, and he immediately felt better. He started ordering it by the case, straight from the manufacturer. He became a believer, and he fed it to his friends, his female acquaintances, and his favorite horses. Reasoning that if a bottle made him feel good, then many bottles would make him feel marvelous, he eventually downed about 1,400 bottles of Radithor by 1931.

There is a problem with ingesting a significant amount of radium. Radium shares a column on the Periodic Table of the Elements with calcium. This means that both elements have the same outer electron orbital structure, which means that they form chemical compounds in basically the same way. With this chemical similarity between radium and calcium, when the human body finds radium in its inventory, it will use it for rebuilding bones.[17] Byers's range of beverage intake did not necessarily include calcium-rich milk, and when his metabolism demanded material for repairing that hairline fracture in his arm, it found plenty of radium on hand.

Bones may seem like hard, immutable structures made of an inorganic calcium compound, when actually they are constantly being torn down and rebuilt. The bones with the most material turnaround are the jaws, which are under tremendous stress from having to support the teeth and chew food. It is surprising how much effort goes into constantly shoring

---

17 Strontium is also in this column. This makes radioactive strontium-90, a fission byproduct with a half-life of 29 years, one of the major contamination concerns when a reactor core comes apart or a nuclear weapon is exploded above ground. It is a pure beta minus emitter, and it is just two disintegration hops away from stable zirconium-90. The interim isotope is yttrium-90, with a 64-hour half-life, emitting a beta minus and a gamma ray that is so weak it can be ignored. Although its danger potential is far less than radium, it is known to cause bone cancer.

up teeth, which seem barely alive but are also under constant maintenance. By the time Byers stopped taking Radithor he had accumulated about three times what is now known as the lethal dose, and it went straight to his bones and teeth. An x-ray machine was not necessary to take a cross-sectional picture of his teeth. They would light up a photographic plate with their own radiation output.

On February 5, 1930, the Federal Trade Commission filed an official complaint, claiming that Dr. Bailey had advertised falsely by claiming that Radithor could be beneficial and would cause no harm. Bailey took umbrage, proclaiming "I have drunk more radium water than any man alive, and I have never suffered any ill effects."

Byers could not make such a claim. He started to complain about unusual aches and pains. He had lost "that toned-up feeling," and he was losing weight. Maybe it was just age catching up with him, who had just passed his 50th birthday, or perhaps it was just a bad case of sinusitis, as his doctor opined. He started having blinding headaches and then toothaches. Soon, his teeth began falling out. Starting to panic, Byers consulted a specialist, Joseph Steiner, a radiologist in New York City.

To Steiner, the problem looked like "radium jaw," a newly identified occupational disease that had been seen in watch-dial painters. The common factor was, of course, the presence of radium in the person's life. It goes straight to the jaw. Frederick B. Flinn, the radium expert from Columbia University, was called in, and he grimly confirmed Steiner's suspicion. There was absolutely no known cure or even a treatment of the symptoms. Although Dr. Moyer, Byers's personal physician, refused to accept the diagnosis, the patient's health was steadily declining. The once solid hunk of man wasted down to 92 pounds but remained alert and lucid. His bones were splintering and dissolving.

The Trade Commission, seeing this as a further indictment of Bailey's work, called on Byers to testify in September 1930. He could not make the trip, so attorney Robert H. Winn was sent to his Southampton mansion to take a deposition. His written description of Byers says it all:

> A more gruesome experience in a more gruesome setting would be hard to imagine. We went to Southampton where

> Byers had a magnificent home. There we discovered him in a condition which beggars description. Young in years and mentally alert, he could hardly speak. His head was swathed in bandages. He had undergone two successive jaw operations and his whole upper jaw, excepting two front teeth, and most of his lower jaw had been removed. All the remaining bone tissue of his body was slowly disintegrating, and holes were actually forming in his skull.

Byers was moved to Doctor's Hospital in New York City. He died on March 31, 1932, at 7:30 A.M. The Trade Commission had shut down Bailey's operation with a cease-and-desist order on December 19, 1931.

Eben Byers had been rich and well known, and the *New York Times* took his death as important and worthy of front-page reporting. The headline was crude, but it still has a compelling effect on even the casual reader: THE RADIUM WATER WORKED FINE UNTIL HIS JAW CAME OFF. On April 2 the subtitles started to unwind the story: DEATH STIRS ACTION ON RADIUM "CURES." TRADE COMMISSION SPEEDS ITS INQUIRY. HEALTH DEPARTMENT CHECKS DRUG WHOLESALERS. AUTOPSY SHOWS SYMPTOMS. MAKER OF "RADITHOR" DENIES IT KILLED BYERS, AS DOES VICTIM'S PHYSICIAN IN PITTSBURGH. FRIENDS ALARMED TO FIND MAYOR HAS BEEN DRINKING RADIUM-CHARGED WATER FOR LAST SIX MONTHS.[18]

This intense tabloid journalism pushed all the right buttons. The reading public was justifiably horrified, and the dangers of radioactive materials came into sharp and sudden focus. The Federal Trade Commission, feeling empowered, reopened its investigation, and the Food and Drug Administration began a campaign for greater enforcement power and new laws concerning radioactive isotopes. Sales restrictions on radiopharmaceuticals, still in place, date back to the Byers affair, and the market for over-the-counter radiation cures collapsed immediately.

William Bailey was not one to be discouraged by federal intervention, and he went on to market a radioactive paperweight, the "Bioray," acting as a "miniature sun" to give you the benefits of natural,

18 James J. "Jimmy" or "Beau James" Walker was mayor of New York City at the time, and he at first refused to give up his radium water because it made him feel so good. Later that year Walker fled to Europe in fear of prosecution for several crimes while in office. Seldom mentioned is Byers's girlfriend, who also perished of radium poisoning from the Radithor that Byers insisted that she drink.

environmental cosmic rays even as you sat in your dismal office space. The Great Depression was killing sales, but he kept on, selling the "Adrenoray" radioactive belt clip and the "Thoronator," a refillable "health spring for every home or office" designed to infuse ordinary tap water with health-giving radon gas. Fortunately, none of his new products were noticeably radioactive.

Meanwhile, a second death-blow to the popularity of radium was developing on the other end of the ladder of success. Thousands of young women were working in factories using radium paint, and they were beginning to die off in horrible ways. They lacked the celebrity of Eben Byers, but there were so many of them that the problem became impossible to ignore. This highly successful industry was making profits in Canada, Great Britain, and France, with particular enthusiasm concentrated in the United States. Its only goal was to make things glow in the dark.

The quest to make a paint that would glow by itself goes back as far as 1750 in industrial Europe. The first concoction to be advertised and sold was probably called "Canton's Own," manufactured by a Professor Tuson in London in 1764. By 1870 there were competing formulae, and luminous paints were selling briskly. Most used strontium carbonate or strontium thiosulphate. It had been found, probably accidentally, that strontium compounds would seem to store sunlight and would then give it back after the sun went down. We now know this phenomenon as a "forbidden energy-state transition" in a singlet ground-state electron orbital. The strontium, like everything else, absorbs and then returns a light photon that hits it, but in this case the return is delayed. The strontium atom, excited to a higher energy state by the absorption of light, "decays," as if it were radioactive, reflecting the light back with a half-life of about 25 minutes. After four hours of glowing, the strontium compound needs to be re-charged with light. By 1877, phosphorescent paint was all the rage. The insides of train cars were painted with it, so your reading would not be interrupted when the train passed through a dark tunnel. Glow-in-the-dark wallpaper was sold.[19] Street signs and souvenir postcards used it.

19 This product was particularly popular in gunpowder factories, where liberal use of gas lighting was liable to end in disaster.

The discovery of radium and its radioactivity in 1898 put an entirely new spin on the concept of phosphorescent paint. It did not take long to figure out that a speck of radium carbonate, invisibly small, mixed with zinc sulphide would glow not for four hours, or four years, but forever. There was never a need to charge the material with light. All the excitation of the zinc came from the radiation emitted by the radium. This world-changing discovery was made not by a crack team of physicists, but by Frederick Kunst, a gemologist working at Tiffany & Co. in New York City. He teamed with Charles Bakerville, a chemistry professor at the University of North Carolina, and they came up with the formula for luminous paint, receiving a U.S. patent in 1903.[20]

Ignoring the concept of patent rights, the Ansonia Clock Co. of New York began selling timepieces with self-luminous dials in 1904, supposedly using their own formula. The concept was an immediate success. Limits to the availability of the active ingredient, radium, kept sales from going out of control, but the glowing clock-dial was an attractive novelty.

In 1914 came the First World War. It was a new type of war, fought partly in the dark of night. All critical equipment needed to be lighted, but not so brightly that it would reveal your position to the enemy. Radium paint was the logical, perfect solution, and demand on both sides of the conflict became huge. Everything needed to glow: gun sights, compass cards, elevation readouts on cannons, land-mine markers, and, most urgently, watch dials. Watches were used widely, to synchronize night charges by mile-wide fronts of men emerging suddenly from their trenches. In 1913 the United States Army began equipping every soldier with a luminous-dial timepiece, and there were 8,500 in use. The radium-dial watch became a necessary part of

20 Bakerville could never grasp the theory of transmutation of elements due to radioactive decay, and he announced the discovery of two new elements, "carolinium" and "berzelium," while studying curious properties of radium. He was thus lauded in the *New York Times* in 1904 as "The Only American Who Ever Found a New Element." America at the time was feeling as if Europe was running away with all the glory discoveries of science. The feeling was not unjustified. Carolinium and berzelium are not to be found in the current Periodic Table of the Elements, but carolinium shows up in H. G. Wells's atomic bombs in his novel, *The World Set Free*. Wells left the second "i" out of carolinium as a poke at the Americans, who had accidentally left the second "i" out of aluminium.

living during that war, and the need came home with the surviving troops. By 1919, the number of glowing timepieces turned out in the U.S. had grown to 2.5 million. The Ingersoll Watch Co. alone made a million radium watches per year, and the demand for the rare metal put a maximum load on the mines in Colorado and Utah. Hospitals began to protest that the supply of radium was drying up, and thousands of cancer patients would be denied the only effective treatment for some forms of the disease. The sum of all radium inventories in the world amounted to only hundreds of milligrams, and any application had to be judicious and without waste.

There were many names given to the glow-in-the-dark product: Luma, Marvelite, Radiolite, and Ingersollite. The most memorable was probably "Undark," sold by the Radium Luminous Material Corporation of New York City. By 1917 it was being used for doorknobs and keyholes, slippers, pistol sights, flashlights, light pulls, wall switches, telephone mouthpieces, watches, and house numbers. The advertising slogan, showing the address numbers on a darkened front door, was "I want that on mine." With things self-illuminating, you would never have to light a match to find them in the dark. In Manhattan, "radioactive cocktails" were served in the best bars, and a musical named *Piff! Paff! Pouff!* celebrated the wonders of ionizing radiation with *The Radium Dance*, written by Jean Schwartz. Uranium mining was stepped up all over the world, in Portugal, Madagascar, Czechoslovakia, Canada, and even in Cornwall, England, as the demand outstripped supply.

Into this madness stepped the American entrepreneurial spirit, building factories from West Orange, New Jersey, to Athens, Georgia, to cover the numerals on watch dials with self-luminous paint. Young women were hired to do the meticulous manual work of applying the paint, as male workers were thought incapable of sitting still for hours at a time to do anything useful. The workers were paid generously, at $20 to $25 per week, when office work was paying $15 a week at most. By 1925, there were about 120 radium-dial factories in the United States alone, employing more than 2,000 women.

Painting the numbers on a watch face was not easy. The 2, 3, 6, and 8 were particularly difficult. You had to have paint mixed to the right viscosity, a steady hand capable of precise movement, and good eyesight. One woman did about 250 dials per day, sitting at a specially built desk with a lamp over the work surface, wearing a blue smock with a Peter Pan collar. The brush was very fine and stiff, having only three or four hairs, but it would quickly foul up and have to be re-formed. All sorts of methods were tried for putting a point on the brush. Just rubbing it on a sponge didn't really work. You needed the fine feedback from twirling the thing on your lips. Some factory supervisors insisted on it, showing new hires how it is done, and some factories officially discouraged it while looking the other way. Everybody did it, sticking the brush in the mouth twice during the completion of one watch dial. The radium-infused paint was thinned with glycerin and sugar or with amyl-acetate (pear oil), so it didn't even taste bad.

The practice of tipping a paint brush started contaminating everybody and everything in a watch dial factory. Painters noticed that after sneezing into a handkerchief, it would glow. You could see the brush twirlers walking home after dark. Their hair showed a ghostly green excitation, and they could spell out words in the air with their luminous fingers. Some, thinking outside the box, started painting their teeth, fingernails, eyelashes, and other body parts with the luminous paint, then stealing away to the bathroom, turning out the lights, and admiring the effect in the mirror. There was no problem finding gross radium contamination in a factory. There was no need for a radiation detection instrument. All you had to do was close the blinds. Everything glowed; even the ceiling. Most workers were each swallowing about 1.75 grams of radioactive paint per day.

By 1922, things started going bad in the radium dial industry. In the next two years, nine young radium painters in the West Orange factory died, and 12 were suffering from devastating illnesses. US Radium, the biggest watch-dial maker in town, strongly denied that anything in their plant could be causing this. No autopsies were performed, and the death certificates recorded anemia, syphilis, stomach ulcers, and necrosis of the jaw as causes. The dead and ailing, however, had dentists in common, and these health professionals had noticed unusual breakdowns of the jaws and teeth in all of these women. It was beginning to look like another case of an occupational hazard, following closely behind tetraethyl lead exposure at General Motors and "phossy jaw" from white phosphorus fumes in the match industry. Could it be the radium?

In 1925 Dr. Edward Lehman, the chief chemist at US Radium, died, and an autopsy showed that his bones, liver, and lungs were heavily damaged by radiation. His skeleton exposed an x-ray plate without the use of an x-ray machine, it was so radioactive. He hadn't even picked up a paint brush. All he had done was to breathe the air in the factory. The Harvard University School of Public Health was brought in by US Radium to examine the factory and give it a clean bill of health. Far from it, the survey found not one worker in the plant with a normal blood count, and the radiation level on the floor was five times above background. The critical report was buried, and a press release issued on June 7, 1928, denied that the study had found any evidence of "so-called radium poisoning."

Sabin A. von Sochocky, immigrated from Austria back in 1913, the inventor of Undark, and the man who started Radium Luminous Materials, was also beginning to feel the effects of occupational radiation. Back in his day, he had been so bold as to immerse his arm up to the elbow in radium paint. Now, his jaw disintegrated, and his hands were coming apart. It was clear that radium was a bone-seeker, leading to no good outcome. Sochocky reversed his attitude, becoming a spokesman against the use of industrial radium and a source of useful admissions. He made available to authorities the vast collection of his papers and company records, and the relationship between luminous paint and death began to clarify. He died at the age of 46 in November 1928 of aplastic anemia, having lost the use of the marrow in his bones.

Finally, a plant worker at US Radium in West Orange, Grace Fryer, decided to sue the company for having subjected her to known health hazards. Five women threw in with her, and the sympathetic press labeled them the "Legion of the Doomed," the "Living Death Victims," and the "Radium Girls," the name that echoes today. They accused the company of subjecting them to illness that would end soon in extremely unpleasant death. Each demanded a quarter million dollars in compensation. The press went viral, and public sympathy surged.

US Radium, still denying everything, talked strategy with their legal team as they delayed the proceedings with everything that could be thrown in the way. They eventually settled out of court for a $10,000 lump sum to each woman plus a $600-a-year pension and coverage of all medical expenses. The plant closed. On August 14, 1929, another worker died just eight days after quitting the Radium Dial Co. in Ottawa, Illinois. Margaret "Peg" Looney, an Irish-Catholic redhead, all of 5 foot 2 inches tall and one of ten children, had worked as a dial painter since graduating from high school at 17. After painting for three years, she started developing trouble with her teeth and an overwhelming weakness. She kept going, needing the income, and her family watched in horror as she started pulling pieces of jaw out of her mouth. She died at age 24 of diphtheria, according to the death certificate, after seven years of radium absorption. She was buried in St. Columba Cemetery.[21] The Radium Dial plant closed shortly afterward, fearing a swell of litigation.[22]

Relentless journalism had made the public painfully aware of the dangers of the radium and its radiation output as no lecture, authoritative text, or a semester of study could. The accounts of horrible disfigurements and lingering deaths suffered by Eben Byers and the Radium Girls still reverberate and became the unfortunate benchmark for the effects of radiation exposure. The federal government became concerned with occupational safety, and labor laws were crafted in Congress resulting from the radium scandals. Radiation tolerance levels were established, and the concept of industrial hygiene for working with radioactive materials was born. The Food and Drug Administration found new powers of enforcement. A fascination with everything radium turned completely around.

This was hardly the best way to introduce the public to the sensitive topic of radiation safety. Isotopes of radium, the first nuclear radiation sources to be commercially exploited, are probably the worst examples

---

21 In 1978 Argonne National Laboratory exhumed Peg Looney and measured the radiation content of her bones, finding 19,500 microcuries of radium-228 remaining after 49 years. That is 1,000 times the maximum allowed level, and, given the 5.75-year half-life of the isotope, it had been a great deal of ingested radium. It's as if she had been drinking the paint. She was reburied in a coffin made of lead.

22 A new factory, renamed Luminous Process Co., opened six weeks later two blocks down the street. It was owned by the same guy, Joseph Kelly, who owned Radium Dial. It was finally closed in 1978 for continual breach of regulatory directives for the safe use of radioactive materials.

out of thousands of radioactive isotopes. Radium has nearly absolute body burden, or a tendency to stay in the metabolism forever, and there are few ways it can escape the biological systems. Its radiations cover a wide spectrum, from alpha to gamma, with unusually energetic rays, and it targets many essential organs. It destroys everything around it, so quickly that cancer doesn't even have time to develop.

Still, there are ironies and unanswered questions concerning this baptism by fire. Radium dial watches were still being made until 1963, when finally they were banned in the State of New York. The US Radium name went away in 1980, when the plant in Bloomsburg, Pennsylvania was renamed Safety Light Corporation, specializing in luminous paints incorporating tritium.[23]

That these radium-dial factories continued operation for decades is not surprising, given the renewed needs for self-luminous equipment during the Second World War, but the persistence of radioactive water for drinking and bathing is astounding. In the 1980s, mineral water became all the rage. It was obviously a better beverage than municipal tap water, which is basically rain-water fortified with fluoride and sanitized with chlorine. Mineral water bubbles up from deep underground, and that (plus its cost) makes it superior to tap water, but we had forgotten why. It is supposedly health-giving because it is radioactive, using the ancient logic of homeopathic medicine. A trace of something that will kill you will only make you stronger. Spring water dissolves soluble mineral substance out of the deep rock, and that would be uranium oxide.[24] Spring water is further fortified with microscopic radon gas bubbles from the radium decay in these same rocks.

This fascination does not end there. Incredibly, the world remains studded with thousands of disease-curing radium springs from Hot

23 Tritium, the heaviest isotope of hydrogen, is still a radioactive substance, but it is not nearly as dangerous as radium. It has a half-life of 12.33 years and it emits a pathetically weak beta ray of only 0.0186 MeV. After 123 years, a tritium sample is effectively all gone. It leaves the body as easily as it comes in, so the burden is slight.

24 Eventually, all the soluble uranium oxide will leach out of the ground by moving water and be washed to the sea, just like sodium chloride, or salt. There is presently an estimated 4,290 million metric tons of uranium in the salty oceans, enough to power the world with nuclear fission beyond the expected lifetime of mankind.

Springs, Arkansas, to the Gastein Healing Gallery in Austria. Japan, a country that seems particularly sensitive to the concept of trace radioactivity in the biosphere, has 1,500 mineral spas. An example, Misasa, in the Tohaku District in Tottori, boasts springs of radium-rich water with radon bubbles. The name "Misasa" means "three mornings," meaning that enjoying an early soak in the magic water thrice will cure what ails you. The town organizes a yearly Marie Curie festival to honor the discoverer of the active ingredient.

Probably no health spa currently does it with more enthusiasm than Badgastein. In 1940 the Third Reich, desperate for wealth, decided to reopen the ancient gold mine running through the Hohe Tauern range in southern Austria. Pickings for gold turned out slim, but they noticed that the enslaved workers were getting healthy working in the hot, radon-contaminated tunnel. This observation was not lost, and in 1946 the *Heilstollen* or Thermal Tunnel was opened up and equipped with small cars to carry bed-ridden patients through the radon surrounded by rock walls, crusty with uranium. By 1980, a million patrons had stayed at least one night at Badgastein. The current brochure for the spa facility, listing diseases that are cured in the tunnel, puts radium advertising copy from 1925 to shame:

> Inflammatory rheumatism; Bechterew's disease; arthroses; asthma; damage to the spinal column and ligament discs; inflammatory nerves; sciatica; scleroderma; paralysis and functional disturbances after injuries; circulatory problems of the arteries; smoker's leg; diabetes, arterio-scleroses; problems with venous blood circulation; heart attack risk factors; infertility problems; premature aging; potency disturbances; urinary tract, gout, and suffering due to stones; and paradontosis.

Even America, home of the Radium Girls, has not lost its love of radioactive water. The use of medicinal springs in the New World dates to prehistory, when the aboriginal residents flocked to the healing fluids, and there are now five towns named "Radium" and three named "Radium Springs" in the United States. One of the most patronized spas in the Western Hemisphere is Radium Hot Springs, in the Kootenay region of British Columbia, Canada.

There is further paradox to the discovery of radiation sickness. William Bailey, the entrepreneur who killed Eben Byers, had ripened to the age of 64 when he died of bladder cancer unrelated to radium in 1949. Twenty years later, his remains were disinterred for study by Professor Robley D. Evans, Director Emeritus of the Radioactivity Center at MIT. A count of radioactivity lingering in his bones proved that Bailey wasn't lying when he claimed to have ingested more Radithor than anyone else. Yet, he had never complained of a toothache, much less died from it. Decades of study suggested that the effects of large radiation loads vary from individual to individual.

Can some people tolerate chronic high radiation better than others? Are certain people better at producing protective hormones such as granulocyte colony-stimulating factor and the interleukins, stimulating the growth of blood cells under radioactive stress?

Hundreds of women are thought to have died or been injured by radium ingestion, but thousands worked at the painting desks. Why didn't they all die? In 1993, when the Argonne National Lab study of radium workers was shut down, there were 1,000 Radium Girls still alive and complaining about the working conditions back in '25. Could we eventually evolve into a race that can withstand high levels of radiation?

Madame Marie Curie, discoverer of radium, died on July 4, 1934, in a sanatorium in Geneva, Switzerland, of a blood disorder for which there was no cure. After many years of sickness, the disease was finally diagnosed as aplastic pernicious anemia. Her bone marrow, contaminated with radium, was unable to produce red blood cells, and the extensive exposure to x-rays during her medical volunteer work in World War I had contributed to the condition.

Her daughter, Iréne Joliot-Curie, had taken up her mother's profession and became a Nobel Prize-winning radiation scientist, working beside her in the Radium Institute. Joliot-Curie was working at her bench in the laboratory in 1946 when a sealed capsule of radioactive polonium exploded in her face. She contracted leukemia caused by her long-term exposure to radiation and the unfortunate large dose she received in the accident at the bench. She died on March 17, 1956, at the age of 58 in the Curie Hospital in Paris.

Chapter 2

WORLD WAR II, AND DANGER BEYOND COMPREHENSION

"It's just like a mule. A mule is a docile, patient beast, and he will give you power to pull a plow for decades, but he wants to kill you. He waits for years and years for that rare, opportune moment when he can turn your lights out with a simple kick to the head."

—Jerry Poole, referring to a nuclear power reactor

By the start of World War II, which in Europe was 1939, the radium scandals had left the public with a strong and somewhat twisted concept of the dangers of radiation. They saw it as deadly in the worst way. It could originate in invisibly small particles of matter, and by the time you realized that you had been dosed with it, it was too late to do anything about it. Swallowing radium was about as bad as radiation sickness could get, but mankind had not seen anything yet. The intense radiation that could be released by a newly discovered phenomenon, nuclear fission, would put radium contamination in perspective. A couple of accidents with fission made it clear that with the discovery of this new way to release energy came novel ways to bring life to an end.

The entire structure of industrial safety had to adjust accordingly. If this new energy source was to be cleaned up for public use, then there would have to be new materials handling procedures, new laws and regulations on the federal level, powerful new government agencies, new controls on every aspect of this prospective industry, and a great deal of secrecy. Unlike with the radium adventure, entrepreneurs, swindlers, amateurs, and fake doctorates would not feel invited to participate. The world had changed, and simple republican democracy was not what it used to be.

Technically, the first public demonstration of nuclear fission by dropping two nuclear weapons on Japan was not an atomic accident, but these events would permanently harden some opinions and perceptions for future nuclear mishaps. The A-bomb campaign was seen as a sure and quick way to bring the war to an end with a minimum number of casualties, but, to be completely honest, it was also a large-scale science experiment. The only hard data that existed concerning the effects of radiation on human beings were studies of the deaths and injuries from radium ingestion. Most scientists working on completion of atomic bomb development speculated that most of the deaths from their new weapon would be from flying bricks and glass as cities were flattened, and not by the radiation from fission or the radioactive byproducts of fission. Yes, thousands of civilians would die, but how was that different from fire-bombing Tokyo, which had killed over 100,000 people? By the end-time, half the capital city was in ashes, with care taken not to bomb out the Imperial Palace.[25]

---

25 It was reasoned that the occupant of this palace, Emperor Hirohito, would be instrumental in issuing an expected surrender. However, on July 20, 1945 a single B-29 strategic bomber dropped a replica of the Fat Man atomic bomb containing 6,300 pounds of high explosive (baratol) from 30,000 feet with the bomb-sight cross-hairs on the geometric center of the imperial residence. It was a clean miss. In the weeks before Fat Man was dropped on Nagasaki, 49 of these "pumpkin bombs" were dropped on Japan, killing an estimated 400 people and injuring 1,200, as practice for the A-bomb mission. With the random aiming uncertainties of high-altitude bombing, the only way to ensure that the Emperor would not be hit was to aim directly at him.

When the atomic bombs were ready to deploy, just about every city in Japan had been bombed to pieces, with a few exceptions. Hiroshima, Kokura, Niigata, and Nagasaki had been purposefully spared. These were the target cities for the atomic bombings, with Hiroshima at the head of the list. It was a little jewel of a city, with 350,000 residents, the Japan Steel Company, Mitsubishi Electric Manufacturing Company, and Headquarters of the Second Army Group, tasked with defending the island of Kyushu from the coming Allied invasion. It was untouched and in perfect condition.[26] There was no sense in dropping the A-bomb on Tokyo, as there was hardly anything left to destroy, but to hit a spared city would yield data as to the destructive power of a single bomb-strike, aimed right at the center. As an experiment, it would end the speculation and guesswork about the effects of fission radiation on human beings and man-made structures, and it would give a calibration for future military operations. The Hiroshima mission consisted of three B-29 heavy bombers: the *Enola Gay*, carrying L-11, or "Little Boy," *The Great Artiste*, carrying the yield measurement instrumentation, and *Necessary Evil*, with the observers and the cameras.

Three instrument pods, having parachutes to slow their descent, were dropped from *The Great Artiste* and synchronized with the bomb-drop from *Enola Gay* with a radio signal. The pods were equipped with radiation counters and barometric instruments, each with a radio channel sending data continuously back to the airplane, where they were recorded. *Necessary Evil* had a Fastax high-speed motion picture camera, shooting 7,000 frames per second, and a still camera recording images of the explosion. A debriefing of the crew, after-action photographs at high altitude, and eventual ground-level evaluations came later. The initial data unraveled by the scientists was sobering, and it took some of the euphoric edge off the celebration.

---

26 The city was unmolested by aerial bombs, but it was not exactly in pre-war condition. That summer before the A-bomb was dropped, school children, aged 11 to 14 years, had been mobilized into a demolition force, tasked with tearing down all the houses or businesses on certain streets. As had been witnessed in other cities many times during this last year, a few B-29s carrying incendiary bombs could wipe out a Japanese city just by starting fires. Japanese houses were notoriously flimsy and made of flammable materials, and multiple ignition points would quickly overwhelm any firefighting effort. Entire streets leveled to the ground were to act as firebreaks, preventing the spread of fire over the entire city by creating zones of nothing burnable.

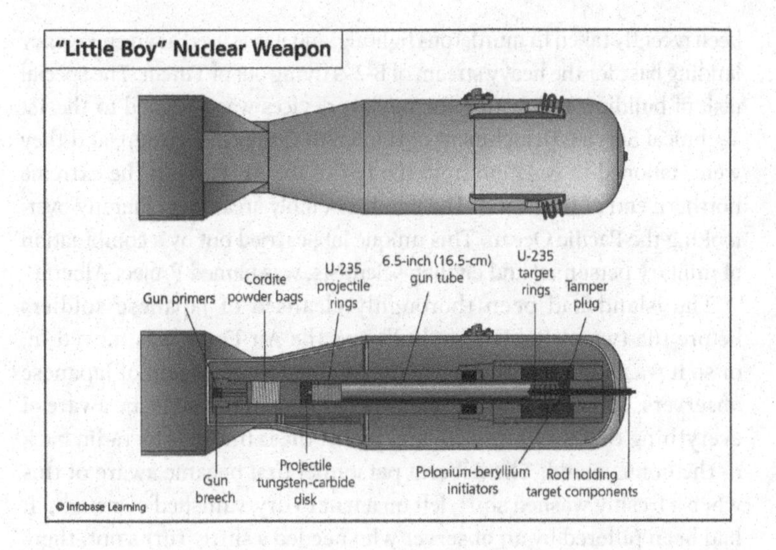

The Little Boy was an "assembly weapon." A cylindrical shell made of a stack of uranium rings was blown against a similar stack of smaller rings held stationary in a block of tungsten carbide, using a smooth-bore 6.5-inch gun barrel. The projectile rings, propelled quickly by three bags of burning nitrocellulose, and the smaller cylinder assembled into a larger, complete cylinder of uranium metal, enriched to 86% U-235. The resulting configuration was hypercritical, and it fissioned explosively.

To maximize the "shock and awe," no leaflets were dropped warning Japan of an impending A-bomb attack, and security was so tight on Tinian Island, the base for atomic operations, that most of the Army Air Force personnel could only guess what was going on.[27] However, the surprise was not as complete as one might think.

Tinian, captured from the Japanese in July 1944, was a sugar-cane plantation just south of Saipan in the Marianas Island Chain. Flat as a pool table, it was an ideal spot for launching heavy bombers against the main island of Japan. Iwo Jima, another small island even closer to Japan, had

27 Once the heavy bombing campaigns started on Japan in 1944, it was standard procedure to drop leaflets warning the population to evacuate. This was good military practice, because it was possible to partially empty out a city and send the residents fleeing to the hills. The war-material factories would thereby lose the workforce, and vital production would come to a stop. Given vague warnings of future bombing raids, 120,000 people of the 350,000 population evacuated Hiroshima prior to the A-bomb attack.

been recently taken in murderous fighting, and it was used as an emergency landing base for the heavy stream of B-29s flying out of Tinian. The special task of building and testing the nuclear devices was assigned to the 1st Technical Service Detachment of the 509th Composite Group, and they were stationed in isolation from the rest of the Air Force at the extreme northern end of the island. The bomb assembly areas were literally overlooking the Pacific Ocean. This unique job, carried out by a combination of military personnel and civilian scientists, was named Project Alberta.

The island had been thoroughly cleansed of Japanese soldiers before the two airfields were built and the Air Force was moved in, or so it was hoped. Actually, there remained a contingent of Japanese observers, and their only mission was to remain invisible, be aware of everything that was going on, and report these findings by radio back to the home island. The Alberta personnel first became aware of this when a freshly washed shirt, left on a tent to dry, vanished overnight. It had been pilfered by an observer who needed a shirt. Turns out, there was a high area in the middle of the north end of the island, about 440 feet above sea level, consisting of coral cliffs, pocked with caves and tunnel entrances. At night, the observers would quietly come down out of the caves and into the 509th area to take notes.

These detailed examinations were useful. The next morning, Tokyo Rose, an English-language radio variety show originating somewhere in Japan, would casually mention details about what was going on at the north end of Tinian Island, broadcasting to the entire Allied force. She apparently knew more than the average sailor, and, grappling for an explanation, some seriously credited the charming radio announcer with clairvoyance. The Japanese, from the Imperial Emperor on down, knew that some special weapon was being prepared. It would take few planes to deliver it, and they even knew which planes would fly the mission and when they took off. Was it a new form of nerve gas? Perhaps it was a powerful anesthetic to be delivered by airplane, and the Americans planned to put everyone on the island of Honshu to sleep, then just walk ashore and take over.

Colonel Paul W. Tibbets, the man in charge of the bombing operation, grew concerned at the accuracy of the radio programs, and he had the markings on his plane, the *Enola Gay*, changed at the last minute. Before the paint had dried, Tokyo Rose announced it to the rest of the listening world, describing the upward arrow in a circle on the tail. Her omniscience could be spooky.[28]

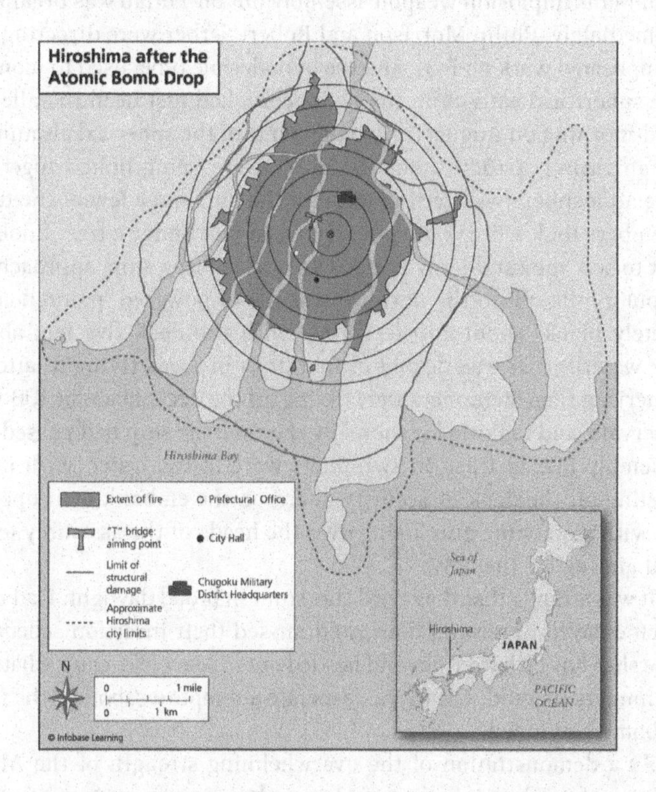

At the end of World War II, Hiroshima was a compact Japanese city with several munitions plants, army storage depots, and an army headquarters. Even though most strategically important cities in Japan had been bombed, Hiroshima had been left untouched. One bomb destroyed its industrial capability and wiped out all communications, power distribution, and transportation systems.

28 Tokyo Rose was a generic name given to any of about a dozen English-speaking women on the NHK propaganda channel, transmitting popular American music, radio skits, and carefully slanted battle news. Listening between the lines, the average soldier could gauge how badly it was going for the Japanese forces by the daily news from Rose. This particular announcer was possibly Iva Toguri D'Aquino, an American citizen who was caught in Japan at the beginning of the war. Convicted of treason in 1949, Toguri was pardoned by President Gerald Ford in 1977.

After Hiroshima was annihilated on August 6, 1945, the Japanese knew better what was going on, and a commando raid on the F-31 "Fat Man" implosion weapon assembly hut on Tinian was organized immediately. Philip Morrison and Robert Serber were directing the complicated work on F-31, and the hypodermic tube, used to monitor the subcritical activity in the bomb core, had just been installed at mid-morning on August 8. Two segments of the spherical aluminum bomb casing, Y-1560-6 and Y-1560-5, were being bolted together. The atmosphere was getting tense in the hut, and a few of the team members took a break outside, trying to rest under a tree. Looking out to sea, they suddenly noticed an odd-looking ship, approaching about a mile off, to the north. It was diesel-powered, painted completely black, about 150 feet long, with the deck five feet above the waterline. It was devoid of markings but was flying a tattered American flag. Swimmers were diving off the deck, at about 100-foot intervals, and making for shore. By the time the ship had passed the assembly hut, at least 30 swimmers were in the water, with more peeling off the deck. A security guard on the embankment opened up with a machine gun, firing over the heads of the assembly techs and aiming for the bow.

It was strange that they tried this stunt in broad daylight. Had they been delayed by several hours and missed their insertion schedule? The ship hove a hard right and headed out to sea, picking up what few swimmers it could. Clearly, a desperate attempt to sabotage the next A-bomb had failed.[29]

As a demonstration of the overwhelming strength of the Allied invasion force bearing down on Japan, dropping a uranium bomb on Hiroshima was unsurpassable. The mechanics of the A-bomb explosion have been thoroughly studied, and here is a summary:

The nuclear fission explosive uses the fact that a uranium-235 or plutonium-239 nucleus can split into asymmetric fragments when it

---

29 The only record I can find of this action is in a book written by team member Harlow W. Russ, *Project Alberta: The Preparation of Atomic Bombs for Use in World War II.* Los Alamos, NM: Exceptional Books, 1984. I won't say that Mr. Russ is a stickler for details, but he wrote down the contents of every meal he had on the way to Tinian Island, and his 1945 New Mexico fishing license is faithfully copied into the appendices.

encounters a loose neutron. This unusual reaction releases about 200 MeV of energy, which on the atomic scale is a great deal. Also emerging from this mini-explosion are two extra neutrons. These neutrons, traveling at high speed, crash into other nuclei in the tight matrix of a bomb core, which consists of a metallic mass of the fissile material. The first reaction thus accelerates into two reactions, and each generation of reaction leads to twice as many subsequent reactions. In fewer than 90 such generations, every nucleus in a 50-kilogram uranium bomb core will experience the fission stimulus, and the combined reactions release the energy equivalence of exploding a million tons of TNT high explosive. Given the speed of the flying neutrons, the size of a bomb core, and the response time of a uranium nucleus, these 90 generations take place in about one millionth of a second. The short time in which this much energy lets go provides the condition for a hell-on-earth explosion.[30]

Most of the energy from this explosion, 85 percent, is released in the form of heat. The heat radiates as light energy, from infrared to ultraviolet. The remaining 15 percent of the energy release is radiation of nuclear origin, but only five percent is immediately involved. Residual radiation, ten percent of the bomb's energy, is released on a falling exponential rate over thousands of years after the instant of detonation.

The World War II bombs, the only nuclear devices ever used as weapons so far, were airbursts, detonated at about 1,900 feet above the ground.[31] The air surrounding the bomb instantly heated to incandescence. This feature is called "the fireball." This rapidly expanding sphere translated a percentage of the thermal energy into blast energy, or a destructive wave of compressed air moving outward at high speed, capable of knocking over concrete buildings.

30 This one-megaton bomb is a theoretical device used by Samuel Glasstone in his definitive work, *The Effects of Nuclear Weapons* (1957), in his detailed analysis of an atomic bomb explosion. The actual bomb that destroyed Hiroshima was smaller, 16 kilotons, but the effects scale down only slightly.

31 The airburst tactic did two things: it maximized the radius of destruction, and it minimized the resulting radioactive fallout. The explosions kicked up a lot of dust, but the only radioactive material to be spread off-site was the bomb itself, which was about 9,700 pounds of metal. The bomb debris consisted of fuel that failed to fission, fission products, and various metals in the structure of the device, a portion of which were neutron-activated to radioactivity in the explosion.

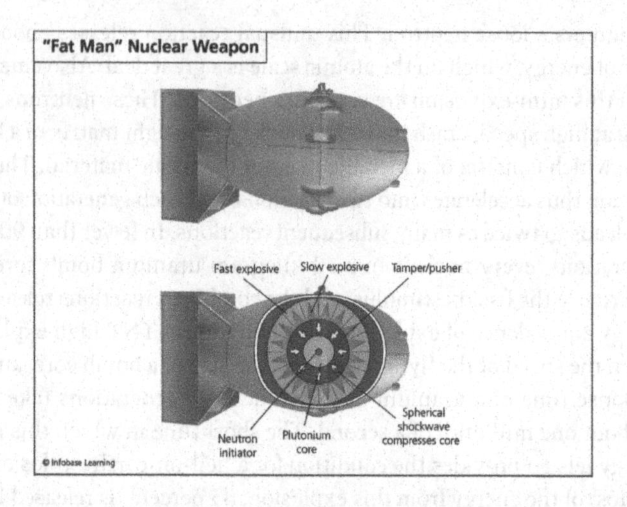

Fat Man was completely different from Little Boy in the method it used to create a hypercritical mass and the fissile material used. A ball of plutonium metal the size of a navel orange was momentarily compressed to the size of a table tennis ball using a powerful explosion turned inward. Although the high explosives surrounding the plutonium ball exploded outward like an ordinary bomb, the inward force of the same explosion was carefully directed into a spherical shock wave. The inter-nuclear distances in the plutonium were shortened by the shock wave, and the resulting hypercritical mass fissioned explosively.

The first thing hit by this airwave was the ground directly underneath the bomb, or "ground zero." This was a hard thump, and it resulted in an earthquake-like shock energy traveling outward through the ground. The total energy from the detonation was thus distributed as 50 percent blast and shock, 35 percent thermal radiation, 10 percent residual nuclear radiation, and 5 percent initial nuclear radiation. The scientists had not been wrong in predicting small damage due to nuclear radiation, but they had been way off in considering the damage done directly and indirectly by the intense thermal energy. The burns that injured many survivors of the A-bombs were not caused by gamma or beta rays, but by light. Simply being caught standing behind a light-shield when the bomb detonated could be life-saving, providing you weren't struck down by the shield as it was blown away seconds later in the air blast. The temperature at the center of the explosion was far outside human experience, probably millions of degrees, approaching the conditions in the center of the sun, and the air pressure produced was on the order of millions of pounds per square inch. Everything flammable within 12 miles caught fire. Some people were vaporized in the fireball, tens of thousands were crushed in the air blast, and tens of thousands more were severely burned by the flash of light. The death-toll would eventually reach about 83,000 people, as some would die decades later from radiation-induced cancer.

The heat and initial nuclear radiation portions of the event were over in about 60 seconds, but the bomb effects continued to develop for 6.3 minutes. The rapidly expanding fireball created a large vacuum in midair, and as the heat dissipated, air from the surrounding territory started to be sucked in. The blast thus blew air both ways: first outward, a pause, then inward, back toward ground zero. This effect is called the "afterwind." Meanwhile, the residual heated air rose in a strong updraft, like a hot-air balloon. Solid material on the ground, now pounded to dust, was drawn up into the rising column, making a dirt-cloud.

In thirty seconds, the cloud reached a height of three miles. When the ever-rising cloud reached an altitude where its density matched that of the surrounding air, at the base of the stratosphere, the cloud started to spread out horizontally. The sight of this feature became an icon, a dreaded emblem of the atomic age—the mushroom cloud.

On August 9, 1945, the Strike Centerboard operation, carrying the Fat Man plutonium implosion device in a B-29 named *Bock's Car*, dropped the second weapon on Nagasaki, and World War II was over except for the shouting.[32]

To develop these science-fiction-level devices into things that could fall from an airplane required a crash program of unprecedented speed and complexity. Not only was the nuclear reactor invented, prototyped, powered up, and operated for three months, but a huge reservation was built in Washington State so that several reactors could be run 24 hours a day at high power, experimental reactors were built and operated in

32 Atomic bomb trivia: Stenciled in black on the bright orange nose of the Fat Man bomb were two things: a profile of a fat man with a capital F at his back, and the cryptic inscription "JANCFU." It meant "joint army navy and civilian fuck-up." The assembly team had been unable to use the heat-tempered armor plates for the bomb's airframe. The plates had been warped in the process of hardening them, and in desperation they resorted to using the mild steel plates left over from the pumpkin series of practice bombs. Once they got it all together, they realized that one of the safety plug sockets had been wired wrong. With no time to take the thing all apart and re-wire the socket, they modified the safety plug to match the error, so this bomb was not completely built-to-prints.

Tennessee and Illinois, massive plutonium and uranium purification plants were built and run, and risky physics experiments were conducted in New Mexico, all without a single fatal accident or even a radiation injury. Thousands of people worked on this project, some in hazardous conditions and most without a clue as to what they were building. The effort was constantly plunging ahead into the unknown, and the potential for disaster was always close; but due to heightened vigilance and a touch of luck, nobody got hurt. There are no atomic accidents in the Manhattan Project on which to report, right up until the last bomb was dropped. There were, however, some close calls that could foretell later problems.

About 25 miles west of Knoxville, Tennessee, was a sparsely populated 60,000 acres of land near the Blackoak Ridge. Blackoak runs north-south and connects two bends of the Clinch River, and it is part of a sequence of five ridge/valleys on the southeast side of the Appalachian Mountain Range. The Cherokees claimed it as a hunting ground, but by 1800 the Treaty of Holston had ceded it to the United States and several farming communities took root in the area.

In 1902 the local mystic, John Hendrix, 37 years old and thought by some to be not right in the head, was enjoying a typical day by lying in the woods on the ground clutter and gazing up at the sky through the trees. His attention was grabbed by a loud voice, telling him to remain there asleep for 40 nights so that he could be shown visions of what was in store for the surrounding acreage. Being given an account of the future by an external source, he repeated this information many times to anyone who would listen. His predictions were positively eerie:

> And I tell you, Bear Creek Valley someday will be filled with great buildings and factories, and they will help toward winning the greatest war that ever will be. And there will be a city on Black Oak Ridge and the center of authority will be on a spot middle-way between Sevier Tadlock's farm and Joe Pryatt's place. A railroad spur will branch off the main L&N line, run down toward Robertsville and then branch off and turn toward Scarborough. Big engines will dig big ditches, and thousands of people will be running to and fro. They will be building things, and there will be a great noise and confusion and the Earth will shake. I've seen it. It's coming.

Hendrix went on to inhabit a mental institution, and in October 1942 Brigadier General Leslie Groves, head of the Manhattan Engineer District and assigned the task of developing an atomic bomb, chose a spot between the Tadlock and Pryatt farms in east Tennessee as his headquarters.[33] It was remote, cut off from the world, and yet blessed with a great deal of surplus electrical power. The Tennessee Valley Authority, set up by President Roosevelt as a make-work project in the throes of the Great Depression, had gotten a little too enthusiastic and peppered all of east Tennessee with hydroelectric plants. Groves would quickly put them to good use.

Nobody among the Axis Powers that were trying to take over the world had ever heard of the place, and it wasn't even on a map. It was perfect for top-secret work. In a couple of years, it would be known as the Manhattan District HQ, the Clinton Engineering Works, or simply as Oak Ridge, and its population would explode into 75,000 people. Tracks were laid for a rail spur off the L&N, right where Hendricks had said, and in Bear Creek Valley was erected an enormous industrial complex of seven buildings comprising the Y-12 site. The largest building in the world, the half-mile-long K-25 gaseous diffusion plant, was built at a bend in Poplar Creek, about 10 miles southwest of Y-12.[34] Construction began at a furious pace, making an instant city. Housing for workers, thoughtfully made of asbestos

33 The first written account of the John Hendrix story I can find is in Robinson, George O., *The Oak Ridge Story; the saga of a people who share in history*. Kingsport, TN: Southern Publishers, 1950, published some eight years after the supposed prophecy came true. There is some question as to when Hendrix died. This book says 1903, and others say 1915, but this volume includes photos of his once home and his gravestone. He did what the Voice told him to do and found a good place to sleep in the woods, fitfully, for 40 nights. During many of those nights, it rained on him.

34 K-25, using the gaseous diffusion process, was used to "enrich" uranium for bombs, research reactors, submarines, and power plants for the next 40 years. The other two methods for uranium enrichment, the thermo-columns at site S-50 and the electromagnetic calutrons at site Y-12, were torn down quickly after the war ended. Two large diffusion plants were built in Paducah, Kentucky, and Portsmouth, Ohio, to increase production during the Cold War, and these were copies of the Oak Ridge facilities. However, K-25 did not add to the highly enriched uranium used in the Little Boy bomb dropped on Hiroshima. The diffusion process was slow, and the uranium fed in at the mouth of the process did not have time to reach the end-product stage by the time the war ended. Using the highly inefficient Y-12 process, by August 1945 we had just enough uranium-235 for exactly one bomb.

to prevent fires, was a priority. Eventually the city had ten schools, seven theaters, 17 restaurants, 13 supermarkets, a library, a symphony orchestra, churches, and its own Fuller Brush salesman.

This bustling metropolis was built from scratch for exactly one purpose: to take mined uranium, which was nearly all worthless uranium-238, and purify it down to the rare and precious ingredient, uranium-235. The atomic weight of this isotope, 235, was an odd number, and that made its heavily overloaded nucleus touchy and likely to explode if a random neutron were to blunder into it. Just any neutron would do, but it was particularly sensitive to slow neutrons, beaten down to move no faster than any common molecule at room temperature. At Y-12, K-25, and S-50 various concentrations and chemical forms of uranium-235 were stored, moved, stacked up, bottled, boxed, and formed into piles. Only the few top administrators and some of the on-site scientists knew what the stuff was for and had a vague sense of the ultimate danger of working with it. There were 12,000 workers in the K-25 building alone, and none of them was made aware of exactly what they were doing.

For the needs of military security, there was nothing better than absolute ignorance. It was impossible for a worker to spill the beans to an Axis spy, even on purpose, and this massive, continent-wide industrial effort to build atomic bombs remained unknown to the enemy powers. However, there were dangers in this enterprise that had never before visited the human race. If one happened to stack up enough of this weird material in one place, it would start to generate heat, and this energy release would increase exponentially until the stack lost its initial configuration. The conditions under which this disaster could happen were varied across multiple dimensions. The "critical mass" condition depended on the purity of the uranium-235 in the material. Uranium fresh out the ground had only 0.73 percent of the active isotope in it, and an infinite stack of it would not approach the energy production threshold. Start increasing the percentage to, say, 3.0 percent, and the probability changed. If the uranium were dissolved in ordinary water, as it often was in the stages of processing it, then the hydrogen in the water would slow down the trigger particles, the free-range neutrons, using collision dynamics. Just like a high-speed neutron hitting a hydrogen atom in water, if you crash your car into one parked, your car stops cold and the one you hit bounces excitedly into whatever is in front of it. In the same way, a high-speed neutron crashing into a slow-moving hydrogen nucleus, which is of similar mass, will kill the speed of the incoming particle. Having uranium dissolved in water, even if it's only slightly enriched, makes a runaway fission situation quite possible.[35]

Another factor is the shape of the stack. The less surface area your stack has per volume of the stack, the better is the probability of causing an energy-release incident. The worst you can do is to stack bottles of enriched uranium oxide dissolved in water, which is a curious green color, in a rounded mound on the floor. In that configuration the surface area is minimized, so fewer neutrons, which bounce around in completely random directions, are likely to escape the stack without causing a fission. Next worse is a neat cube. The best way to stack it is in a straight, single-file line. The same number of bottles can either be benign containers of mineral water or a glowing inferno, depending on how you stack it.

Bricks of pure uranium metal are another matter of concern. Power-producing fission is possible using high-speed neutrons as triggers, freshly minted in the fission process, but the probability is lower. In pure metal, neutrons were not slowed down to desirable speed just by running into atoms. Hitting a uranium nucleus was like running your car at high speed into the side of a bank building. You might make the building move, but not by much, and your car bounces off in the opposite direction with most of its initial speed. It takes more enriched uranium mass in pure metal form to make it go nuclear than in a water solution, but that's not to say it does not happen. Stack up enough enriched uranium metal in a shape that will encourage fission, and you have it melting through the floor.

35 Actually, uranium oxide dissolves in water, and not pure uranium metal. Another possibility is to have natural, out-of-the-ground uranium oxide dissolved in heavy water, or deuterium oxide. Using ordinary water, the plain hydrogen in it can occasionally absorb a neutron, and this is a neutron that misses the opportunity to trigger a fission. Deuterium, which is hydrogen that already has that neutron in the nucleus, doesn't absorb neutrons, and for that reason heavy water encourages fission more than ordinary water. You could stack up bottles of plain uranium in heavy water and cause a meltdown. The dynamics of fission are that sensitive.

Technically, this type of potential accident builds an impromptu nuclear reactor, and not a nuclear weapon. There is no way to stack up pure uranium bricks fast enough for them to explode as a bomb, because an entire explosion takes place in about a microsecond, much faster than anyone can lay down blocks. But such a situation of stacking enriched uranium bricks would still be extremely dangerous, as it would make a basic nuclear reactor without bio-shielding and without even rudimentary controls. It runs wild until the heat generation is sufficiently severe to wreck the stack and make it subcritical by virtue of shape.

The behavior of the water-solution stack and the solid-metal stack are significantly different. In a water stack, the power generation is dependent on neutrons slowed to thermal speed. Not only are the neutrons slowed down, they are separated by distance from uranium nuclei in the dilute water solution. The reactor is spread out over a large volume, the size of a garbage can. The metal reactor has more uranium in it, but it is extremely compact. In the optimum configuration, a sphere, it is the size of a grapefruit, and it is extremely sensitive to its environment. If you have a barely subcritical sphere of uranium-235 sitting on a tabletop, not fissioning or causing any radiation to speak of, then simply walking by it or waving a hand over it will cause it to go supercritical, raising its temperature and spewing out radiation in all directions, increasing exponentially.[36] This happens because your body consists of about 70 percent water. Random neutrons, born of spontaneous fissions and escaping off the surface of the sphere, will hit your hand occasionally and slow down in your water component. Those occasional neutrons are knocked in all directions. A scant few wind up drifting back toward the sphere from which they came. They re-enter the fissile material, and this extremely slight increase in the number of available slow neutrons can set off a chain reaction. The neutron population bursts into high production, and it's off to the races.

---

36 This problem is only hypothetical. During the war, there was never enough uranium-235 existing at Oak Ridge to make a purely metallic, highly enriched critical mass. As soon as a piece of it the size of a silver dollar accumulated, it was sent off to Los Alamos, where every scientist knew of the potential hazard of collecting the metal in one place.

Incredible as it seems, the difference between the subcritical neutron population in a uranium mass, making no fission, and supercritical, making wild, increasing fission, is a very small number of available neutrons out of trillions: all it takes is just one neutron.

A similar possibility of accidentally assembled reactors existed at the Hanford Works, built a year after the Oak Ridge facility out in the desert in the middle of Washington State. It was another instantly derived city, a bit larger than Oak Ridge, having 50,000 people. Its product was plutonium-239, an artificially produced isotope made by subjecting uranium-238 to neutron bombardment. The fissile material was nearly 100 percent pure, and having low enriched material was never a problem. In water bottles or stacked in bricks, it was as problematic as pure uranium-235 and much more plentiful.

At Oak Ridge in 1944, batches of enriched uranium began to accumulate, and a memo arrived at Los Alamos from a plant superintendent, expressing concern about the possible peril of having bottles of uranium-water solution neatly stacked in a corner. Would it be advisable to install a special fire extinguishing system? This memo set off alarms on multiple levels, and J. Robert Oppenheimer, head of the scientific mission to develop the A-bomb, dispatched Emilio Segré to Tennessee to assess the situation.

Segré was a typical worker at Los Alamos, in that he was a brilliant physicist and a recent immigrant from fascist Europe, having been driven away by the enforcement of official anti-Jewish regulations. He would eventually win the Nobel Prize in physics and discover two new elements and the antiproton, but in 1938 he was a refugee stuck in a $300-per-month job as Research Assistant at Ernest Lawrence's Berkeley Radiation Lab in California. When Dr. Lawrence, who believed strongly in fiscal responsibility, figured out that Segré had nowhere else to go, he dropped his salary to $116 per month. The talented Segré felt fortunate to have been grabbed by the U.S. Government to work on the bomb program in New Mexico. As head of the experimental division's radioactivity group, Oppenheimer thought he could spare him for a few days to see what was going on in Tennessee.

Examining the situation at Oak Ridge, Segré found that no workers knew that they were making an explosive, much less that it was a very tricky one, and only a few top officials were aware of the problem of bringing together a critical mass. They had been given the talk, but it had mentioned only the problem of stacking metal bricks, and they had no idea that water diluting the active substance only made it *easier* to produce a runaway reaction. The accumulating stores of wet uranium at Oak Ridge were on the verge of disaster. Oppenheimer responded to Segré's grim report by dispatching his best man, Richard Feynman, immediately to the scene.

Feynman was only 27 years old, the youngest group leader in the mass of heavy thinkers gathered at Los Alamos. Working under the director of the theoretical division, Hans Bethe, he was one of the few natural-born Americans on the T-section payroll. He grew up in Far Rockaway, New York, and earned his physics degrees at MIT and Princeton. He had quickly established a reputation as a quick mind with brilliant insights and an ability to find the problem in any aspect of the complex bomb development. He also gained fame by an apparent ability to crack any combination security lock at the lab. Everyone was impressed, particularly Oppenheimer, who was not disappointed by Feynman's sharp analysis of the problem.

It was even worse than Segré had reported. There were storage drums of different sizes stored in dozens of rooms in many buildings on site. Some held 300 gallons, some 600 gallons, and some an eye-opening 3,000 gallons of uranium oxide dissolved in water, in a range of uranium-235 enrichments from raw, natural uranium to nearly critical concentrations. Some were on brick floors, which was fine, but some were on wooden floors. Wood is an organic compound, and it contains hydrogen, which would moderate the speed of leaking neutrons and reflect them back into a drum, enhancing the conditions for fission. In some cases large drums were segregated into adjoining rooms, but if two drums were backed up against the same wooden wall, the two subcritical nuclear reactors were capable of coupling into one critical assembly, using the wall as a neutron-moderating connection.

Atop all those problems, the very shape of a drum encouraged fission. Drums were made to minimize the amount of metal needed to build a container of a given size, so the volume-to-surface-area ratio was optimized. Feynman examined the floor layouts of the agitators, evaporators, and centrifuges used in the sequential processing of the uranium. From the blueprints of the buildings, he could tell that the architects did not have nuclear physics in mind when they drew the floor plans. The entire industrial complex of the Clinton Works was a disaster under construction. The potential meltdown, in which a nuclear reactor could be unintentionally assembled and run up to power, was given a name: *the criticality accident*.

There was no set rule for how much uranium water could be stored in one location, or how close two drums could be located. There were simply too many variables at work to be able to look in a room and say, "Put one more drum in here, and it will take an aerial photo to see the entire crater."[37] Feynman relished the task of mathematically solving the impossibly complex interactions of bricks of metal near steel drums scattered in random locations in connected rooms, but the problem boiled down into one fact: the workers in this production facility could not be kept unknowing of what was going on. Some raw knowledge was the key to preventing a nuclear disaster. Oppenheimer gave him the go-ahead. The rabid security measures were now working against the project, and this would have to be an exception to the total ignorance policy, or the uranium and plutonium production could self-destruct. Feynman prepared a series of lectures for workers and supervisors, starting with the simple basics of nuclear physics. This action probably saved many lives, but in the next few decades the lesson would have to be learned over and over, continent by continent.

Aside from the usual industrial accidents and hazards of using dangerous chemicals, working at a nuclear facility under war footing was remarkably safe. There were no radiation injuries.[38] However,

---

37 A blatant exaggeration, but you get the point. As will be chronicled in a later chapter, this accident has happened on many occasions. When the contents of the drum go supercritical, the water boils vigorously. Just the boiling action changes the configuration enough to throw the reactor into subcriticality, and the reaction stops. The radiation burst of an uncontrolled approach to criticality in an unshielded assembly is deadly to any nearby organism.

38 Well, maybe a few. On June 6, 1945, the exact critical mass of a uranium-235 bomb core was still unknown, and an experiment using 35.4 kg of 79.2% enriched uranium metal was devised. Blocks of the uranium were built into a pseudosphere surrounded by paraffin. The assembly was placed in a large tank of water. As the tank was being filled, the thing went unexpectedly critical, and there was no way to scram it. Someone lunged for the drain valve, which was 15 feet away, and finally brought it under control. Three scientists received "significant" radiation exposure, but they seemed unscathed.

there was a reactor explosion that destroyed a building. It was not in the Western Hemisphere, and, as would prove the case in many future nuclear accidents, the reactor was nowhere near running on fission. The problem involved water.

Werner Heisenberg, a respected German theoretical physicist, had made a name for himself well before the war started. He was famous for having expanded quantum mechanics with his uncertainty principle and his matrix spin operator, and he won the Nobel Prize in physics in 1932 for "the creation of quantum mechanics," which was a bit overstated. With the German universities cleansed of nuclear talent by Nazi anti-Jew policies in the 1930s, Heisenberg, a Lutheran, was almost all that was left for mounting a nuclear weapon project. The necessary tasks were parsed and funded by the Reich Research Council. Nicholaus Kopermann was in charge of uranium production. Paul Harteck got heavy water production. Walther Bothe drew nuclear constants measurement, and Georg Stetter was given transuranic elements. Heisenberg was assigned the core problem, to prove the validity of the chain-reaction concept and then use the resulting nuclear reactor as a neutron source for further experimentation and data collection. Oddly, a separate uranium-enrichment task was spun off for Manfred von Ardenne, a German television pioneer, funded by the German Post Office.

Truth be known, Heisenberg was a brilliant theorist but not so good as an experimentalist, and his task involved building a nuclear reactor, which was heavy on the experimental side. He was grateful to be assigned Robert Döpel, professor of radiation physics at the University of Leipzig, to assist. It was Döpel and his wife, Klara, who decided that deuterium, the hydrogen isotope in "heavy water," would be the ideal neutron moderator in a reactor using natural uranium. The construction of the first reactor, the L-I *uranmaschine*, was completed in August 1940. It was far subcritical, but it did accomplish neutron multiplication, producing more neutrons than were being injected from an external source, and it indicated that they were moving in the right direction. It would have to be rebuilt, larger, using heavy water, which was a precious material available sealed in 20-milliliter vials.

The Manhattan Project was doing basically the same thing in 1941, with a slightly different approach. Deuterium was indeed a fine neutron-moderating material, but, unlike the Third Reich, the United States had not captured a heavy-water-production plant in Norway. Instead, chemically pure synthetic graphite was used, delivered by the ton from Union Carbide. Enrico Fermi, a refugee nuclear scientist from Italy, headed the project, starting with a small pile of uranium and graphite in the corner of a lab at Columbia University. It was subcritical, but it multiplied the neutrons from a source. They were also going in the right direction, but they were one year behind the Germans.

By June 23, 1942, Heisenberg and Döpel had constructed L-IV, a bigger, more sophisticated version of their reactor in a dedicated laboratory building at the University of Leipzig. A large, circular pool of water was sunk into the middle of the floor in the lab. At the bottom was a frame, made of steel girders bolted together. Held off the bottom by the frame was a hollow sphere, one meter in diameter, of cast aluminum, three quarters submerged in the water. A flange around the circumference of the sphere was holding the upper and lower hemispheres together using 22 bolts. Four lifting lugs were cast into the flange with steel cables attached to a hoist above, and a long chimney emerged from the top, bolted to a flange on the upper hemisphere. On the inside surface of the sphere was a layer of uranium metal held in place by another, smaller flanged aluminum sphere. The inner space was filled with heavy water surrounding a still smaller sphere having another layer of uranium inside. A last aluminum sphere at the very center was filled with heavy water, and the chimney extended down through the center of it. Four neutron counters were arrayed on the top hemisphere.[39]

---

39 This incident is fairly well known, but there is some confusion about the exact configuration of the reactor. The uranium is often described as being in powdered form, which makes no sense. If it were powdered, the uranium metal would have reverted to uranium oxide before the sphere was loaded, and metallic uranium was thought to be essential. It was most likely uranium metal formed into marbles, so that they could be packed between aluminum shells. To machine hollow hemispheres of uranium would have been unnecessarily difficult. An alternate, more plausible configuration of L-IV is in a drawing captured by the Alsos Mission in 1945. It shows flat plates of alternating uranium metal and deuterated paraffin stacked inside the aluminum sphere. With ten plates of uranium (551 kilograms) and ten plates of paraffin, only the top hemisphere was filled. The plan was to load two plates at a time while monitoring the neutron multiplication, stopping when criticality was achieved.

The plan was to lower a fixed neutron source consisting of a mixture of radium and beryllium powders down the chimney to the center of the reactor. The neutrons would be slowed by the heavy water and hit the first hollow sphere of uranium from all interior directions. High-speed neutrons from the fission reactions in the uranium would fly into the second layer of heavy water, slow down, and impinge on the outer layer of uranium, causing a chain reaction and sending a portion of the resulting neutron burst back through the heavy water and into the inner uranium shell. The water immersion in the pool was supposed to keep the assembly from melting when criticality was achieved. They could not have been overly optimistic, as they had no particular plan for what to do if the thing sprang to life as a supercritical reactor, with the heat exponentially increasing. The sphere was sealed up tightly, with a gasket separating the two halves, because the metallic uranium would react chemically with any water leaking in, jerking the oxygen right out of the H_2O.

That morning, Döpel had noticed something odd about L-IV. Bubbles were coming out of the sphere and bursting on the surface of the water in the pool. They had been experimenting with it since June 3, and it had seemed complacent, even dull and unresponsive, but now it looked angry. As he stood and tried to figure out what was wrong, the bubbles stopped. Nothing to be concerned about. Döpel struck a match over the last bubble as it surfaced, and it popped with a bang. Yep. The gas leaking out was hydrogen, or it could be deuterium. Somehow, water was getting to the uranium.

After lunch, Döpel and Paschen, the lab mechanic, winched the thing out of the water and started to loosen the bolts. A gasket must have failed and it would require replacement. As soon as the seal broke, the sphere made a sudden hiss. A vacuum had developed inside, and air was rushing in. They stood frozen for a second. It was quiet, then suddenly flames started shooting out around the flange, followed by molten uranium, scattering all over the lab. Döpel doused it with water as Paschen tried to re-tighten the bolts, and the flames seemed to subside.

Heisenberg was summoned. He did not know exactly how to handle this situation. A nuclear reactor had never caught fire before. He ordered the ball to be lowered into the pool. At least that would cut off the oxygen and keep it cool. Nothing burns, he thought, under water. He left it to Döpel and went to the adjacent building to hold forth at his weekly nuclear physics seminar.

At about 6:00 Döpel barged in, saying "You must come at once!" Heisenberg spun around to upbraid him for interrupting, but he saw a look of cold terror in his face. "You've got to come look at the thing!"

They hastened to the lab, and Döpel pointed down into the central pool. Steam was rising from the sphere. It looked as if it were . . . expanding? It gave a little shudder. Both scientists spun in unison and lunged for the door. The L-IV exploded with a roar, sending flaming uranium against the 20-foot ceiling and setting the building on fire. For two days it burned, with no amount of effort able to extinguish the burning remnants of the reactor, and it finally settled down into a gurgling swamp of radioactive debris.

Although the project was supposed to be a secret, the explosion was not, and Heisenberg had to endure winks and hearty congratulations from associates on his success with his atomic bomb. By the time the story had leaked across the ocean to the United States, it had grown considerably. An entire room full of German scientists had perished in a nuclear weapon test. For Heisenberg, it was no success at all. The metallic uranium in their pathetically subcritical assembly had simply caught fire. It was a setback. By December, the Americans had caught up with the Germans and passed them with a self-sustaining chain reaction. Security was so tight, the Germans did not even know they had been beaten. At the end of the war, their only accomplishment had been the world's first nuclear reactor accident, caused by water leaking past an inadequate gasket.

The war ended with Emperor Hirohito's "Jewel Voice" recorded radio announcement to the people of Japan on August 15, six days after the final atomic bombing run. It was over, and to the Manhattan Project the shock was deep. After this intense effort and all the frantic war research and industrial production in the United States, to have all activity stop suddenly was not exactly possible. There would have to be a short wind-down, before nuclear weapon development would rebound. The design of the plutonium implosion bomb was under constant modification and improvement, even as the Fat Man was dropping on Nagasaki, and reasons would be found to continue the work.

The model Y-1561 bomb, while successful, left much to be desired, and work was underway to increase its efficiency, as if a 20-kiloton blast was not big enough. The nuclear explosion occurred when a barely subcritical ball of plutonium metal, 3.62 inches in diameter, was crushed down to the size of a large marble by an explosive shock wave, turned inward. The nuclei of the plutonium were forced closer together than normal, and the chances of being hit with a flying neutron and fissioning were increased accordingly. The subcritical sphere became supercritical, at least three times over, and the uncontrolled chain reaction grew with devastating speed.

The little ball of plutonium was plated with 5.0 mils of nickel to prevent it from spontaneously catching fire as it was exposed to air. Around the fissile ball was assembled a "tamper" shell, 8.75 inches in diameter. Its purpose was to keep the plutonium ball together as long as possible as it was exploding to ensure that a maximum number of fissions could occur. With the fission rate doubling 90 times in a microsecond, the once-solid ball would become a superheated plasma, trying to expand from an inch in diameter to hundreds of feet in diameter as quickly as possible. Reasoning that even an atomic blast could not accelerate matter from rest instantly, the scientists decided to make the tamper shell of uranium metal, depleted of its fissile isotope. Aside from plutonium, it was the heaviest element available, and therefore it would provide the most inertial resistance to sudden expansion.

It was a touchy design. The plutonium component was built so close to criticality, the material that would immediately surround it had to be chosen carefully.[40] There was reason to believe that substituting tungsten carbide (WC) for the uranium in the tamper would up the yield by a kiloton. There was one question that could not be answered

40 Georgia Tech physics professor Nesbit Kendrick, my faithful source of atomic bomb stories, told me about the problem of inserting the plutonium "peach-pit" into the center of an Mk-4. You had to hold it connected to the end of a T-wrench and carefully insert it through a hole that was opened up in the high explosives (HE) that surrounded it. The explosives were complex organic compounds containing hydrogen, which is a very effective neutron moderator, and as the pit passed through the hole, slow neutrons would bounce back into it and it would go supercritical on a slow period! One did not linger in the midst of the HE, lest the fission neutrons boiling off the plutonium sphere burn the skin off the hand gripping the wrench handle.

by theory: Exactly how much WC could surround the plutonium ball before the carbon atoms would reflect enough neutrons back into it to make it cross the line and go supercritical?

Improbable as it now seems, the answer to that question was to experiment standing over a plutonium bomb core with some bricks made of WC, stacking them up until the thing was on the verge of a runaway chain reaction. A plutonium ball on a workbench was not a plutonium ball crushed by an explosive shock wave, and there was no way to make it go off as a bomb, but it could be the world's smallest, most simple nuclear fission reactor. Change its situation slightly, like by reflecting some stray, spontaneous neutrons back into it, and it could "go critical," a condition in which it was producing exactly as many neutrons by fission as were being lost by leakage or absorption. "Supercriticality" could be slight, in which the energy-release rate increases slowly, or it could be great, depending on the degree with which it was perturbed. "Tickling the dragon" involved the skill of making an eight-story house of cards. You had to be focused, alert, and stone sober.

Haroutune "Harry" Krikor Daghlian, Jr., was born in Waterbury, Connecticut, on May 4, 1921, to Haroutune and Margaret Daghlian, immigrated from Armenia. He earned a Bachelor of Science in physics at Purdue University. In the autumn of 1943, recruiters from the Manhattan Project found him working on the cyclotron at Purdue, trying to produce 10-MeV deuterons, and by 1944 he was working in Otto Frisch's Critical Assembly Group at the Omega Site, Technical Area 2, at Los Alamos.

The Omega Site was stuck in a canyon, out of shrapnel range of the administrative and theoretical offices, so that only the technical class would be wiped out if an experiment were to go suddenly awry. By the end of the war, Daghlian had tickled the dragon so many times, he was at that very dangerous point where experience and confidence were so extreme, there was no need to be careful. Unlike the Oak Ridge workers, as a nuclear physicist he did not have ignorance as an excuse for not being terrified of his tasks.

All day on August 21, 1945, six days after Japan gave up, Daghlian worked on the WC loading for a 6.2-kilogram Mk-2 bomb core, standing over a low steel assembly table in the 49 Room at the Omega

Site.[41] There were workbenches on all four walls of the room, a desk for the SED security guard on the east wall exactly 12 feet away from the assembly table, and in the southeast corner was a special vault, made to store bomb cores isolated from each other and from any radiation source.[42] A stack of WC bricks of different sizes and shapes was piled on a rolling dolly to his left, and he would try various configurations against the ball of plutonium, always aware of the radiation counters ticking in the racks to his right. He had two fission chambers running numerical counters, each sounding a click in a loudspeaker every time a neutron hit, and a BF3 chamber indicating the neutron count rate visually with a strip-recording milliammeter.[43] An experienced lab technician could tell easily if a criticality was imminent just by hearing the ticking sound become frantic, or at least mildly excited. In a specially machined brick, a 5-millicurie Ra-Be fixed neutron source sat against the ball, providing rogue neutrons to be multiplied by the plutonium and indicate its approach to criticality.

Daghlian was using rectangular WC bricks, 2.125 by 2.125 by 4.250 inches, and he found that the ball went critical when surrounded by five layers of bricks arranged as a cube with two bricks on top. He tried stacking the bricks differently, experimenting to find the minimum amount of WC that would cause the plutonium to take off. He logged out of the room at the end of the day after returning the sphere to the vault, scheduling another experiment with the bricks for the next day.

After dinner, he wandered over to the evening science lecture at theater no. 2, but something was bothering him about his last stack of bricks. He could not get it off his mind, and when the lecture broke

---

41 The information in this account comes from the classified "REPORT OF ACCIDENT OF AUGUST 21, 1945 AT OMEGA SITE." The report was unclassified on August 28, 1979, and publicly released on January 28, 1986.

42 SED means Special Engineer Detachment.

43 The "fission chamber" is an ion-chamber tube lined with uranium-235. If a neutron wanders into it, a fission will likely occur in the uranium, and this ionizing event causes the gas in the tube to conduct electricity, which is countable as one neutron interception. The BF3 chamber is not quite as sensitive. The ion chamber is filled with boron trifluoride gas, and if a slow neutron is captured by the boron, the resulting radioactive decay of the activated boron will also ionize the gas and register as an encounter with a neutron.

up at 9:10, he went back to Room 49 in the canyon, arriving at 9:30. It was against regulations to perform a criticality experiment without an assistant, and it was certainly forbidden to do it after hours, but there was something he had to try or he could not sleep that night. Lights were on in the building.

Daghlian walked into the room, stood over the assembly bench for a second, then crossed the room to the plutonium vault to recover the ball. Sitting at the desk was SED guard Private Robert J. Hemmerly, reading a newspaper. There had to be a guard on duty 24 hours a day in the room where bomb cores were. Daghlian looked nervous and apprehensive for some reason. Hemmerly said "Hi, Harry," and returned to reading.

By 9:55 Daghlian had built his five-layer brick house around the bomb core, holding the brick that would seal the top in his left hand. Slowly he lowered it toward the pile, and the neutron counters started chattering madly. He had passed the critical line, barely, but the sudden radiation was startling. His left arm jerked upward to get the brick away from the pile. It slipped out of his hand.

Hemmerly was still sitting with his back to the assembly table, but he heard the rash of counts over the loudspeaker and then the clunk and the WC brick fell across the top of the plutonium ball, centered perfectly. The neutron detectors overloaded and the speakers went quiet as the wall in front of him lit up with a blue flash, and he twisted around.

Daghlian had caused a problem, and every instinct told him to immediately erase the problem. With his right hand he knocked the WC brick off the top of the assembly, glowing a pretty blue, and he noticed the tingling sensation of direct neuron excitation. He then stood there, arms limp by his sides, coming to grips with what had just happened.[44] He decided to dismantle the pile of bricks, and he calmly told Hemmerly what had occurred. Joan Hinton, a graduate student, happened to have just arrived at the Omega Site, and she drove the stunned scientist to the Los Alamos hospital as Hemmerly alerted

44 A characteristic that all criticality accidents seem to have in common is the blue flash. It is caused by the sudden blast of radiation from the uncontrolled chain reaction ionizing nitrogen gas in the air and causing it to glow a characteristic color. In this first case, the glow was about two inches deep around the box made of WC bricks.

Sgt. Starmer. Starmer was in the Omega Site office, which was separated from the 49 Room by a five-foot-thick shielding wall.

Daghlian's right hand had endured a high dose of x-rays, gamma rays, and high-speed neutrons. There was no direct way to record the dose to his palm, used to brush aside the WC brick, but it was probably 20,000 to 40,000 rem. His left hand took a hit of 5,000 to 15,000 rem as the brick hit the pile. His body absorbed about 590 rem.[45]

The first symptom of Daghlian's radiation exposure observed at the hospital was the swelling and numbness in his right hand. Unrelenting nausea started 90 minutes after the accident, and continued for two days with a break only for prolonged hiccups. After 36 hours, a small blister appeared on his ring finger. Shortly after, the circulatory system in his hand collapsed and it turned blue, beginning with the nail beds. The blistering spread to the palm and then the back of the hand, and the hand essentially died. He was given opiates and ice packs in an attempt to control the pain.

After two days, he was feeling better and he was hungry. His arms, face, and body were turning red and skin was starting to come off, but he ate well and seemed to be improving. On the tenth day, the severe nausea returned, and he was no longer able to keep anything down. He started losing weight. He was given a blood transfusion, large doses of penicillin, vitamin B_1, and quinidine sulfate. No treatment was reversing the condition. After 25 days, he slipped into a coma. He died at 4:30 P.M. on Saturday, September 15, 1945. His obituary in the *New York Times* said that he had perished from chemical burns. Harry Daghlian was the first person to die accidentally of acute radiation poisoning. It was history's first mini-disaster involving nuclear fission out of control. The bomb core, not in its assigned role, had inadvertently become an unshielded nuclear reactor, suddenly achieving supercriticality and with no automatic shutdown system in place. There was nothing that could have been done medically to save his life.

The other victim, Private Hemmerly, had been exposed to the same radiation burst, but from a distance of 12 feet. He was confined to a

45 Roentgen Equivalent Man. The rem is an obsolete measure of radiation dose, taking into account the unique sensitivities of the average human body. Doses in the range of 1,000 rem are usually fatal. The current measure of dose is the sievert. To convert rem to sieverts, divide the value in rems by 100.

bed for two days, with his only complaint that he felt tired. His blood samples showed increased leukocytes, but this condition was only temporary, and he was released after three days and returned to active duty. He went on to father two more children, and he died at the age of 62, showing no medical evidence that he had ever been exposed to a naked nuclear reactor. The difference between him and Daghlian was apparently the distance to the radiation source. In informed retrospect, if Daghlian had recoiled, jumping back from the assembly table when he dropped the brick instead of bending over to brush it off the pile, he would have survived. If he had been standing on the south side of the table instead of the north side, as was the case with Heisenberg and Döpel, he could have been out the door in three desperate bounds, with Hemmerly right behind him.

But, what would have happened to the supercritical plutonium ball? After a few seconds of power increase, the immediate temperature rise would have shut it down as the sphere expanded in the heat. The supercritical condition in a metal reactor of this size is so sensitive to perturbation, just a slight increase in the distances among plutonium nuclei is sufficient to stop the fissions. It would have then sat there with the heat diffusing slowly to the surface of the ball and radiating out into the room. As soon as it had reached room temperature, it would again become supercritical, and the cycle would start again, hosing the room once more with radiation.[46] After a few cycles, the movable WC bricks would be nudged to the sides by the expanding ball enough to no longer encourage another supercritical excursion, and the assembly would be stable but dangerous. A technician would enter the room behind a lead shield and dismantle the pile using a 20-foot metal pole, and Daghlian would have never been allowed again in the 49 Room.

This shocking event should have been a strong lesson learned, with measures implemented immediately to prevent its further occurrence.

46 The "supercritical" condition is necessary for this to be a disaster. Any reactor can be exactly critical, conducting self-sustaining fission, while generating no detectable power. The only way to bring an operating reactor up to a useful power-level is to temporarily add some reactivity, making it slightly supercritical. When the power level is achieved, the reactor is leveled off at exactly critical. To lower the power level, the reactor is rendered slightly subcritical on a temporary basis.

But, it wasn't. Louis Alexander Slotin, an expert at assembling bomb-core experiments, was one of three investigators who submitted the accident report on August 26, 1945, five days after the Daghlian incident.

Slotin was born in 1910 to Jewish refugees who had fled the pogroms of Russia to make a life in Manitoba, Canada. He grew up on the north end of Winnipeg in a tight cluster of Eastern European immigrants, and he proved to be academically exceptional. He entered the University of Manitoba at age 16, earning a Bachelor of Science degree in 1932 and a Master of Science a year later, both in geology. Further study at King's College London led to a Ph.D. in chemistry in 1936 and a wealth of dubious exploits. Later in life he would claim to have test-flown the first jet plane developed in England, despite lacking a pilot's license. He captivated those listening with tales of having volunteered for service in the Spanish Civil War just for the thrill of it, although there was some confusion as to which side he was on.[47] At King's he won the college's amateur bantamweight boxing championship. His first job out of school was testing rechargeable batteries for the Great Southern Railways in Dublin, Ireland.

Back home in 1937, Slotin was turned down for a position with Canada's National Research Council. He wangled a job as a research associate at the University of Chicago, where he worked on a cyclotron under construction in the Old Power Plant building. The pay was pitiful, but with help from his father to buy food he stayed on for a few years, using the new particle accelerator to make carbon isotopes for biological studies. It was claimed that he was present at Enrico "The Pope" Fermi's CP-1 reactor startup in 1942, but nobody remembered him being there. He was caught in the sweep of the Manhattan Project draft and wound up at the Clinton Works in Oak Ridge.

At Oak Ridge he gained a reputation as someone who would step over the safety line and take chances that should not be taken. One Friday afternoon, young Louis wanted the X-10 graphite reactor shut down so that he could make adjustments to his experiment at the bottom of the fuel pool. It was a tank of water under the floor at the back of the reactor where hot, very radioactive fuel was dumped to cool off.

---

47 Slotin's brother, Sam, in a later interview revealed that Louis had gone on a short walking tour in northern Spain and had participated in the revolution only in spirit.

The head of health physics, Karl Z. Morgan, nixed the idea. The pile could not possibly be shut down. It was being used as a pilot plant for the plutonium production reactors being built at Hanford, and the thing had to run 24/7, balls to the wall. Every few days, new fuel was pushed into the front face of the reactor, and the burned-up fuel would fall into the pool. The bottom had to be heavily contaminated by now.

When Morgan returned to work the following Monday, he discovered that Slotin had stripped down to his shorts, dived into the pool, and made his adjustments. Morgan was appalled. Slotin was reassigned to Los Alamos, where daring was better appreciated, in December 1944. He quickly earned respect for a natural ability to assemble the complicated implosion bomb without excessive worrying and hand-wringing. He expertly put together the bomb core for the Trinity test in New Mexico in July 1945. His unofficial title was Chief Armorer of the United States. The only reason he was absent on Tinian Island when the Fat Man was assembled was his lack of U.S. citizenship.

Slotin was shocked and saddened when Daghlian, his assistant and fellow dragon tickler, died in the criticality accident, and he spent days at his bedside in the hospital. This tragedy, however, did not affect his supreme confidence. He brushed aside advice that he should automate the critical assembly experiments, even when the very wise Fermi warned him that he wouldn't last a year if he kept doing that experiment. The central problem pointed out by Daghlian's death was approaching criticality from the top, where gravity could accidentally complete the operation. It would make more sense to assemble from the bottom. If anything was dropped, it would fall away from the plutonium sphere instead of into it. Slotin discounted the advice as an unnecessary complication.

The next atomic bomb explosion was to be a test in the middle of the Pacific Ocean at Bikini Atoll, designed to demonstrate that a navy flotilla could survive a nuclear attack and proving that the water-borne armaments had not been made obsolete by this recent innovation. The date for the Able shot in Operation Crossroads was set for July 1, 1946. The implosion bomb was under constant improvement, and the WC tamper had been replaced by a beryllium tamper. It was machined into a pair of concentric shells, 9.0 and 13.0 inches in diameter, and split into hemispheres to fit around the plutonium bomb core. The beryllium would act as a secondary neutron source during the explosion, hopefully increasing the number of fissions as the core destructed. A hole was bored into the top hemisphere so that the "initiator" modulated neutron source could be inserted into the core without dismantling the entire bomb. The criticality experiment for this revised tamper design was moved to a new building in Pajarito Canyon, and it would be performed using the same ball of delta-phase plutonium that had killed Daghlian.

On May 21, 1946, Slotin was training his replacement, Alvin C. Graves, who had a Ph.D. in physics from the University of Chicago. Slotin had grown weary of the bomb work, and was planning to bail out and go work in biochemistry back east. This would be his last bomb core.

It was about 3:15 in the afternoon. The experiment was set up on the low assembly bench, with the bomb set up near the edge and the 5 millicurie Ra-Be neutron source placed a few inches in front of it.[48] Radiation detectors of several types were set up on and near the bench and warmed up, giving continuous recordings and audible clicks. There were seven men in the room, which was unusual for a criticality experiment, but this one was informal and was not scheduled. Two men had been working on initiator tests on a bench on the east side of the room, and the sensitivity of the required radiation counts had delayed them several times as the background counts were perturbed by tests outside the building. Slotin's demo for Graves would also interrupt them, but it would be interesting to see him do the now famously dangerous criticality test. The SED guard was present, as were two other scientists, and they were all fascinated by watching the skilled Armorer at work. Three were standing directly in front of the bench. The room was brightly lit with overhead fluorescents and low sunlight through the windows.

The formal test called for wooden spacers to hold the top nine-inch tamper hemisphere off the bottom hemisphere, and it was to

48 Declassified documents differ on the type of neutron source used. One says it was a radium-beryllium source, and one says it was plutonium-beryllium. Usually not mentioned was an extremely active 30 curie polonium-beryllium source, three months old, located about seven feet east-northeast of the assembly table. These seem minor points, but one strives for as much accuracy as possible.

be gradually lowered onto the core by changing out the spacers, one at a time, with smaller ones until the assembly was very close to criticality. Both the 13-inch and the 9-inch tamper hemispheres were installed only on the bottom of the assembly. The tamper pieces would then be sent back to the shop to have some metal removed, and the test would be done over until the assembled bomb was stable and very near the critical condition.

Slotin discarded the spacers and used a big-bladed screwdriver instead. With the blade under the lip of the tamper, he could lever it up and down, impressing Graves and his audience by making the neutron count rate zoom in and out on the loudspeaker. Graves was close behind him, looking over Slotin's right shoulder. Slotin's left thumb was through the access hole on top of the tamper, with his fingers on the curved side, adding to the downward tension of the movable tamper-half. He pulled the screwdriver handle up, increasing the angle between the bottom hemisphere and the straight blade, with the top hemisphere riding up, increasing the gap and making the neutron count rate fall precipitously. At an angle of 45 degrees, the screwdriver arrangement became precarious, as the side thrust, pushing the screwdriver outward from the gap, equaled the downward thrust holding the screwdriver down. Ever upward Slotin angled the tool. Beyond 45 degrees, the outward thrust overcame the downward thrust, and the screwdriver suddenly escaped the gap.

Bang.

The top tamper fell squarely on the bomb assembly, and prompt criticality was achieved instantly.[49] The blue flash lit up the entire room, as the neutron counters, ticking merrily along, suddenly jammed and went quiet. Slotin, on pure instinct, jerked the tamper off the assembly and dropped it on the floor. He could feel the tingling in his left hand and he could taste the radiation on his tongue. It had happened again, and there was no ignorance at work here. Familiarity to the point of nonchalance had just claimed another victim.

---

49 "Prompt criticality" is an important term. It means that a state of at least break-even chain reaction was reached using only promptly available neutrons. There was no need to wait for the delayed fission neutrons, which could take seconds, to get a load sufficient to declare the assembly critical. The additional reactivity needed to go from delayed to prompt criticality is exactly one dollar. The practice of expressing reactivity in dollars and cents, still in use today, was coined by Dr. Louis A. Slotin.

Slotin had a body dose of 2,100 rem of mixed radiation, or twice the dose of guaranteed lethality, and he died the same way Daghlian had, only faster, nine days later. The same radiation pulse hit Alvin Graves, standing an inch from Slotin, but he was partly shielded by the Armorer's body. He stayed in the hospital a few days and was released. The other men in the room showed minimal effects from the incident.

The Crossroads tests went on as planned, with the Able shot using the bomb core that had killed two scientists. The bomb, dropped from a B-29, was affectionately named *Gilda*, and had a picture of Rita Hayworth painted on the side. The yield was 23 kilotons, or 3 kilotons more than the device dropped on Nagasaki a year earlier with a uranium tamper. The target flotilla consisted of 95 vessels of all types, from a captured Japanese battleship to a floating drydock.[50] All were sunk, lost, damaged beyond repair, or made dangerously radioactive except one, the U.S. submarine *Dentuna*, which was refurbished and returned briefly to naval service. The ships were manned by 57 guinea pigs, 109 mice, 146 pigs, 176 goats, and 3,030 white mice. Some lived through the air blast and the radiation pulse, with the most famous survivor being Pig 311, who was found swimming in Bikini lagoon after it stopped raining battleship fragments. He lived out his life at the Smithsonian Zoological Park in Washington, D.C., on a government pension.

There would never be another manual bomb assembly experiment, anywhere or any time. There was still a need to test-assemble the core parts to find unwanted critical conditions, and even an application of a chain-reacting naked plutonium core, to produce the specific radiation spectrum of an atomic bomb explosion. All further work was done at a distance of a quarter of a mile, using remote controls, television cameras, and a quick shutdown capability. The practice of bringing very small, bare metal reactors to the power-production point

---

50 The Japanese battleship was the *Nagato*, carefully placed in the array so that it would be certain to sink. It had been the command ship from which the Pearl Harbor attack was directed back in 1941, and to sink it with an A-bomb was symbolic. It was a well-built ship, and two nuclear devices failed to send it to the bottom. The target point was the battleship *Nevada*, a survivor of Pearl Harbor. It was painted red so that the bombardier could see it (not likely in warfare), but they missed it by 710 yards, putting the bomb on top of a lowly transport ship, the *Gilliam*.

was still an extremely ticklish, sensitive action, but at least nobody could get hurt. That was the intent, at least, but fissionable materials always seemed capable of finding a flaw in the best intentions. All the remote-controlled assemblies in the United States were named "Godiva."[51]

The first Godiva to go out of control was on February 1, 1951. The bomb designers at Los Alamos were working on the Mk-8, a lightweight bomb similar to the one used on Hiroshima, and two sections of highly enriched uranium, the "target" and the "projectile," were suspended by poles in a water tank to see how close they had to be in a moderating medium to reach criticality. The poles ran on motorized tracks, so the distance between the two uranium pieces could be controlled. There were three ways to scram the experiment: the target could be withdrawn using a pneumatic cylinder on its pole, the water could be drained out of the tank, and a cadmium sheet could be dropped between the target and the projectile. When the assembly barely reached criticality, all three scrams were put into action.

The cadmium screen, absorbing neutrons with a vengeance, dropped. The water started draining, and the target started pulling out of the tank as quickly as possible. Just then, the TV camera whited out from steam coming out the top of the tank, and the neutron detectors jammed. If they had had a color camera, they would have seen the water vapor turn blue. To the amazement of the experiment crew, the thing had gone prompt critical.

This was not a life-threatening situation, because the experimenters were far removed from the incident, but still it was another criticality accident, and it once again rammed home the fact that a metal-on-metal reactor was tricky beyond theory. All the minds at Los Alamos had yet to outfox it. The recognition of this flaw led to further design work and improvements. There was not going to be another Slotin incident at a national lab.

51 Not really. At the Los Alamos Lab there were also the Topsy and the Jezebel assemblies. These specialized devices may have been better designed, as no unplanned criticalities were ever reported for them. The French were conducting similar experiments at about the same time, first at Saclay with the PROSERPINE and ALECTO assemblies and then after 1961 at the Valduc Research Centre using the CASTOR and POLLUX machines (rig B and rig D). No accidents were recorded.

Analysis and re-creation of the accident found the problem. As the target was jerked out of the tank, the back-wash from the swirling water had banged the two pieces together, flexing the poles that were holding them and overcoming the reaction-dampening effect of the cadmium sheet. Who would have predicted it? Further criticality mistakes occurred at Los Alamos on April 18, 1952, February 3, 1954, February 12, 1957, and June 17, 1960.[52] Causes were one too many uranium disks added to a stack, a piece of neutron-moderating polyethylene left too close to Godiva, and pieces that were supposed to slide past one another not doing so. There were criticality accidents with Godivas at Oak Ridge on November 10, 1961, Lawrence Livermore on March 26, 1963, White Sands Missile Range on May 28, 1965, and the Aberdeen Proving Ground on September 6, 1968.[53]

It is interesting that all of these Godivas jumping out of control were loaded with metallic uranium or uranium alloy. Not one criticality accident using plutonium was logged, despite the fact that very few bomb designs used uranium cores. The safety measures were laudable, as not one worker was exposed to any radiation, and machine destruction was never something that could not be fixed in three days.[54] Meanwhile, our Soviet counterparts were doing the same thing, experimentally assembling fissile components to find

52 The uranium assembly built in 1952 was named "Jemima." The Mk-8 uranium bomb was already deployed by then. Jemima was probably the W-9 280mm artillery shell, being handed over to the Army that month.

53 Lockheed Nuclear Products received the go-ahead to build a Godiva II on an aluminum railcar for use at the Georgia Nuclear Aircraft Laboratory in 1957. This plutonium "pulse reactor" was supposed to simulate a nuclear weapon explosion below the nuclear-powered bomber as it flew away after a drop. Would the radiation pulse confuse the engine instruments and cause a scram in the aircraft? I find no record of its use nor accident reports.

54 There was also a "mini-Godiva" for portable use by weaponeers. An aluminum Halliburton case was filled top and bottom with paraffin, with cutouts shaped to fit the target-piece for a uranium assembly weapon. A BF3 neutron detector and a fixed neutron source were also embedded in the paraffin, with a cable connection for the counter electronics in another Halliburton. The target was made of little uranium discs that would screw together into a cylinder. In the field it could be shimmed to the proper level of activity by adding or subtracting discs. The specialist would lay the cylinder in the paraffin nest, close the case, and push a button on the neutron counter. After counting neutron hits for a minute, the activity level of the target was evaluated using the results.

that exactly subcritical configuration that would be stable yet triggerable in a bomb. Their experiences would prove to be even more dramatic than ours.

The old town of Sarov in Nizhny Novgorod Oblast in eastern Russia disappeared off all maps in 1946 when it became the home of the All-Union Scientific Research Institute of Experimental Physics, or the Soviet equivalent of Los Alamos, New Mexico. It was renamed Arzamas-75 to confuse, indicating that it was 75 kilometers down the road from the town of Arzamas, which it was not. The name was changed to the more accurate Arzamas-16 when it was realized that Arzamas wasn't on a map either.

In Building B were two "vertical split tables," which were very well designed machines that would conduct critical assemblies by remote control. The top half of a bomb was supported on a steel table, and the bottom half was pushed slowly into position from underneath using a motorized jack-screw. Not being ones to assign cute monikers, Soviet scientists named them FKBN and MSKS.

The building was well designed for safety, but it was not perfect. The FKBN was located in a concrete room with seven-foot-thick walls and a vault door. There was no straight crack around the door where radiation could escape, and the control room was in an adjacent space. The MSKS was in a long chamber with rails on the floor, across the corridor from the FKBN control room. It was shielded with five feet of concrete, and the split table and a separate neutron source trolley could roll back and forth on rails to vary the mutual distance. The MSKS control room was across the corridor from the source trolley, and it was set up better than the FKBN control room, in that it had a back door. Dash out of the FKBN control room, and you were right in front of the door to the MSKS. The rules were that before you could set up an experiment on the MSKS, you had to first prove that it would not run away on the heavily shielded FKBN.

On March 11, 1963, the chief of operations and the head engineer were setting up an approach-to-criticality experiment on the MSKS without bothering to try it first on the FKBN. The assembly was a boosted implosion device with a delta-phase plutonium core, 135mm in diameter. The core was surrounded by a tamper shell, 350mm in diameter, made of lithium deuteride. A fixed neutron source was installed in the middle of the core, instead of the initiator source, throwing about a million neutrons per second into the assembly. The neutron detectors were turned off, so there was no automatic scram system working, and the two supervisors were trying to adjust the lift mechanism on the fully loaded split table, bumping it up and down. It was sticking. On the last try, the two halves clapped shut.

The room lit up with a blue flash. There was no audible count-rate or anything else to indicate that something was wrong, but the two experienced nukes had an excellent idea of what had happened. The ball of plutonium had gone prompt critical, which can happen when two organic neutron-moderating reflectors are kneeling at the thing, jockeying the controls. The two lunged for the door and scrambled down the hall, turning left into the control room. The chief hit the down button for the lift.

They did not die. With doses of 370 and 550 rem, they were just under the lethal limit, although they were definitely injured and spent months in the hospital. One lived another 26 years, and the other was still alive in 1999. They were guilty of gross violations of the MSKS operating procedures, even though one of them had written the manual.

Another interesting mistake was made at Arzamas-16 on June 17, 1999. A highly respected experienced scientist wanted to recreate an experimental assembly he had made back in 1972. He first made the Daghlian error, working alone and not having completed the paperwork, in a new building made to house an improved split table, the FKBN-2M.

This device was shielded by nine feet of concrete on all four sides and the ceiling, with the control room outside the south wall. The lower works would move up and down with a hydraulic lift, but the fixed upper portion of the assembly could be rolled back on rails to give you room to build up the bottom half of your bomb experiment. A sensitive automatic scram would gravity-drop the bottom half quickly if it went supercritical.

The experimenter opened his old logbook, looked up the dimensions of his original assembly, and started stacking components on the lift, with the top half rolled out of the way. It was an unusual bomb, built using an imploding uranium-235 core with a copper tamper. His second error was that in his log he had written the wrong diameter of the reflector. It had been 205mm, but he had written 265mm. He scrounged up the right nested copper bowls to build up his reflector. "Like a *matryoshka* doll," he thought.[55] He built up the bottom reflector using four bowls, then dropped in the uranium ball with a hundred-thousand-neutrons-per-second fixed source inside. He wanted to build up two layers of reflector on top, then roll the top assembly over it, retire to the control room, and slowly assemble his experiment into a critical mass. He dropped the first layer of copper bowl over the core.

Oops. The assembly went prompt supercritical, instantly spiking at over 100 million watts. A blue flash, of course, over-lit the room. Obviously, there was too much reflector under the core. The assembly scrammed, dropping to the floor, but there was nothing to drop away from. All the reactivity was present on the lower half of the assembly, and the top of the machine had been moved out of the way. The experimenter, knowing what he had done, ran out of the room, closed the vault behind him, told two guys in the control room what had happened, and died of severe radiation poisoning two days later. His radiation dose from neutrons alone was several times the lethal level.[56]

The assembly heated up to 865°C, expanded, and settled down to a stable power level of 480 watts, fissioning away for six and a half days until the emergency crew was able to position a vacuum gripper on it and pull off the copper tamper-piece on top.

In 1957 an additional atomic city was built in the Chelyabinsk Oblast in the Urals district of Southern Russia. It was named Chelyabinsk-70, home of the All-Russian Scientific Research Center of Technical Physics, or the VNIITF. After the end of the Cold War it was reassigned the name Snezhinsk, which was easier to pronounce. The extensive research facilities included an FKBN vertical split table, just like the one at Arzamas-16.

---

55 Perhaps. I actually don't know what he was thinking, but the metaphor was too thick not to use it.

56 Technically, he absorbed 4,500 rad of neutron radiation and 350 rad gamma rays. The rad is an obsolete measurement that does not take into account the "Q," or the radiation quality factor and how it affects humans. Assume a Q of one, and he got a body dose of at least 4,850 rem. It was probably closer to 40,000 rem or 400 sieverts, which would drop an elephant.

On April 5, 1968, two very knowledgeable, experienced criticality specialists were experimenting with a special reactor setup on the split table. The goal was to make a tiny reactor to be used in pulse-mode to investigate the effects of the radiation spike from a nuclear weapon detonation. All day they had tried different configurations. At the center of the reactor was a hollow sphere of 90% enriched uranium, or 43.0 kilograms of uranium-235 in a 47.7 kilogram ball, 91.5mm in diameter with a 55mm cavity inside. The reflector halves were natural uranium, making a hollow sphere 200mm in diameter. In the last configuration they tried, the uranium sphere had nothing but air in the center. They had lowered the top reflector half onto the ball using an overhead electrical winch, then retired to the control room, closing the shielding vault door, and slowly drove the lower reflector up toward the assembly until it went critical. Satisfied with the result, they then drove the bottom reflector down until the assembly went subcritical, which was with the southern hemisphere 30mm below the stop.[57]

It was late and after hours. The health physicist and the control room operator had gone home. The two specialists had tickets to the theater, and they were in a hurry to leave, but there was one last thing they wanted to try. Not bothering to turn on the criticality alarm, they used the winch to lift off the top reflector half, removed the top core half, and inserted a polyethylene ball in the empty cavity. For some inexplicable reason, these two experts did not expect a hydrogen-containing moderator at the center of the reactor to change anything, but they just wanted to make sure. One operated the control box for the overhead winch while the other steadied the heavy, 308-kilogram hemisphere as it came down on the core-ball at 100mm per second.

Blue flash! With his hands on the reflector, one felt a shock, as if the thing had been struck with a mallet. Both were hit in the face by the wave of heat as the system's reactivity flew past the prompt critical level. When the power level hit one kilowatt and rising, the

57 If you are amazed by some of the detailed information available concerning Soviet nuclear work, read an example of *glasnost* in the Proceedings of ICNC'95, Vol. 1, pp. 4.44-4.47, "Criticality Measurements at VNIITF Review," V.A. Teryokhin, V.V. Pereshogin, and Yu.A. Sokolov.

scram activated, and the bottom of the assembly fell away, but it was too late for the specialists. Before they left the control room the lower reflector should have been lowered to the bottom stop, but it was kept at the level that was barely subcritical for the assembly with a hollow center. The one with his hands on the uranium absorbed between 2,000 and 4,000 rem, and he died three days later in the Bio-Physics Institute in Moscow. The man who was holding the winch control only received something between 500 and 1,000 rem, and he managed to cling to life for 54 days.

These two men suffered from the same supreme confidence in what they were doing that had killed Louis Slotin. They had violated many rules, including the most important one: *Every unmeasured system is assumed to be critical.* It is the same as finding a pistol sitting on a table. Assume that it is cocked and loaded.

The nuclear age had arrived with a pronounced bang, and by 1947 two experts had died trying to achieve zero-power criticality in the simplest possible reactor configurations. It had become obvious that an extraordinary level of caution would be needed to do anything practical with this new discovery, this new, novel, and dangerous way to heat the old cave. Be careful, or the innocent-looking ball of metal could pin you to the wall like a mule with a long-festering grudge. And a radioactive one at that.

Nuclear reactor systems were about to get a lot more complicated, with more moving parts, pumps, valves, controls, indicators, and data recorders, and a great deal of plumbing. The heart of the system, the reactor core, was going to be covered up by layers of safety-ensuring machinery and made abstract by the interpretive instrumentation; but we must never forget that at the center of it all, danger still lurks. Remain alert, capable of terror, and never so familiar with the routine that you are certain that nothing could happen.

Chapter 3

A BIT OF TROUBLE IN THE GREAT WHITE NORTH

"A scientist need not be responsible for the entire world. Social irresponsibility might be a reasonable stance."

—advice given to young physicist Richard Feynman by mathematician Johnny von Neumann

THE DECADE OF THE NINETEEN-FIFTIES is often cited as a dull period of time, lacking the excitement and colorful excesses of the following decade, the sixties. The sixties exploded with John Kennedy, the Beatles, recreational pharmaceuticals, space travel, and hippies. What did the fifties give us? Dwight Eisenhower and black-and-white television?

Deeper research indicates that this comparison of two decades is upside down. The utter wildness of the nineteen-fifties, a decade in which 100 new religions were formed, psychedelic drug experimentation was on an industrial scale, and vast scientific experiments outstripped science fiction, makes the sixties a wind-down.[58]

Eisenhower, the subdued old Republican who liked to play golf, reversed everything that his predecessor, Harry Democrat Truman, had worked so hard to nail down. He stopped Harry's Korean War in mid-advance. He played a clever game with the Soviet Union, forcing them to be the first to orbit a satellite that passed over the United States, thus setting the international precedent for down-looking reconnaissance from space. Most surprising, he opened the files of the Manhattan Project, insisting that every document, scientific finding, and gained expertise that did not relate directly to the weapons be declassified and released to the entire world. Truman, seeing this knowledge as proprietary property of the United States government, had denied access to our most trusted allies. Even the British and the Canadians, who had participated in the development work, were allowed no access. Eisenhower wished to give all the world enough knowledge to pursue civilian-owned nuclear power. He railed at the "military-industrial complex," warning of its desire to make profits from developing new, more advanced weaponry.

Simultaneously, this tranquil administration oversaw the rapid development of the hydrogen bomb, a weapon 1,000 times more powerful than those used to wipe out entire cities in Japan with single drops, and the exotic hardware to deliver it. Nuclear rockets capable of sending a fully equipped colony to Mars in one shot were designed. Most of the nuclear power research effort went into submarine propulsion, with civilian electrical plants a minor sub-topic. Enormous scientific and engineering development efforts, such as the nuclear-powered strategic bomber and earth-moving by atomic bombs, call

58 A prime example of a wide, government-sanctioned program of psychedelic drug experimentation on uninformed civilians was project MK-Ultra, enthusiastically administered by the Central Intelligence Agency beginning in 1953. Experiments were performed at 44 universities and 36 other venues, including hospitals, prisons, and pharmaceutical companies. Citizens were subjected to drugs, hypnosis, sensory deprivation, torture, and abusive language. Today, they would not let us do any of this to a goat. (Don't ask me how I know.) Records of this project were declassified in 2001.

into question the enthusiasm of this ten-year span. Some projects were so insanely reckless, the public perception of anything nuclear was permanently damaged.

A case in point is Castle Bravo, the code name for the first test of a practical H-bomb at Bikini Atoll in the Marshall Islands archipelago. The concept of a nuclear fusion weapon had been resoundingly confirmed on November 1, 1952, with the explosion of the Ivy Mike thermonuclear device on what used to be Elugelab Island in the adjacent Enewetak Atoll. That bomb weighed 82 tons, sat in a two-story building, and required an attached cryogenic refrigeration plant and a large Dewar flask filled with a mixture of liquefied deuterium and tritium gases. It erased Elugelab Island with an 11-megaton burst, making an impressive fireball over 3 miles wide, and the test returned a great deal of scientific data concerning pulsed fusion reactions among heavy hydrogen isotopes, but there was no way the thing could be flown over enemy territory and dropped.[59]

The Castle Bravo shot on March 1, 1954, tested a lighter, far more compact H-bomb named "Shrimp." It used "dry fuel" or lithium deuteride as the active ingredient, and it needed no liquefied gases or the cryogenic support equipment, yet it gave the same deuterium-tritium fusion explosion in an "F-F-F" sequence: first a RACER IV plutonium implosion bomb (fission), followed by a large deuterium-tritium compression event (fusion), and finally a fast-neutron chain reaction in the uranium-238 tamper (fission). Sixty percent of the power from this and subsequent thermonuclear devices came not from the hydrogen fusion, but from the fission of the humble uranium tamper, a mechanical component with a lot of inertia intended to keep the bomb together for as long as possible while it exploded.

---

59 Or, at least you would think that a rational military-industrial complex would have ruled it impractical. The Sandia Committee in New Mexico proposed the design of a weaponized Ivy Mike device designated TX-16/EC-16, to be carried aloft in a B-36 "Peacemaker" 10-engine strategic bomber. The "EC" in the designation referred to "emergency capability" version, meaning that it was to be used only in the dire situation of an enemy attack, which was anticipated to occur at any time. The problem with the liquid hydrogen isotopes slowly boiling off and being lost was solved by installing large Dewar flasks in the airplane with piping to replenish the cold liquid as it gassed off into the atmosphere. Only five of these monsters were built, and it was never tested. It looked as big as an Airstream house trailer, and it was named "Jughead."

The tritium used in the fusion event was made during the explosion from the lithium component of the dull gray lithium deuteride powder.[60] The light isotope of natural lithium, lithium-6, captures a surplus neutron from the explosion of the RACER trigger device and immediately decays into tritium plus an alpha particle, or a helium-4. This tritium plus the deuterium nucleus in the same molecule fuse, being caught between the severe x-ray pressure front from the fission explosion and a plutonium "spark plug" in the center of the fusion component.[61]

The explosive yield of this arrangement was predicted to be 5 megatons, with no possibility of exceeding 6 megatons. It could not be as efficient as the Ivy Mike device using liquid hydrogen isotopes, because the lithium was not all lithium-6. Natural, out-of-the-ground lithium is only 7.5 percent lithium-6; the rest is lithium-7. Lithium-6 has an enormous neutron activation cross section, or probability of capturing a neutron and exploding into tritium plus helium. Lithium-7 has an insignificant cross section and would not participate. With great effort, the bomb makers were only able to enrich the natural lithium to 40 percent lithium-6, and the rest would be inert and wasted.

In the week before the Castle Bravo test, the wind was blowing consistently north. That was good. Any fallout kicked up by the explosion would be blown out over a large Pacific range, empty of islands and inhabitants. Early in the morning of the test, the wind shifted, blowing east. That was bad. From 60 to 160 miles east of ground zero were inhabited islands that could be hit with a load of radioactive debris. Delaying the detonation until the wind direction improved was

60 At 8 grams per mole, lithium-6 deuteride is the second lightest compound known to chemistry. Lithium-6 hydride is the lightest.

61 This extreme force, capable of bending atomic structure, exerted by an x-ray front is completely outside human experience. Imagine sitting in the dentist chair and being blown through the wall of the building when the technician pushes the button on his x-ray machine. There are other forces at work in an H-bomb, such as the gamma front, the neutron front, the beta front, the alpha front, and, of course, the shock wave caused by the explosion, but the x-rays are the first out of the box. The x-rays are caused by accelerating electrons originating in the loosely coupled outer orbitals of the atoms. Before the atomic nuclei have time to fully react, the electrons have been bounced and are sending off x-rays at the speed of light in an extremely dense mob.

debated, but the operations director vetoed it. There were too many time-dependent experiments set up, and it would cost too much to interrupt the tight schedule. The countdown continued.

The Shrimp was set up on an artificial island on the reef next to Namu Island, and at 6:45 local time it was detonated, becoming the first nuclear accident involving a weapon test. We will never know exactly how powerful the Castle Bravo was, because all the measuring equipment, close-in cameras, and recorders were blown away in the blast, but it is believed to be between 15 and 22 megatons, making it the biggest explosion ever staged by the United States, and much larger than what was planned for. In one second it made a fireball four and a half miles in diameter, visible on Kwajalein Island 250 miles away. The top of the mushroom cloud reached a diameter of 62 miles in ten minutes, expanding at a rate of four miles per minute and spreading radioactive contamination over 7,000 square miles of the Pacific Ocean. Did they do anything like that in the sixties? Not even close.

All hell broke loose. The Rongelap and Rongerik atolls had to be evacuated. Men were trapped in control and observation bunkers, sailors suffered beta burns, and fallout rained down on Navy ships in the area. The bomb had cleaned out a crater 6,500 feet in diameter. The coral in the reef was pulverized and neutron-activated to radioactivity, mixed with radioactive fission debris, and in 16 hours spread into a dense plume, 290 miles long and heading due east in the wind toward inhabited islands. Permanently installed testing facilities at the atoll were knocked down, and radioactive debris fell on Australia, India, and Japan. Circling the world on high-altitude air currents, the dust from the test was detected in England, Europe, and the United States. American citizens were alarmed when warned of milk contaminated with strontium-90, a major product of the uranium-238 fissions in the tamper.

What happened? The expectation of no action from the lithium-7 component of the lithium deuteride was incorrect. The neutron density in a thermonuclear bomb explosion is inconceivably large, and in this condition it does not really matter how small the activation cross section is. Neutrons will interact with the lithium-7, producing tritium, and helium-4, plus an extra neutron. All of the lithium deuteride was therefore useful in the explosion, and the yield was three times the expected strength. Not only was more energy released in the deuterium-tritium fusion, but the unexpected neutron excess increased the third-stage fission yield in the tamper, made of ordinary uranium. While the fusion process was considered clean, producing no radioactive waste products, the uranium-238 fission was unusually dirty.

A complicating problem was the choice of the Director of Operation Castle, Dr. Alvin C. Graves. As you recall from the previous chapter, he was standing close behind Louis Slotin when he made his fatal slip with a screwdriver and a plutonium bomb core went prompt critical. Graves caught 400 roentgens right in the face. He could have died easily from the acute exposure, but he lived on to rise in the ranks at Los Alamos.[62] Graves therefore could see no particular problem putting men close to atomic blasts in several experiments, from the Marshall Islands tests to the above-ground explosions in Nevada.[63] This peculiar tendency is similar to the case of Bill Bailey and his Radithor, noticing no ill effects from his elixir while subjecting Eben Byers to a horrible death. Both men, Graves and Bailey, endured later condemnation for exposing so many people to so much radiation.

A medical study of Marshall Island residents, Project 4.1, was put together hastily to document the radiation injuries. The investigation found that 239 Marshallese and 28 Americans were exposed to

62 Alvin C. Graves, Ph.D. physics, became head of the Test Division at the Los Alamos National Laboratory. The accident with Slotin made him temporarily sterile, and his eyesight was never the same. He appeared in the documentary motion picture *Operation Cue*. Graves died of a heart attack in 1965 at the age of 56.

63 Subjecting soldiers and village people to radiation from above-ground nuclear weapons tests just to see what would happen was not unique to the United States. In September 1954, large-scale human response tests were performed at the Totskoye Military Range in Orenburg Oblast, Russia. In the military exercise "Light Snow," about 45,000 Soviet soldiers and officers were exposed to the radiation from an above-ground detonation of a 30-kiloton nuclear device, dropped from a Tu-4 bomber for realistic effect. This was a complete surprise to the test subjects, who, unlike their American military counterparts, were given no protective gear or a hint as to what was going to happen. Deputy Defense Minister Georgy Zhukov observed from a safe distance in an underground bunker. The medical records of the thousands of people affected seem to have disappeared.

significant but non-fatal levels of radiation. The final report was classified SECRET, "due to possible adverse public reaction."[64]

Over-yield of the Castle Bravo device was frightening to many who worked on it, but the real tragedy unfolded far west of the test site, in Japan. It is called the "Lucky Dragon Incident," and its everlasting effect on the public's perception of nuclear radiation was outside the control of the test program. It would mark in history the first and last record of a death caused by a United States nuclear weapon test.

The *Daigo Fukuryū Maru*, or the Lucky Dragon 5, was a wooden 90.7-ton Japanese fishing boat with a 250-horsepower diesel engine and a crew of 23. On March 1, 1951, she was trawling for tuna where the fishing was good and competing with 100 other Japanese fishing boats in the general area of the Marshall Islands. There had been vague warnings from the U.S. earlier that year, defining a rectangular area of hazard around Bikini Atoll and hinting at nuclear weapons tests, but no dates had been specified. The Dragon got as close as it could to the western edge of the rectangle, within 20 miles of the boundary. Tuna liked to swim near the Marshalls.

At 6:45, the sun seemed to rise in the west. The crew stopped their preparations for the day's fishing and stared at the fireball lighting up the sky. Seven minutes later, the shock wave, reduced by distance to a mean clap of thunder, rolled over the boat. Still, the men fished. In a few hours, it began to snow, and the boat, the fishing equipment, and the men started to become covered with white flakes of coral, blasted to a fine ash by the explosion of the Shrimp over in Bikini. For three hours it fell, beginning to form drifts against the wheelhouse and impeding movement on the deck. The men started scooping it into bags with their bare hands, initially unaware that it was fallout, infused with a fresh mixture of radionuclides, but starting to get the

64 It has been alleged that the victims of the Bravo test were purposely exposed to fallout radiation as "guinea pigs" in a radiation experiment. Micronesian Representative Ataji Balos charged that Castle Bravo was directed purposely at the inhabited Marshall Islands because the Marshallese were considered to be expendable and of marginal status in the world at large. There is nothing to support this charge. The engineers and scientists in charge of the Bravo test could not have bent the wind east if they had wanted to, and they were honestly surprised at the power of the explosion. Project 4.1 did not exist until four days after the test.

dreaded feeling that they had witnessed a *pikodon*—Japanese for atomic bomb.[65] They had to get out of there fast, but first the money-maker had to be reeled in. It took several hours to recover and stow the trawling net, with the men wiping the calcium snow out of their eyes. Thirteen days later the Dragon chugged into its home harbor in Yaizu, Japan, filled with radioactive fish.

The crew was suffering from nausea, headaches, burns on the skin, pain in the eyes, and bleeding from the gums—all symptoms of radiation poisoning, and as their boat was unloaded and their catch put on ice the men were sent to the local hospital. Several were obviously sick. For some reason the radio operator, Aikichi Kuboyama, who should have been inside and not on the deck, was in the poorest condition. The men were scrubbed down several times, their hair was shaved off, and their nails were clipped, all to remove the radioactive dust that was ground into their surfaces, but the doctors were stumped when nothing seemed to help.

News of the contaminated crew traveled fast. The entire world became interested, and there was explaining to do. In retrospect, the public relations efforts were dreadful. Lewis Strauss, head of the Atomic Energy Commission, first claimed that the fishermen's injuries could not have been caused by radiation, they were inside the no-fish zone, and besides that it was a Soviet spy boat that had gathered classified information on the bomb test and simultaneously exposed its entire crew to radiation just to embarrass the United States. Requests from Japan for an inventory of the radioactive species in the fallout so that treatments could be specified were denied, on the grounds that the nature of the bomb could be derived from this information.[66] The extent of contamination was claimed to be trivial, in parallel with the Food and Drug Administration imposing emergency restrictions on tuna imports. The

65 Literally "flash-boom." First comes the flash and then the boom.

66 It was possible to derive the composition of such a device from analysis of the fallout, which scientists in the Soviet Union proceeded to do without help from the AEC. There was no lack of fallout material to analyze. Measuring the percentage of one nuclide, U-237, in the fallout using a single-channel pulse-height analyzer, it was possible to deduce that Shrimp had been a three-stage F-F-F device with a U-238 tamper.

impression given to the people of Japan, still sensitive about atomic bombs, could not have been worse.

A young biophysics professor in the city university in Osaka, Yashushi Nishiwaki, read about the Lucky Dragon in the paper, and he called the health department to see if any tuna had been shipped there from Yaizu. Yes, tons of it. He took his Geiger counter down to the market and waved it over some tuna. To his alarm, the needle on his rate-meter slid off scale. He was counting 60,000 radiation events per minute. The entire catch was heavily contaminated. Even loose scales and paper wrappings of fish that had been bought and eaten by now reeked of fission products. It was headlines in the evening paper, and mass hysteria took the city, then the region, and Japan. First, the Misaki fish market closed. Fish mongers scrambled for Geiger counters so that they could run them over the fish and prove to buyers that there was no radioactivity, but it did not help. People stopped buying fish. Yokohama closed, and then, for the first time since the cholera epidemic of 1935, the Tokyo fish market closed. It was revealed that fish were banned from the Emperor's diet, and that was it. Prices for tuna crashed, and dealers filed for bankruptcy. It would take years to recover.

Meanwhile, the Lucky Dragon fishermen were recovering, except for Aikichi the radio operator. His liver was failing. His condition worsened and he died on September 23 at the age of 40. "I pray that I am the last victim of an atomic or hydrogen bomb," were his last words, splashed all over the news. The United States government eventually paid the widow the equivalent of about $2,800 and agreed to pay Japan, with the wrecked fishing industry, $2 million for their trouble. From this donation, each crew member was given $5,000.

Out of the disaster came Nevil Shute's great novel, *On the Beach*, later made into a major motion picture starring Gregory Peck, and the entire Japanese monster movie industry, beginning November 3, 1954, with *Godzilla*, a city-wrecking beast mutated by contaminating radiation. The Lucky Dragon 5 was stripped down, decontaminated, and rebuilt. It was sold to the government for use as a training vessel in the Tokyo Fisheries School, renamed the *Hayabusa Maru*, or the Dark Falcon. Today, the Lucky Dragon 5 is preserved for all time, lest we forget, in the Tokyo Metropolitan *Daigo Fukuryū Maru* Exhibition Hall. The other 22 crew members all recovered with no lingering health effects from the fallout contamination.[67] As health physicists always point out, if the men had simply lowered themselves into the water and washed off the gray dust, they would not have suffered any effect from the fallout. It was the fact that it stayed on their skin for so long that caused the trouble. If they had cut loose the nets and headed north at full power, while hosing off the deck, history would be different.

These nuclear shenanigans of the United States in the early 1950s were interesting for how they helped shape the growing public angst, but they were part of a mutant off-shoot of the larger task of taming the atom for use as a power source. The weapons tests were fascinating, almost recreational, but not really helpful from a long-term, scientific perspective. The rest of the world together had a smaller research budget, but progress toward understanding nuclear reactions was being made independently and usually in secret in a few foreign countries. In the beginning, right after the Second World War, England, France, and the Soviet Union were very interested in coming up to speed, but the first nuclear reactor outside the United States was built and tested in the second largest country on Earth: Canada.

With a population about the size of Metropolitan Los Angeles and a million square miles of uninhabitable permafrost, Canada did not seem to have the makings of a nuclear research hub, which required money, a wide-ranging technical manufacturing base, hundreds of highly specialized scientists and engineers, and yes, still more money. But Canada did have a portion of the scientists involved in the Manhattan Project, the largest and most pure deposit of uranium ore on Earth, and Chalk River.

The Chalk River Laboratories, in some ways similar to the Oak Ridge facilities in Tennessee, were built in an isolated rural setting northwest of Ottawa in Ontario Province during World War II. It started out as an independent Canadian/British effort to develop an atomic bomb independent of the United States in a house belonging to McGill University in Montreal. It was near the end of 1942, and the

67 Why was the radio operator the only one who died, and of what did he perish? Aikichi may have been a person who was more sensitive to radiation exposure than the rest of the men. Even given that, the radiation load would not have been high enough to kill him in seven months. A quiet conclusion was that he died of infectious hepatitis from the many blood transfusions he was given at the Tokyo University Hospital.

Manhattan Project was still fairly scattered and not looking too successful. The British deeply wanted an atomic bomb project, but they wanted it somewhere besides Britain, where there was no assurance that the Germans would not take it over in an invasion. Canada, as part of the ever-untwining Empire, was the logical choice, and a group from the Cavendish Laboratories at Cambridge shipped over.

There were complications. The Cambridge group had actually originated in Paris, and only one of the six senior members was British. To the security-conscious Americans, the initial research staff seemed questionable. It included a Frenchman with jealously guarded patent rights to nuclear systems, a potential defector to the Soviet Union, a possible spy, and a Czechoslovakian. The laboratory director, Hans von Halban, a French physicist of combined Bohemian-Jewish-Austrian-Polish descent and a convinced secularist, lacked certain management skills and tended to irritate the National Research Council of Canada, his sponsor.

Experiments toward a bomb began with attempts to create a self-sustaining nuclear fission reaction. To that end, the group stacked cotton bags filled with uranium oxide powder interspersed with bags of powdered coke in the corner of a room.[68] Performance of this first pile was disappointing, as it seemed to just sit there and not make any attempt to fission. Obviously a much larger stack of bags would be necessary to achieve any sort of success. The group needed a larger working space and more bags.

In March 1943 the lab moved to a new building at the Université de Montréal, originally intended for a new medical school, and they expanded to a staff of 300, half of whom were Canadians. By June, the level of enthusiasm had reached a low point. Walls and floors of the building were black and filthy with a mixture of uranium and coke powders that had escaped bagging, morale was low, few fission neutrons were produced, and the Canadian government considered closing the project down.

As it turned out, the Americans had a heavy-water plant in Canada, barely over the border in British Columbia in the town of Trail. DuPont

68 No, it is not something to be inhaled or poured over ice. In this case "coke" is the porous, carbonaceous solid made by heating bituminous coal. Bituminous coal has a minimum amount of sulfur and other impurities, and the coke derived from it is almost pure carbon.

Chemicals was directing the work extracting deuterium from fresh water using an electrolysis process, not exactly because the Manhattan Project desperately needed heavy water, but because the Germans had a heavy-water plant, and maybe they knew something that we did not. Ergo, a heavy-water plant had to be acquired. The Vermork hydroelectric plant at Rjukan, Norway, a fertilizer factory, had been producing high-purity heavy water for no particular reason since December 1934, and the Germans had taken over operations and had been sending barrels of the stuff back to the Fatherland since April 1940.[69] The fear was that they were working on an advanced form of nuclear reactor, possibly more sophisticated than anything the Americans had come up with.[70]

On August 19, 1943, an Anglo-Canadian-American understanding had been officially reached. This "Quebec Agreement" was drawn up to ensure that this close, English-speaking component of the Allied forces would be working together on the atomic bomb and not duplicating

69 The last time I saw any genuine Norsk Hydro heavy water was in April 1989. Reproduction of the Pons and Fleischman "cold fusion" experiment was all the rage in nuclear research at the time, and I was briefed on parallel work ongoing at the Oak Ridge Laboratory. It was proving difficult to get positive results from the "fusion in a bottle" experiment, which involved electrolysis of pure deuterium oxide using palladium as a cathode. The researchers wanted a perfect setup, so instead of using the 99.8 percent deuterium oxide from Sigma Chemicals, no. D-4501, like everybody else, they dusted off a box of flame-sealed glass vials of heavy water sitting on a shelf in a World War II-era stockroom. It was straight from Vermork, 1942, found aboard surrendered German Type XB submarine U-234 soon after VE Day, apparently being shipped to Japan. It was old and exotic and for paranormal reasons seemed the better electrolyte for an experiment that was out on the fringe anyway. The cold fusion still didn't work.

70 The fear was so intense, the Allied Forces attacked the Norsk Hydro plant any way they could, beginning on October 18, 1942, with Operation Grouse, manned by four Norwegian members of the Special Operations Executive (SOE). This attempt at sabotage was unsuccessful, followed in November by the all-British Operation Freshman, which also failed. In February 1943 Operation Gunnerside was more successful, managing to blow up the electrolysis machinery and waste 500 kilograms of heavy water. In November an American bomber raid failed to drop explosives on anything of significance, but on February 20, 1944, the Norwegian resistance managed to sink the ferry boat HYDRO, carrying with it several large barrels of partially enriched deuterium oxide and dozens of innocent Norwegians. Strangely, the German High Command did not interpret all this attention being paid to deuterium production as indicating that we were terrified of their nuclear work, and the atomic bomb effort never gained a sense of priority.

efforts because of excessive secrecy. From this new sharing of information came news that the Americans had already achieved a successful critical nuclear assembly using graphite back in December 1942, and there was no need to prove it again. The Canadians were encouraged to see what they could do using heavy water as a neutron moderator, trying to duplicate whatever the Germans were doing. They could have all the heavy water being produced in the Trail plant, and they should build an even bigger deuterium-extraction operation somewhere else.

Either graphite or deuterium oxide (heavy water) was a usable moderator for use in building the feeble nuclear reactors of the time. The only known isotope that would fission was U-235, and it was a rare component of mined uranium, being only 0.73 percent of the pure metal. It was possible to build a working reactor using such diluted fuel, but all conditions had to be carefully optimized. The speed of the neutrons that were born in fission, which were necessary to cause subsequent fissions and make the process self-sustaining, was too high. The neutrons have to be slowed way down to "thermal" speed, or the speed of ordinary molecules bouncing around at room temperature. The way to slow them down was to allow them to crash into bits of matter that were standing still, thus transferring all the energy by billiard-ball action. Think of a neutron as the cue ball on a pool table. When the cue ball hits another stationary ball, it stops cold and the ball that was hit takes off at the original speed of the cue. Not only does this action slow the neutron down to fission speed, it also transfers the energy, or the heat, from the frantic neutrons to another medium. The material used to slow down the neutrons and absorb the heat is called the "moderator."

The perfect moderator consists of tiny balls that are almost exactly the same mass as the neutrons. That would be a fluid consisting of protons, which happens to be hydrogen, which is conveniently included in the common material, water. Using water as a moderator would seem ideal, because it can be pumped through holes in the reactor core, slowing down the neutrons while actively cooling the metal to keep it from melting and transferring the heat to some useful application. One direct hit of a neutron against a hydrogen nucleus, or proton, and the enthusiastic particle has decreased speed from 2

MeV to 0.025 eV.[71] A reactor moderated with ordinary tap water would therefore be very compact, not requiring a long chain of repeated contacts to slow down the neutrons. Hit a stationary polo ball with a billiards cue ball, and it does not come to a complete stop with one impact, but only gives a fraction of its energy to the target.[72]

Unfortunately, given the natural uranium that was available in the early 1940s, water was not quite good enough. On rare occasion, a neutron would stick to a proton instead of bouncing off it, thus taking a neutron out of the pool of fission-producing particles. The maintenance of criticality, or the ability to produce as many fission neutrons as were lost, is so sensitive to the number of available neutrons, just losing one out of trillions can shut the process down. That is, in fact, why we always call it "criticality." By capturing a neutron, an ordinary hydrogen atom instantly becomes deuterium, but the incident is too rare to make any difference in the composition of the moderator.

The usable moderating material had to be one that would slow neutrons down to thermal speed after some reasonable number of collisions while having an extremely low probability of capturing a neutron and taking it out of the race. For at least a first experiment in criticality, graphite was ideal. It had an extremely low neutron capture cross section, and as a solid material it would also double as the structure of the reactor. No metal tanks, girders, struts, or nuts and bolts, each a potential neutron absorber not contributing anything to the nuclear reaction, would be necessary to build a working reactor. The first reactors in the United States were therefore large cubical heaps of graphite blocks. They were referred to then and for decades afterward not as reactors but as "piles."

71 The speed of a neutron emerging from nuclear fission has a mean energy of 2 million electron volts. To put this in conventional terms, this is 20,000 kilometers per second. As a rule, any neutron traveling faster than 1 MeV is considered "fast." A thermal neutron is going about 2.2 kilometers per second. Neutrons born from hydrogen fusion are booking at 14.1 MeV, or 17.3 percent of the speed of light, and at this speed even U-238 can be fissioned readily.

72 Please note that every collision between a flying neutron and a standing proton is not head-on, and most encounters are a glancing blow with only a partial loss of speed. However, the hydrogen in water is still a terrific moderator. Much as it hurts to do so, I have to simplify explanations or I fear I would lose readers in the drone of details.

There was another possible moderating material. Heavy water, the compound made of two atoms of deuterium, the isotope that weighed twice as much as plain hydrogen, plus an oxygen, looked, tasted, and poured just like tap water, and it also had a very low capture probability for neutrons. Each deuterium nucleus was a proton that had already captured a neutron and was unlikely to need another one. Although it lacked the slowing-down power of pure hydrogen water, it had other advantages of water. Uranium held in a matrix and immersed in it would transfer its fission energy to this moderator, and it would be an ideal coolant. Unlike the graphite, it could be pumped into and out of the reactor space, being a mobile material. Graphite was stationary, and energy deposited into it had to be removed by another moving material, such as air, helium gas, or even water pumped through channels in the structure. Heavy water, on the other hand, had the advantage of not absorbing any neutrons, just as graphite, but it was also a mobile energy transfer medium. Perhaps this was why the Germans were so intensely interested in the Norwegian heavy-water plant?

The Americans suggested that the Canadians should try the alternate moderation scheme of using heavy water, and the Brits were keen on the idea. It would be excellent for their purposes to have a unique set of reactors working, hidden away in rural, uranium-rich Canada. At that time the purpose of nuclear piles was not seen as a power-generating technique, but as a way to make an alternate fissile isotope. Over 99 percent of mined uranium is uranium-238, which is inert for the purposes of fission but is important for the production of the new, man-made element, plutonium. Uranium-238 captures neutrons and becomes uranium-239, which quickly beta-decays into neptunium-239, which in a few days beta-decays into plutonium-239. Plutonium-239 is fissile and is appropriate for making atomic bombs. Uranium-235 is also a bomb material, but separating it from the uranium-238 as mined is an extremely difficult, time-and-energy-dependent process. The Pu-239 doesn't have to be isotope-separated from anything, and it is immediately usable. The Canadians took the challenge with enthusiasm.

This commitment to heavy-water moderation became a Canadian trademark, following them for decades and into the next century, starting with a primary agendum, to make plutonium for the Brits. The traditional use of natural uranium in Canadian reactors has its advantage and its disadvantage. The upside is that no expensive U-235 enrichment is required, as it is for all American power reactors. You just use the uranium as it comes out of the ground. The downside is that this natural uranium contains so little usable isotope, you have to constantly remove the expended fuel and load in new fuel, as the reactor is running. This feature becomes complicated and the radiation hazards in a reactor that is open at both ends are considerable, but the fuel juggling is essential.

There is a powerful, secondary advantage to this need for constant refueling. A percentage of the U-238 component of the fuel is converted into Pu-239, the bomb ingredient, as happens in all uranium-fueled reactors. In a reactor built to the American pattern, using enriched fuel, the refueling is once every few years. In that protracted time among flying neutrons, a percentage of the Pu-239 is up-converted to Pu-240, and this ruins the plutonium for use in atomic bombs. In the Canadian-style reactors, the fuel does not stay in the core long enough to contaminate the plutonium with Pu-240.[73]

By 1944 the Anglo-Canadian nuclear effort to build a heavy-water reactor, named NRX, was moved to Chalk River in a newly built facility. The Brits, after having been cleansed of most nuclear assets by the Manhattan Project, contributed one of their last treasures to the project, the eventual Nobel laureate, Dr. John Douglas Cockcroft OM KCB CBE FRS. Cockcroft, the son of a mill owner, was born in the English town of Todmorden in 1897, and he began his journey through knowledge at the Todmorden Secondary School in 1909. Continuing his education at the Victoria University of Manchester, he moved to the Manchester College of Technology after an interruption by the First World War, in which he served as a signaler for the Royal Artillery. After two years studying electrical engineering, he moved on to St. John's College, Cambridge, in 1924 and enjoyed the privilege of working with Lord Ernest Rutherford unwrapping the structure of the atom.

---

73 For several decades the United States produced bomb material using the graphite-moderated reactors at the Hanford Laboratory in Richland, Washington, and using Canadian-style heavy-water reactors in the mirror-plant, Savannah River Laboratory, in Aiken, South Carolina. Any uranium-fission reactor, even the fast-in/fast-out designs, produces a small percentage Pu-240 residue. The problem with Pu-240 is that it tends to fission spontaneously instead of waiting for the bomb trigger. It makes the bomb design easier if you can minimize the percentage of Pu-240 in the plutonium.

In 1932 Cockcroft and the notable Irish physicist Ernest Thomas Sinton Walton stunned the scientific community by reducing lithium atoms into helium by bombarding them with accelerated protons. This accomplishment was made possible using their invention, the Cockcroft-Walton "ladder" voltage multiplier, which on an unusually dry day could produce an impressive 700,000 volts at the top electrode. They jointly won the Nobel Prize in physics for this work in 1951.[74]

Cockcroft took the helm of the Montreal Lab, which was in the process of moving to Chalk River, from von Halban in late April 1944. By July he had assessed the situation and saw the value of quickly building a small, zero-power heavy-water reactor to gain knowledge and experience for designing the ambitious NRX. In August, project approval and a new British engineer, Lew Kowarski, shipped over from the home island. Von Halban, feeling somewhat miffed, traveled to Paris to celebrate its recent liberation from German control, and he was suspected of having given his French compatriot, Frédéric Joliot-Curie, a run-down on the nuclear progress being made in North America. This action was strictly forbidden by the Quebec Agreement.[75] Kowarski was put in charge of the small reactor project, with Charles Watson-Munro as second in command. Cockcroft named it the Zero Energy Experimental Pile: ZEEP.

The design chief, George Klein from the National Research Council, was under pressure to design a 1-kilowatt pile, but he was careful to keep it to essentially zero power, or one watt. The purpose was to make ZEEP as versatile as possible, so that they could change the fuel configuration every which way to find an optimum

---

74 It turned out that the famous Cockcroft-Walton voltage multiplier, which has since been used in everything from Xerox machines to television sets, was invented in 1919 by the Swiss physicist Heinrich Greinacher. Impress your friends by referring to the cascade voltage doubler as the "Greinacher multiplier."

75 Much as one would like to think that Britain and Canada were managing their own plunge into nuclear science and engineering, they were not. Control of everything, including the work of these friends of the United States, was ultimately under one man, General Leslie Groves of the Manhattan Engineer District. When Groves formed the vaguest suspicion that this Frenchman had spilled some information, he was jerked out of the project immediately and was unable to travel or find work until the war was over. A bit of nuclear apocrypha has General Groves holding forth to Robert Oppenheimer in a hallway, when in mid-breath he peeled off his tunic and handed it to the colonel standing next to him with instructions to have it dry-cleaned. Without batting an eye, Oppenheimer turned to the physicist by his side, pointed to a mustard stain on the cuff of his jacket, and asked him to lick it off.

configuration for NRX, the big reactor. By keeping the power down, there would be very little bio-shielding in the way, and they could fiddle with the internal construction of the reactor core without being in danger of residual radiation exposure from high-power fission.

ZEEP was housed in a metal building, about the size and shape of a hay barn, with the enormous, four-story, brick NRX building being built next door. Final approval for construction was given on October 10, 1944. By that time, the Americans had already built the CP-3 heavy-water research reactor at the Argonne Lab in Illinois, and they were glad to unload some advice and a truck filled with graphite bricks. The graphite was used to build a box-shaped neutron reflector in the center of the building, and the cylindrical reactor vessel, made of aluminum, was installed inside. Aluminum-clad fuel rods made of uranium metal were lowered down into the vessel, which would eventually be filled to a depth of 132.8 centimeters with heavy water. The reactor top could be easily removed for access to the fuel, but during operation it was covered with cardboard boxes filled with borated paraffin, another gift from Argonne. The paraffin was there to prevent neutrons from escaping out the top and bouncing into the control room area. In the basement was a large holding tank for heavy water.

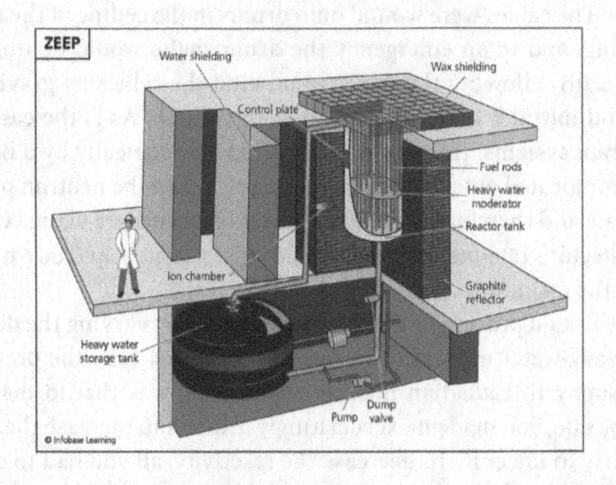

ZEEP was a small experimental reactor, built to test the concept of building a larger reactor using heavy water as the coolant/moderator and natural uranium as the fuel. Cardboard boxes filled with a mixture of paraffin and boric acid were stacked on top of the open reactor vessel, discouraging neutrons from streaming out into the building.

It was the earliest time in nuclear reactor development, and it was wide open to experimentation. There were few set design rules or well-established ways to do things. In those days, the reactor system design was based mainly on anticipatory terror. The fear was that the nuclear fission process, which was normally quite tame and easy to control, could suddenly go skittish, running wild with destructive tendencies, due to an unforeseen mechanical failure or a mistaken move by a human operator. This anxiety was not entirely unfounded, although nothing had ever happened to a natural uranium reactor with an active cooling system. The event of concern was called the "power excursion," in which the rate of energy release from fission would increase rapidly, overcoming the ability of the coolant to remove heat and possibly leading to the dreaded steam explosion.

By 1944 one thing that was standard in the nuclear camp was a "scram" system, or an auxiliary shutdown mode that could kill the neutron flux and quench the fission as quickly as possible to avert a power excursion. For ZEEP, this part of the reactor consisted of heavy plates made of cadmium, known to have a very high neutron absorption cross section, hung over the top of the reactor vessel on cables. The cables were wound onto drums in the ceiling of the reactor building, and in an emergency the drum-shafts would be unlocked electrically, allowing the plates to unwind the cables by gravity and descend into the vessel to stop an energy-climb. As is the case with all scram systems, this one was triggered automatically by a neutron detector located outside the reactor vessel. When the neutron production rate and therefore the reactor power reached a pre-set maximum, the detector's rate-meter circuit would close a switch, and down would come the cadmium.

The fission process in ZEEP was controlled by varying the depth of the heavy-water moderator in the vessel. At that time the prevailing philosophy in Canadian nuclear engineering was that to make the reactor safe, you made it excruciatingly difficult to increase the power and easy to lower it. To decrease the reactivity, all you had to do was open a valve. The heavy water would drain out of the reactor vessel and into the holding tank downstairs, and as the moderation of the fission neutrons decreased, the power level would drop. To increase the power level, the operator had to push forward a spring-loaded slide switch that would turn on a pump and bring heavy water back out of the basement and into the reactor core. With each push, the pump would only run for 10 seconds, then cut off. You had to push the switch again for 10 seconds of pump action. To bring the reactor up from cold shutdown was quite arduous, requiring the operator to push his damnable switch 1,000 times. ZEEP first went critical at 3:45 P.M. on September 5, 1945, only 11 months after its construction was approved. It was the first reactor to run successfully outside the United States.

Oblivious to this safety measure or because of it, an interesting incident occurred at ZEEP in the summer of 1950.[76] The reactor was shut down so that two physicists could insert metal foils into a few fuel rods. These foils would be activated by the neutron flux, and the resulting radiation count at a later time would identify hot spots in the reactor core. They had cleared off the paraffin boxes and were standing directly over the naked reactor core.

The operator knew that he would have to restart the reactor after the physicists were finished on top, so to save some time he started filling the reactor vessel with heavy water. To save his thumb, he had jammed a chip of wood into the pump switch so that it would run continuously.[77] The phone rang. Unfortunately, the telephone was on the wall at the opposite end of the building. The operator rose from his chair, hustled to the phone, and answered. Getting immersed in the conversation, he forgot about having left the moderator pump running. The water rose slowly in the vessel. After a while, it reached 130 centimeters. The two physicists on top of the core were now down

---

76 The official word in "A Review of Criticality Accidents" from Los Alamos says that this event occurred "late 1940s or early 1950s." The date is unknown, because this accident was never reported, and it was a dark secret until 1992, when the participants were interviewed for *Nuclear Chain Reaction: The First 50 Years*, a book published by ANS. They couldn't remember what year it was. By back-tracing the operating log summaries of the ZEEP and the NRX, I was able to pin it down. The excursion caused a scram in the NRX next door, so it had to be running, and NRX was first started in 1947, at which time ZEEP was down, because its heavy water load was needed for NRX. ZEEP was brought back online in April 1950, and it was busy with experiments until August.

77 So goes the story, but it does not ring exactly true. Just keeping the switch in the ON position should not have kept the pump going. The operator had to have jammed the 10-second timer switch in the ON position as well.

on their knees, fiddling with the foils, and they did not notice as the moderator level slowly crept up. 131 centimeters. The operator was leaning against the wall, getting stiff from standing so long and talking on the phone.[78] 132 centimeters. 133 centimeters. Wait for it. . . .

SNAP! The two busy men froze, then looked up. The scram reels had tripped loose, and they were spinning as the cadmium plates slithered down and into the core. Uh oh. Both physicists, who had left their lab coats downstairs with their radiation dosimeters clipped to the pockets, were being painted with a blast of broad-spectrum fission radiation as ZEEP went supercritical. Although the power threshold was set at three watts, there was no telling how the power was still climbing as the cables unwound. Next door at the very large NRX reactor, another snap, and the entire staff jumped. The gamma radiation beaming from the top of ZEEP went through the physicists, through the roof, reflected off the cloud cover, and back through the ceiling at NRX. Its scram system, ever vigilant, interpreted the sudden radiation increase as a fission runaway and slammed in the emergency controls. The world's first power excursion had scrammed two reactors at once. The staff at ZEEP was far too embarrassed to admit that they had done anything so careless, so this historic incident went unreported, and the NRX staff was at a loss to explain why their reactor had scrammed. With nothing reported, the only lesson learned was the importance of not saying anything about it. The magnitude of the dosage received by the scientists and how they expressed their disappointment with the operator's performance are unknown. The ZEEP continued on for a distinguished and uneventful career and was decommissioned in 1973. It is now on static display at the Canada Science and Technology Museum in Ottawa.

Just two years later, Canada scored another two milestones in the history of nuclear power, within yards of the ZEEP, at the new NRX reactor. After nearly three years of development and construction, the National Research eXperimental pile, or NRX, became operational on July 22, 1947, literally overshadowing the smaller ZEEP. For a while,

78 An alternate account has the operator responding to a shout from one of the physicists to bring a tool up to the reactor top, so there were three men on the opened reactor. The telephone story is straight from the Canadians, and I believe it is true.

it was the most powerful general-purpose research reactor in the world, initially running at 10 megawatts. By 1954, improvements in the cooling system allowed its thermal output to be increased to an impressive 42 megawatts. The squat, flattened design of its reactor vessel and the use of heavy water for a moderator made it purposefully wasteful of neutrons, yet capable of great power. This unusually large flux of extraneous neutrons was put to use, developing radiation therapy to treat cancer, neutron scattering measurements of materials, medical isotope production, and testing of other reactor design aspects under constant neutron bombardment.

The reactor vessel was a cylindrical aluminum tank, named the calandria, about three meters tall and eight meters in diameter. Inserted into the top of the calandria in a hexagonal lattice were 175 aluminum tubes filled with uranium metal, and around each fuel tube was a larger aluminum cooling tube, filled with ordinary water. The moderator was 14,000 liters of pure heavy water, with the depth adjustable as a means of controlling the fission reaction. In 1952 the power level had been increased to 30 megawatts, which was carried away by 250 liters of water per second from the Ottawa River flowing through the cooling tubes. The versatility of the NRX was increased by allowing any cooling tube to be blown clear of water, using outside air as the coolant instead.

A blanket of inert helium gas was kept on top of the heavy water to prevent corrosion, and it was kept at a pressure of about 3 kilopascals above atmospheric pressure. The top of the calandria was sealed and connected by a pipe to an external tank holding about 40 cubic meters of helium. As the level of heavy water in the calandria changed, the top of the helium tank was free to move up and down on greased tracks, maintaining a constant pressure.

In retrospect, the controls for the NRX may have been overly complex, or, as W. B. Lewis, Director of Research at Chalk River, put it, there were 900 ways to keep the reactor from operating and only one to make it work. Twelve of the 175 rods stuck in the calandria, called "shut-off rods," were filled not with uranium but with boron carbide, meant to absorb neutrons and not release them.[79] If as few as seven

---

79 There was also a 13th rod which could be run in and out of the core with an electric motor, used as a vernier to make fine adjustments and keep the reactor at just critical. Its maximum reactivity, achieved by withdrawing it all the way, was the equivalent of adding 10 centimeters of heavy water moderator.

of these tubes were fully inserted into the reactor core, then no self-sustaining fission reaction could occur. Normally during operation these rods were kept out of the reactor, locked in position by active electromagnets. If everything failed, including the electrical power in the control room, then the electromagnets would lose their grip and the shut-off rods, each weighing 29 pounds, would simply fall by gravity into the core and kill the power production. There were also controls that would put compressed air into the top or the bottom of the shut-off rod tubes, forcibly running them into or out of the reactor core. With compressed air on top, you could drive the rods halfway down the 10-foot run in half a second. With only gravity, it would take as long as five seconds.

Operation of these shut-off rods was the subject of a complex set of startup rules and safeguards. The 12 rods were arranged in six banks, with Bank 1, the "safeguard" bank, consisting of four rods. There were four important push buttons on the control desk. Push button 1 raised the safeguard bank out of the core. Push button 2 raised the remaining eight rods in an automatic sequence. Push button 3 temporarily increased the current in the electromagnets to make sure that the rods were held tightly at the top of travel. Push button 4, mounted on the wall separately from 1, 2, and 3, put compressed air on top of all the rods and drove them in. Although it is not entirely clear why, when you pushed button 3, you had to be pushing buttons 1, 2, and 4 also, and this was awkward. The full-up position of each rod was indicated by a red light on the control panel. With all the shutdown rods fully withdrawn from the reactor, there was no "neutron poison" to prevent fission in the core.

On Friday afternoon, December 2, 1952, most of the senior staff at Chalk River had gone home, and the last experiment of the day was to find the reactivity of a newly installed air-cooled fuel rod with the reactor running at low power. To make it easier to shuffle some fuel rods for test purposes, several were disconnected from the main cooling water circuit and instead were connected to flexible hoses. The reactor was going to be operated at very low power for this experiment, so the quality of the cooling was not a concern.

It was 3:00 P.M. An operator was in the basement, routinely checking to make sure everything looked right before they started up. Being unusually vigilant, he noticed a bank of compressed air bypass valve handles that seemed to be in the wrong position, and he proceeded to correct them. The reactor supervisor, sitting at the control desk upstairs, noticed the red lights starting to come on, indicating that the shutdown rods, which should have been completely in the core, were hitting the full-out stops. He blinked once, grabbed the telephone, punched the basement button, and told the operator to stop doing whatever in the hell he was doing. Just to make sure, he got out of his chair and hustled downstairs to see for himself.

Arriving quickly, he was horrified to find that the operator had applied compressed air to the underside of four shutdown rods, blowing them clean out of the core. Fortunately, someone had removed the handles from the other valves, so the reactor had not gone supercritical. The supervisor returned the four valves to their correct positions, letting the rods drop back to full-in position by gravity, and received word over the phone that the red lights had indeed gone out. Unfortunately, only one of the rods had actually gone back into the core, and the other three had fallen just far enough to clear the light switches and were hung in the tubes.

Still on the phone, the supervisor then told his assistant at the control desk to press buttons 4 and 1, just to make certain that the shutdown rods were firmly seated in the full-in positions. The assistant immediately proceeded to carry out this instruction, but to do so he had to put down the phone and use both hands. The supervisor, realizing almost instantly that he had meant to say push 4 and 3, tried screaming his correction into the phone receiver. He could not be heard with the control room phone sitting on the desk. By pushing button 1, the assistant had pulled the entire safeguard bank out of the reactor, overriding button 4. With three other rods stuck in the out position and the core full of heavy water, the reactor was now supercritical, with power doubling every 2 seconds. It was 3:07 P.M.

Red lights were coming on, indicating that the shutdown rods had come out, instead of going in, and the assistant found this surprising. After 20 seconds of contemplation, the power level had risen to 100 kilowatts. He reached forward and hit the red scram button with his palm, which was supposed to kill the power to the electromagnets that were holding up the shut-off bank and drop them into the core. Two of the red lights remained on. All four should have gone out. It turned out that only one of the four rods had dropped, and it took one and a half minutes to do so. Two were stuck in the full out position, and one had slipped down just far enough to clear the light switch.

Reactor power was indicated by a Leeds and Northrup Micromax galvanometer recorder with a lighted spot moving across a scale, and it seemed to indicate that power was still rising. The power level was now 17 megawatts. The reactor had been set up for very low-power operation, and the rubber hoses could not run water through the cooling tubes fast enough to handle the megawatts of heat. The water boiled furiously.

The boiling action of the plain-water coolant caused two problems. Moving water can absorb heat and carry it away. Steam cannot. The voids caused by steam bubbles reduced the cooling effect to zero, and the temperature in the fuel rods climbed quickly. In a heavy-water moderated reactor such as NRX, the loss of water from the cooling tubes was beneficial to the fission process. With no tap water in the core to occasionally absorb a neutron, the power level took off too fast for the power recorder to follow it. The instruments ran off scale.[80]

By now there were two physicists, an assistant reactor branch superintendent, and a junior supervisor in the control room, and they were all getting frantic. The only thing they could do now was to drain the heavy-water moderator into the holding tank downstairs and stop the fission action. The superintendent gave the word, but one of the physicists was already lunging for the dump switch.[81] It was 44 seconds since the assistant had pushed button 1, and it took five seconds for the dump valve to fully open.

---

80 Later analysis of the accident found that the reactor power had peaked at 80 megawatts, far outside the designed power level. Most nuclear reactors have a digital simulator for training operators and investigating odd operations without the risks of using the actual system. In the case of NRX in 1952, no digital simulator existed, but they did have an analog electronic simulator built using vacuum-tube operational amplifiers. The Micromax recorder from the control room was connected to the simulator, and the accident was duplicated so as to give the instrument the same response it had given during the power excursion, running off scale. From this analysis along with the physical damage to the reactor, the 80-megawatt maximum power was confirmed.

81 Oddly, the superintendent said "Dump the polymer!" I would have thought it correct to say "dump the deuterium oxide" or perhaps "dump the heavy water," or even "dump the moderator," but "dump the polymer"? Was "polymer" a code word for still-secret heavy water?

As the heavy water drained, it occurred to the men that the helium tank would not be able to keep up with the sudden loss of heavy water, and the suction could implode the aluminum calandria. The helium pressure gauge went off scale low, confirming a possible problem, and the superintendent reached for the dump switch and turned it off. Unexpectedly, the domed lid on the helium holding tank continued to go down, indicating . . . what? After thinking about it for a few seconds, he turned the dump valve back on. To everyone's relief, the power level dropped to zero and all the instruments returned to scale. The assistant superintendent declared the reactor power excursion over, having run out of control for about 62 seconds, and he left the control room to tell his boss, the reactor superintendent, the bad news and then the good news.

The incident was not quite over. The moving top of the helium tank, having lost all its gas, fell to its lowest possible position, cocked slightly, and jammed in place. In the basement, the door to the base of the reactor had been left open, and an operator could see water gushing down a wall and rolling through the doorway. It was radioactive. The metallic uranium fuel, deprived of cooling water, had melted, burning through the aluminum tubes which defined the fuel rods and the surrounding water-cooling jackets. The entire coolant inventory was draining into the basement, taking with it whatever fission products in the molten fuel could be dissolved out.

The assistant superintendent returned to the control room only to meet an operator at the entrance with interesting news. The floor had trembled and a low rumble came from the reactor, with water spurting out the top. Just then, the radiation alarms started going off. A hand-held "cutie pie" ion chamber showed 40 mr/hr in the control room and 90 mr/hr at the door to the top of the reactor.[82] Opening the door made the detector jump to 200 mr/hr. None of this was normal, by

82 "mr/hr" means millirems per hour, or the rate at which a man is absorbing radiation. "rem" means roentgen equivalent man, or a roentgen of absorbed radiation corrected for the sensitivity of an adult male, and a millirem is 1/1,000 of a rem. One roentgen is 0.01 joules of energy from radiation absorbed per kilogram of body mass. To put this all in perspective, to exhibit any clinical effects of radiation poisoning, a man would have to be exposed to over 25,000 mr/hr for an hour. Being exposed to 25 mr/hr for 1,000 hrs for some reason does not necessarily give the same effect.

any stretch. The phone rang. It was the chemical extraction plant next door. Their air activity monitor had gone off scale, and they wanted to know what was going on. Sirens started going off all over Chalk River, and plant evacuation was called for.

The reactor with its coolant running out and all normal seals broken had sucked out all the helium in the holding tank and was now pulling in air through unplanned holes. The uranium, heated by uncooled fission to beyond the melting point, had started oxidizing, pulling oxygen out of the water, leaving hydrogen gas. Given the new influx of fresh air, the hydrogen ignited, and the sudden oxidation, or the explosion, sent the top of the empty hydrogen tank flying upwards until it stuck in the fully extended position. It had now been 4 minutes since the assistant pushed button 1. The radiation level at the control room door was now 900 mr/hr. The reactor staff donned respirators, to prevent breathing radioactive dust into their lungs, and the radiation level in the basement, near the north wall, reached 8,000 mr/hr, which was extremely serious. The heavy-water holding tank in the basement was full, so the dump valve was closed at 3:37 P.M. There was no fear of a return to criticality, as the heavy-water moderator was now contaminated with tap water, and the reactor core was a shambles. After a few days, the basement had filled with one million gallons of water containing 10,000 curies of various radioactive fission products.[83]

This accident was small, but it was a harbinger of things to come, being the world's first core meltdown and the first hydrogen explosion in a nuclear reactor.[84] These events would happen again, perhaps on a larger and even more dangerous scale, but in numerous ways the NRX accident was typical of nuclear disasters. Not one person was harmed, and the medical histories of personnel involved were studied

---

83 10,000 curies is a great deal of radiation. One curie is 37 billion nuclear disintegrations per second. To own a detector calibration sample of over 10.0 microcuries requires a federal license and a lead vault.

84 Whenever I use the term "meltdown" when writing about something nuclear, some of my colleagues start breathing hard and call me to task, arguing that this word brings on images of the reactor turning to fluid and running out on the floor. They would rather I call it something like "a modification of the fuel matrix geometry." Sorry, guys, but when the fuel temperature exceeds the melting point and it liquefies at least a portion of the reactor core, I call it a "meltdown," just for simplicity.

for decades afterward, looking for health issues that could be attributed to having worked at Chalk River in 1952. The reactor itself was a total loss, and the world's first nuclear accident cleanup would begin shortly.

What did the scientific community learn from this accident? Very little, I am sorry to say, although there were lessons aplenty to glean. The problems at NRX that led to the accidental series of events were operators trying to out-think the system, woefully inadequate instrumentation, and the use of tap water to cool a reactor having a separate, highly efficient moderator. Through the remaining years of the 20th century and into the next, these fundamental problems would destroy nuclear reactors.

The most difficult problem to handle is that the reactor operator, highly trained and educated with an active and disciplined mind, is liable to think beyond the rote procedures and carefully scheduled tasks. The operator is not a computer, and he or she cannot think like a machine. When the operator at NRX saw some untidy valve handles in the basement, he stepped outside the procedures and straightened them out, so that they were all facing the same way. They should not have been manipulated, but why did the valve handles exist if they were never to be touched? Someone in the past had started to correct this by removing the valve handles, but he had stopped with four handles left. They were poorly labeled, if at all. Inadequate labeling of controls is a disaster setup, given that a non-computer mind can try to think around it. The supervisor corrected the operator's error, as he should have, but then he thought beyond the simple repositioning of the valve handles. He stepped outside the problem, wanting to improve the situation even further with an extra effort. He called up to the control room and made a simple error in his instruction to the junior man, who, unfortunately in *this* case, did not think beyond what he was told to do and, like a computer, did exactly as he was told. The telephone as a communication link failed at this time, due to its position on the console, and this was similar to the telephone problem at ZEEP. The assistant did not receive the counter-instruction.

When the moderator was dumping out of the calandria, the men thought beyond the task of stopping the fission. The heavy water was draining too fast, and the atmospheric air pressure outside could buckle the reactor vessel. They stopped the flow. This train of thought, trying to prevent stress to the reactor vessel when in reality the core was melted and there was nothing left to save, is not unusual in times of nuclear stress. The same short decision chain, thinking outside the box and trying to reduce the physical damage to a minimum, would take out the entire Fukushima Daiichi power plant in 2011.

At this point in the NRX accident, the inadequate instrumentation was the problem. The red lights were supposed to indicate that the shutdown rods were in the up position, completely out of the core. One was to assume that if the rods were not out of the core, then they were in the core. In reality, these lights did not give a clue as to the position of the control rods. The best one could surmise was that the rods were not touching the switches at the top of travel when the lights extinguished. This lack of position information from a very important mechanism combined with an operator's out-thinking the system would cause another meltdown 27 years later at the Three Mile Island power plant in Pennsylvania.

Cooling a heavy-water- or graphite-moderated reactor with tap water because it is inexpensive remains in effect today, even though when the water goes absent the reactor goes supercritical. Canadian CANDU reactors, still using heavy-water moderation, are in use all over the world, and the latest designs use tap-water cooling in the core. Certain Soviet-era Russian reactors, such as the infamous RBMK-1000 at Chernobyl in the Ukraine, are graphite-moderated with in-core tap-water cooling, and this feature, the "positive void coefficient," helped lead to the worst nuclear reactor disaster in history in 1986. There are 11 of these plants still running.

It took two years to get the radiation spill cleaned up and the NRX back to working condition. The reactor was pulled out of the building and buried somewhere in the yard outside. The remaining water in the coolant tank was emptied into the Ottawa River, and the basement water was pumped into the tank for temporary storage. Chalk River now had a basement with every square inch contaminated with fission products, a tank full of radioactive water, and an empty space where the pile used to be.

The technique for dealing with high-radiation cleanup is to make certain that each worker has an acceptably small cumulative radiation exposure while near the contamination. With the radiation exposure rate fairly high, this means that one man can only work a certain number of minutes. Therefore, the entire job would require a great number of men, each having a security clearance and knowledge of radiation contamination.

In stepped Captain Hyman Rickover, head of the United States Navy's secret nuclear submarine program. Rickover was developing a compact pressurized-water reactor plant that would fit in the cramped engine room of a submarine, giving the submersible weapons platform the ability to remain hidden under water continuously. It was an ambitious project, and, for the strict requirements of this machine, new techniques and materials had to be invented. One of Rickover's brilliant ideas was to use an alloy of zirconium, zircaloy, as a high-temperature, corrosion-resistant fuel cladding and structural material for the inside of the reactor. The behavior of zircaloy under heavy radiation bombardment was unknown, and the only way to find out how it would perform in a submarine reactor was to test a fuel assembly in a high-power research reactor, of which there was none in the 48 states to be borrowed or commandeered.

There was one in Canada, NRX.[85] Although it was technically forbidden by federal law to ship enriched uranium out of the country, Rickover did it anyway, flattering the Chalk River Laboratory with praise and talking them into allowing him reactor time. He shipped a model fuel assembly over the border under guard, labeled "materials test." The time spent in NRX gave valuable results. A black crud built up on the fuel, apparently iron, nickel, and cobalt oxides coming from dissolving stainless steel.[86]

Shortly after the Mark I fuel element was shipped back from Canada, NRX melted, and Rickover in his optimistic mode saw this

---

85 A duplicate reactor, CIRUS, exists in India. It was supplied to India by Canada (heavy water supplied by the U.S.) for peaceful purposes in 1954, but the Indian government used it in secret to produce plutonium, which was used in 1974 to build the Operation Smiling Buddha atomic bomb. Canada was pissed. CIRUS was shut down on December 31, 2010.

86 A nuclear engineering myth has it that the word CRUD is an acronym, meaning Chalk River unidentified deposits, and it came out of the NRX cleanup operation. Once all the contaminating material had been sorted out, the unknown residue was classified as CRUD. In truth, the acronym came out of Rickover's fuel test program, and there is ample reason to believe that the word existed before 1952.

as a bonanza. The United States had a great deal of nuclear expertise, more than all other nations of the world combined, but we had no hard experience cleaning up a major nuclear radiation spill. In 1952 we had not actually spilled anything of significance, and this would be a terrific opportunity to discover the gritty details of what it would take to decontaminate an accident site. He generously volunteered 150 nuclear workers, all security cleared, to Canada to assist with the cleanup. He wanted 50 from the Bettis Lab, 50 from General Electric, and 50 from Electric Boat.[87] Naval officer James Earl Carter, the eventual President of the United States, would lead the Navy personnel.

The 862 staff members at the Chalk River Lab participated in the cleanup, as well as 170 Canadian military personnel and 20 construction workers who were in charge of cranes and digging machines. A pipeline was constructed, leading to a flat, sandy area about a mile from the plant, and the heavily contaminated coolant out of the basement was pumped there and allowed to seep into the ground, where it apparently disappeared. Fission product contamination would not be disposed of in this way now, of course, but this was early in the evolution of nuclear power. Carter and his men spent time scrubbing the floors and walls in the basement, dressed out in full rubber suits and respirators, trying to erase all evidence of an accident. A new calandria was installed, and the improved NRX, having a new set of operating procedures, was up and running for the next 40 years, finally decommissioned in 1992. Officer Carter's impression of stationary nuclear reactors would remain somewhat warped forevermore.

Right next to the NRX in an even larger brick building, eight stories tall, the Canadians built a more powerful reactor, the NRU, the National Research Universal pile. This facility was designed starting in 1949 to run on natural uranium, using heavy water as the moderator, producing 200 megawatts of heat. There were about 1,000 fuel rods in NRU, all installed vertically from the top in a hexagonal matrix of aluminum tubes, sitting in thousands of gallons of very expensive

87 This initial offering adds up to 150 men, and this agrees with Chalk River's official count, but according to the final report Rickover sent 214 men from his naval reactors program to Canada, expending 1,300 man-days of cleanup effort.

heavy water.[88] Each fuel rod was 10 feet long, consisting of a stack of metallic uranium cylinders sealed against moisture and air in a welded aluminum sheath. Its first full startup was on November 3, 1957, and the reactor would be put to use testing materials and techniques for use in the CANDU commercial power reactors. In 1964 NRU was converted to use highly enriched uranium as the fuel, eliminating the need for regular refueling, and the power was dialed back to 60 megawatts. In 1991 it was modified again to use less-expensive low-enriched fuel, and the maximum licensed power was increased to 135 megawatts. NRU is still running. It is the primary source for medical isotopes in the Western Hemisphere, and it supplies these critical diagnostic and treatment materials for 200 million people per year in 80 countries. Eighty percent of all nuclear medical procedures use technetium-99m, and two thirds of the world's supply of this material comes from NRU, the now-antique research reactor at Chalk River.

Although there were advantages to using natural uranium fuel in Canadian reactors in 1958, there remained problems. The fissile content of the natural fuel is so marginal, the fuel can stand no contaminants or dilution, and therefore it has to be pure uranium metal. Unfortunately, in its metallic form uranium will catch fire in air and burn like gasoline. For this reason, most nuclear reactors use uranium oxide fuel. Uranium oxide is already burned, and it cannot possibly burst into flame. Uranium oxide also has a much higher melting point, 5,189°F, than uranium metal, 2,070°F, and it will be the last material to melt in a power excursion.

There is also an inherent problem in making power by nuclear means. You can turn off the fission chain reaction instantly, but you cannot absolutely turn off the heat generation. The process of making high temperature using nuclear reactions has everything to do with changing the identity of elements. Uranium splits into two lesser elements, and this debris left over from a fission event all together weighs less than the original uranium nucleus. This mass deficit converts into pure energy. However, it does not happen all at once. The two nuclei left over after a fission event are neutron-heavy nuclides, and

88 In small quantities, 99.8% deuterium oxide is about a dollar per gram, or a dollar per cubic centimeter.

they are unstable, bound to decay into nuclei of slightly lesser weight and releasing yet more energy from the fission. This nuclear decay is time-dependent, with each decaying species having a characteristic probability of decay, or a half-life. Most decays result in yet other unstable nuclides, which continue to decay in steps until stability is reached. While almost all of the fission energy is released in a few seconds, there is about a one percent residue that takes its time. It can take millions of years for the energy to be completely gone. In theory, it never completely goes to zero. The uranium used in the startup of the ZEEP reactor, wherever it is buried, is still making energy from fissions occurring in 1945.

For a modest research reactor, running at say one kilowatt thermal, the residual heat after shutdown from a long run is about 10 watts, or the heat from a Christmas tree light spread out across the entire machine. For a commercial power reactor running at a billion watts, the heat after shutdown is about 10 megawatts. To put this in perspective, the *Nautilus* nuclear submarine running at full speed used about 10 megawatts. The higher the reactor's operating power is, the higher is the latent heat being generated after shutdown. This inescapable problem would plague nuclear reactors from the beginning of the art, and much engineering has gone into its solution. This, and the fact that uranium metal burns, would work together to form the third great nuclear accident at Chalk River. It was late in the day on Saturday, May 24, 1958, and the refueling crane was busy.

Running on natural uranium, the NRU in 1958 is refueled often, using a semi-robotic traveling crane running on rails above the reactor.[89] To extract a fuel rod, the crane operator punches in an address, like M-25, on a keyboard, and the crossed crane-arms move with electric motors in or out, left or right, to find the correct location and position the fuel flask over the indicated rod.

The fuel flask is a big metal cylinder, tall enough to contain a fuel rod and filled with heavy water so that the rod, still hot from fissioning, remains below melting temperature after being extracted. To

89 How often? I'm not sure. In 1958 the "research" reactors at Chalk River were running constantly, making plutonium-239 by conversion for the secretive British nuclear weapons buildup. Reactor schedules from this period are not readily available.

grab the rod and haul it from the reactor core, a cylindrical tool, small enough to fit in the aluminum tube containing the fuel rod, lowers down until it has the fuel rod in its grasp, clamps down on it, and then gently pulls it up into the flask. The crane then moves over an open pit filled with cooling water, slowly lowers the entire flask assembly into it, releases the fuel rod, and returns the flask to its high position. The crane mechanism is electrically interlocked against any action that it does not consider appropriate, so the operator is blocked from doing anything unusual.

The day before, on Friday, NRU was feeling fatigued after having run full tilt for a week, and the instruments that monitor the fuel were acting erratic. The fission products were building up in the fuel, and the heavy water was becoming polluted with stray radioactive debris. With no further warning, NRU called it quits, scramming itself and blowing all the control rods down into the core with a loud clap and a shudder of the concrete floor. Something in the long list of reasons to scram had irritated the system. With a cursory look, the weary operations staff could find nothing obviously wrong, so they restarted the reactor. WHAP! In went the controls a second time. It really did not want to be started. Loud, rude alarms started going off.

Taking it seriously this time, the operators determined that three fuel locations were extremely radioactive, and this indicated that the aluminum fuel rods had broken open. Fission products were leaking out, at least. They would have to deal with it tomorrow.

Starting Saturday morning, the crane operator went after the first fuel rod thought to be having trouble. He positioned the crane over the hole, sent the tool down, grabbed the rod, pulled it up into the flask, trucked it over to the spent-fuel pool, lowered it down, and released it. It took all morning, but the operation went smoothly. Next on the list was the fuel rod in hole J-18. This one seemed to be not feeling well. It was swollen up like a poisoned dog, and the extraction tool would not fit over the top of it. The crane operator called for a larger tip. It was a pain to change tips, and two workers spent hours installing the larger sized unit. They were so focused on the task at hand, neither of them noticed that a valve had failed open on the flask and all the heavy water had drained out. The fuel flask was dry, without a drop of coolant inside.

With the new tip installed, the operator punched in J-18 and watched as the crane moved over the hole, lowered the tool, and successfully clamped onto the damaged rod. Up she came, easily this time, but just then the operators noticed that the flask was innocent of cooling water. Desperate, they tried to turn on a pump and get some water flowing, but the interlock system prevented one from starting a pump when there was no water. The crane operator tried to re-insert the fuel rod back into the core. It would not go in. It cocked sideways and jammed.

Some of the crew were already decked out in rubber suits and respirators, and at this point they jumped to it, pulling hoses over the top of the reactor to try to hit the fuel rod with cooling water. By this time, the fuel with its aluminum sheath cracked open had been without coolant for nearly 10 minutes. The crane operator reversed the insertion tool to pull the damaged rod back up into the flask. The motor groaned. Something snapped. The operator pulled the flask off the reactor face, and it came up with half the fuel rod still jammed in the core and half of it held by the crane. The fuel in the dry flask caught fire.

Seeing a need to work quickly, the operator tried to send the flask over to where the crew was standing by with the hoses. It would not move. The interlock system had detected something wrong in the flask, and under this condition it would not allow crane motion. Radiation alarms in the building started going off, blaring loudly as the men flipped switches and turned valves, trying to get something to work. Smoke was streaming out the end of the insertion tool. The crane was completely locked up, and the intricate safety system that had been designed to prevent accidents was working against them.

Thinking beyond the procedures manual, a technician took the cover off a relay panel and hot-wired the system using a jumper cable. The crane could now be moved under manual control, and the operator moved it toward the fuel pool. The men in the rubber suits hit the hot flask with high-pressure water from the hoses. As it moved closer, they were able to attach a hose to the top of it and start to fill it with coolant. To their dismay, they found that the valve at the bottom of the flask was still stuck open, and the cooling water washed the fission products out of the now opened, burning fuel rod segment and out the valve. The highly radioactive water splashed onto the floor and down the steps to the basement. They started to see cleanup duty and overtime pay in their future as the radiation monitors slid off scale. Things were starting to get very complicated.

The crane was now moving in the right direction, making for the safety of the fuel pool. The water in the pool would quench the fire, bring the hot uranium down to room temperature, and shield the reactor building from the intense radiation that comes from reactor fuel that spends a week fissioning at full power. To get there, it had to move over the repair pit, a sunken area in the floor. Directly over the pit a piece of uranium metal, three feet long and burning tiger-bright, fell out the end of the tool and hit the floor with a shower of sparks.[90] At that instant, the problem with the stuck fuel rod became a major accident. The entire reactor building was now being contaminated with radioactive fission products boiling off a cylinder of partly fissioned metallic uranium, flaming freely in the air. Directly over the pit, the radiation level was over 1,000 rem/hr, which could mean death for anyone lingering there, looking down at the burning metal. Something had to be done to put out that fire, and spraying it with water was not going to work. They would have to cover it with sand. Quickly.

The steadfast rule of working in a high radiation field still applied: use a large force of men with each individual given a small slice of time under hazard. Everybody in the building was suited up. Bookkeepers, janitors, geeks, forklift drivers, directors—all were given the three-minute instruction on how to breathe through a full-face Scott respirator and lined up along the wall. Each would have to climb an open steel stairway to a catwalk, walk quickly across until the burning fuel was straight below, empty a bucket of sand, and then continue moving and exit out the other side of the building.

The first up was an accountant. He climbed the dizzying stairs, not looking down, trotted to the point specified, dumped the load, and exited in haste. In those few minutes, he absorbed his entire radiation dose allowed for one year working at Chalk River. Everyone working at the lab site had an assigned dosimeter always attached to his clothing,

90 Some accounts say that the segment was only one foot long. The wide extent of resulting contamination would indicate the larger piece.

and his cumulative dose had been tabulated daily. In 15 minutes of sanding, the fire was out and the source of radioactive vapor was covered. The entire building was contaminated, because the ventilation system had been jammed in the open condition, and the air in the highbay had circulated all over, even soaking the outside. It was just before midnight, and the cleanup operation started immediately. It would take three months of round-the-clock work by more than 600 men to restore the NRU to operation. They were careful, and nobody was injured by radiation exposure.[91]

What was learned from this accident? This was a case that argued against the danger of having operators trying to outthink the system. The operators found the fuel-handling equipment locked up and unable to move because of the safety rules rigidly wired into the electrical circuits. They reasoned their way out of this predicament, modifying the system as wired and making the crane do something that it was never allowed to do, moving a broken fuel rod. Without the ability of men to think beyond the designed parameters, this accident could have been a disaster.

Not really. In fact, if the operators had not rigged the circuit to move the crane, the burning fuel rod would have remained over its hole in the top of the reactor, where it was supposed to be, and the flaming segment would not have fallen into the repair pit and spread contamination all over the facility. If the operator had lowered the open bottom extension of the flask back down onto the reactor face, the fuel would still be burning, but slowly. It would lack the free air circulation afforded

91 So says the official report, but there may well have been at least one victim of radiation contamination. Bjarnie Hannibal Paulson was a corporal in the Royal Canadian Air Force, and in 1958 he was transferred to Camp Petawawa to participate in the NRU cleanup. He began to suffer from cancerous carcinomas in 1964. Backtracking these outbreaks, it looks as if Paulson's three-layer rubber suit had been covered with an alpha-emitting dust during a cleanup procedure. Removing his respirator with his right hand, he seems to have first rubbed off a surface coating of contaminated material with his palm. He then inserted his hand under the respirator mask and across his face to remove it, burying the dust in hair follicles. Alpha radiation is extremely difficult to detect in sub-surface conditions, and his whole-body count did not show it. He then showered, but the dust was not immediately on the surface and the dustmotes remained in place for years, eventually giving him radiation-induced skin cancers. At this writing, AECL admits no responsibility for Paulson's cancers, and he has been unsuccessful in applying for a military disability payment.

by being hung in mid-space. The highly radioactive ash would have fallen straight down, back into the heavily shielded reactor. Eventually the crews would have been able to spray water onto the overheated flask from hoses, and the cleanup operation would have been limited to one of a thousand fuel locations. The problem was that the staff was more worried about disabling the reactor than they were about a dangerous, three-month cleanup operation. Cooling the fuel flask with tap water running out the bottom could have diluted the deuterium content in the reactor moderator, and letting a burning fuel segment fall into the core might have contaminated the moderator with fission products. In retrospect, this course would have been better than what happened, and the crane was correct in disallowing motion under the circumstances. The greater problem was caused by a recurring human need to reduce the problem immediately and make it go away.

These examples of technical adventures in the 1950s, a period that gave us thermonuclear weapons, Scientology, and the Urantia Book, would not be repeated in exactly these ways, but the pioneering accidents in Canada would outline fundamental problems that would continue to bug the nuclear power industry. By the end of the decade, the art of fissioning uranium to make heat was in its adolescence, only 18 years old, and the release of mayhem had only begun.

Chapter 4

BIRTHING PAINS IN IDAHO

"Everything went quite well as long as a mechanic from Augsburg and an engineering school professor were permanently on hand."

—a note from an early diesel engine owner
to Rudolph Diesel, 1898

CANADA WAS SHOWING GREAT INITIATIVE, independence, and creativity in their evolving quest for nuclear power at Chalk River in the 1950s, but whatever they were doing, the United States was doing it larger, faster, and in dazzling variety. Experimental projects, trying new and previously unheard-of ways to build atomic piles, were underway in Illinois and New Mexico, while plutonium and enriched uranium were being produced by the ton in Washington State and Tennessee. Europe was behind but gathering speed in England, France, and the Soviet Union.

In a rare fit of post-war brilliance, the government of the United States decided to relieve the military services of research, development, manufacturing, testing, and ownership of any nuclear weapons or power systems, beginning January 1, 1947. Everything having to do with nuclear reactions, including facilities, access roads, employees, and waste dumps, was transferred from the Army Corps of Engineers to a new federal agency, the Atomic Energy Commission, or the AEC. The Navy would own the nuclear submarine *Nautilus*, but the engine and the fuel would belong to the AEC. If you kicked up a rock in your back yard and it was carnotite, a uranium ore, it belonged to the AEC. David Lilienthal was appointed chairman. Lilienthal had proven his mettle by keeping both hands around the throat of the Tennessee Valley Authority as it evolved from a make-work project under President Franklin Roosevelt into a major source of electrical power. Piloting this new agency was no easier than controlling the TVA. The AEC was tasked with promoting world peace and improving the public welfare while developing weapons that could return civilization to low-population Stone Age conditions in a few seconds.

In 1949 the AEC decided that the public welfare could be improved by moving all of the experimental nuclear reactor developments to the middle of nowhere in Idaho. Montana had been considered for the site, but there was plant life and some animals scattered around in it. Roughly in the middle of Idaho, between Arco and Idaho Falls, was a flat desert resembling the surface of the moon. It even had craters, left over from its tenure as the Naval Proving Ground during the war.[92] Frankly, if a reactor experiment happened to go rogue and self-destruct, there was not much there to be harmed, and it was good practice to concentrate all the dangerous stuff in one place. The new facility was named the National Reactor Testing Station, or the NRTS, and its headquarters was built in Idaho Falls.

With more than 50 prototype reactors and some large-scale chemistry experiments, personal safety at the NRTS was not bad. There were a couple of meltdowns and three steam explosions that were considered to be accidents. Two of the blow-ups were staged events

92 The US Navy tested everything explosive at the Proving Ground, from machine guns to the largest battleship cannons. A favorite at the Proving Ground was testing blast shields, boxcars, and explosive bunkers to see how they would perform when subjected to 250,000 pounds of TNT going off all at once. Although the Navy no longer owned the facility, they were allowed to test 16-inch guns from the USS *New Jersey* battleship as late as 1970. Unexploded ordnance was still showing up on the ground in 1990.

that happened to have larger effects than were expected, but one, the infamous SL-1 incident, was the strangest and most tragic reactor accident in American history. Lessons were learned and stringently applied in most cases. The causes of these unfortunate incidents were all over the map, and their listing does not necessarily help us pin down anything we saw in Canada as the universal reason for problems using nuclear fission.

The first accident at the NRTS involved the Experimental Breeder Reactor One, or the EBR-I, in 1955. EBR-I was an odd piece of atomic enthusiasm. It was designed in 1949 at the Argonne National Laboratory in suburban Chicago with the encouragement of Dr. Enrico Fermi, the Italian who conceived the CP-1 pile back in '42 and the *pontifex maximus* of reactor physics. The design team was led by Walter Zinn, a nuclear physics Ph.D. from Canada. When the CP-1 was started up at the University of Chicago on December 2, 1942, Zinn was responsible for withdrawing the safety control rod, called the "zip." By 1946 he was the head of the Argonne Lab, in charge of all the nuclear reactor research in the United States.

The importance and omnipotence of Walter Zinn's monopoly was felt by the other labs still standing after the war. The Singing Oak Ridge Physicists expressed their feelings with this song at the 1947 Christmas party, to the tune of *Deck the Halls*:

Pile research is not for us'uns.
Fa la la la la, la la la la.
Leave it to our Argonne cousins.
Fa la la la la, la la la la.
Engineering is for we'uns.
Fa la la la la, la la la la.
We're a bunch of dirty peons.
Fa la la la la, la la la la.

Zinn was comfortable at Argonne, but just as the breeder reactor design started to gel, this and every other reactor construction project was moved to Idaho, which some people came to know as "Argonne West."

At the time, there was not a single electrical-power-producing reactor in the entire world. All the piles were used either for plutonium production or as neutron sources for research. Uranium mining out west was a small enterprise, usually involving old vanadium mines, and the thought of relying on Canada or the Belgian Congo for the fuel to run the industrial economy of the United States was not attractive. There simply was no promise of enough uranium to do anything but make bombs and run a few submarines. Rickover's nuclear sub fuel was expensive, exotic 50-percent enriched U-235 from the still-top-secret gas diffusion plant at Oak Ridge, and there was no way that commercially competitive power could be produced by burning it in a privately owned pile.

Backing up at Oak Ridge by the hundreds of tons was an inventory of depleted uranium-238, the waste exhaust from the uranium enrichment process. In theory, this useless stuff could be converted to fissile plutonium-239 by fast neutron capture. This was possible in the graphite-moderated converter reactors, because the fissioning U-235 was co-located with the U-238 in the natural uranium fuel. Before it had a chance to escape the fuel rod and be moderated down to thermal energy, a freshly born neutron running at ramming speed had a chance of being captured by a nearby U-238 nucleus. The uranium nucleus, heavier by one captured neutron, would then be U-239, which would beta-minus decay into neptunium-239 with a 23.47-minute half-life. The neptunium would beta-minus decay in its own sweet time with a half-life of 2.365 days into almost-stable plutonium-239, a very useful nuclide. Once it made it out of a fuel rod and into the graphite, a neutron's value changed from being able to convert U-238 to being able to cause fission in another U-235 nucleus as it slowed down to thermal speed. Walter Zinn postulated a reactor that could run without a moderator. All the neutrons would be fast, causing fissions at top speed rather than slowed down, and therefore neutrons not used for fission would be able to convert the wasted U-238 into useful plutonium. Each fission released more than two hot-running neutrons, and only one was required to trigger another fission.[93]

93 Probability factors into all sub-atomic processes. In the case of neutron release, every 100 fissions in U-235 at thermal speed produce eventually an average of 243 free neutrons. Some of those neutrons are delayed, and may wait hours to release. In Pu-239, you get 290 neutrons per 100 fissions, but when high-speed neutrons cause fission in U-235, you also get nearly 290 neutrons.

Zinn's concept was a reactor that burned the artificial nuclide Pu-239 to make power. Manufactured plutonium was almost 100 percent pure, containing no worthless isotopes, and it did not need the advantage of thermalized neutrons to fission efficiently in a fast-neutron environment. The fission process in pure Pu-239 was not marginal and barely able to make it, as was the natural uranium fission scheme. The plutonium reactor core, bleeding fast neutrons from every external surface, would be surrounded by a blanket of pure U-238. Wasted fast neutrons escaping the fission process would be captured in the uranium shield, converting in a few days into more Pu-239. Calculations showed that such a pile would produce more Pu-239 than it burned in the fission process. It seemed utopian. It was a power source that produced more fuel than it consumed. There was enough U-238 piling up from the weapons production to provide the total energy needs of civilization for thousands of years into the future, without drawing down the weapons material or the submarine fuel. This gorgeous concept would be known forevermore as the "fast breeder reactor." There were some minor complications to be worked out.[94]

A power reactor must have some means of transferring energy from the fission process to the power-generation process. In Rickover's submarine reactor, being assembled for testing just over the horizon from the EBR-I site, the medium of transfer was water. The water absorbed heat from the fission by direct contact with the fission neutrons, which were born going 44 million miles per hour. Zinn's breeder had to be different. There would be no neutron moderation, no slowing them down to thermal speed. The coolant in the breeder had to be heavy, metallic, and liquefied, so that it could absorb heat from the reaction by conduction to metal surfaces in the reactor structure without slowing anything down. Neutrons hitting heavy metal nuclei would simply bounce off and retain their speed. Both transfer methods, water and metal, kept energy from accumulating in the machine and causing it to melt. The fluid medium

94 Recently the phrase "fast breeder reactor" has been changed to "Integral fast reactor." I'm not sure why. A sign on Highway 20 leading into the NRTS cautioned travelers, "WARNING: DO NOT DISTURB BREEDING REACTORS." It was a real knee-slapper at the time.

would flow into the bottom of the reactor core, picking up heat and exiting through a pipe at the top. For the EBR-I, the logical choice of coolant was a mixture of sodium and potassium metals, pronounced "nack."

By mixing sodium with potassium, the melting point of the metal could be brought down to about room temperature. It was a very sluggish fluid, but it would flow, particularly when heated to 900 degrees Fahrenheit with the reactor at full power. Unlike water it was, of course, completely opaque. There was no way to visually inspect the reactor internal structures with the coolant in it, the way it was possible using water moderator. The sodium tended to activate under the neutron bombardment of fission, becoming sodium-24, throwing beta and gamma radiation with a 15-hour half-life. The potassium was a lesser activation risk, but it would contribute some to the induced radioactivity. The worst characteristic of nack is that it would react vigorously with air, water vapor, or particularly liquid water, becoming sodium and potassium hydroxides. In their undiluted form, these are nasty chemicals. Step in a puddle of it, and it will first dissolve your heavy leather boot followed by the foot inside the boot. Anything aluminum touched by it, like an airplane or a Pullman railway car, turns to white powder. There were entire reactors made of aluminum, but this one would have to be stainless steel.

Zinn found that plutonium-239 was impossible to obtain for reactor experiments. By 1951 the AEC was turning out plutonium-based implosion bombs on an industrial scale, and surplus material was simply unavailable at the time. It was easier to obtain 97-percent enriched uranium-235 for the EBR-I fuel, and it was an adequate substitute for the plutonium.[95] At Argonne, the 52 kilograms of bomb-grade uranium was fabricated into 179 stainless steel-clad fuel rods

95 The use of plutonium in a fast-neutron reactor had already been proven. In 1945, an experimental reactor was built at Los Alamos in a deep canyon, far removed from other activities in case it misbehaved. At the time, the code-name for plutonium was "49," and the scientists involved in the project were called "49ers." You can probably guess the name of the reactor: Clementine. It ran experimentally at a maximum power of 25 kilowatts for six years, until a fuel-cladding failure shut it down for good. The dangerous coolant was pure mercury, which would enthusiastically activate into radioactive Hg-203. The purpose of Clementine was not to advance power-reactor technology, but to test bomb materials under the neutron-soak expected in a nuclear explosion.

with some spares, made pencil-thin for good heat conductivity to the nack. The rods were mounted vertically in a steel tank, separated from each other in the preferred "hexagonal pitch" configuration so that the coolant could flow through them from bottom to top.

Control of a fast reactor is a bit touchy, because a smaller percentage of the fission neutrons are delayed than in the usual thermal reactors. With most of the neutrons born instantly upon fission and no slowing-down time allowed, the smoothness of the controls is diminished. The fission in EBR-I was managed using boron neutron-absorption rods, moved up and down into the reactor core by electric motors. To shut it down quickly, you could bring your palm down on the red scram button and operate the motors at maximum in-speed, or you could hit the reflector release button. Under this condition, the bottom half of the breeding blanket would unlatch and fall away by gravity with a floor-shaking CLUMP. Just losing the blanket and its tendency to reflect some neutrons back into the core was enough to kill the fission process and make the reactor quickly subcritical. In the middle of the control room there was also an alternate scram control. It was a triangular metal handle, hanging by a chain. Anyone in the room could pull that handle at any time and shut down the pile instantly.

In late May 1951, the experiment was ready for a power-up, and dignified guests and visitors, having endured a dreary ride out to the site, were eagerly anticipating success. With all the controls carefully pulled out, the reactor remained serene and subcritical. Zinn remained calm and estimated that the core was so completely dead, he would need another 7.5 kilograms of fuel to make it work. With a hint of reluctance the AEC granted him some more U-235 from military stockpiles, and it took three months to get it fabricated into fuel.

On August 24, 1951, the world's first fast breeder reactor was the first reactor at NRTS to go critical. At zero power it worked fine, but measurements indicated that the fuel assemblies would have to be modified if the thing was going to boil water and make steam, as was necessary for a power reactor. Two thirds of the fuel rods were shipped back to Argonne, where they were made shorter and wider by a hydraulic press. Arriving back in Idaho, they were arranged as a belt-line around the thinner fuel tubes.

On December 20, the reactor was run up to about 45 kilowatts. The hot metal coolant was pumped continuously through a heat exchanger with water on the other side of the heat transfer process. At 1:23 in the afternoon, an operator opened a steam valve connected to the heat exchanger, and steam ran through the power turbine. Four 200-watt light bulbs on the turbine deck began to glow white-hot. At that instant, electrical power was first generated by a nuclear reactor.[96] Zinn took a piece of chalk, wrote his name on the concrete wall of the building, and invited his entire team to do the same. Those names remain on that wall to this day.

Success followed success with EBR-I. Three days later, the crew was able to connect the entire facility to the generator and run everything off reactor power. They ran it at power until early in 1953, when they shut it down for a fuel and breeding blanket evaluation. Representative samples were sent back to Argonne for chemical analysis, and the results came back positive. There was more plutonium made in the blanket than there was uranium lost in the fuel. The breeder concept was verified, and the electrical power future of the world could now be moved beyond the point where the coal runs out. Zinn believed that in the short run, breeders would be the only type of reactor that could compete commercially with fossil-fuel power plants. He pressed on vigorously and started the plans for EBR-II.

Experiments continued, revealing the subtle unknowns in breeder reactor behavior. One characteristic was a particular nag. The breeder had a positive temperature coefficient of reactivity. That is, when the temperature increased, the thing went supercritical, and the operator would have to pull it back down by inserting control rods until it was back to the desired power level. Zinn had a theory

96 So it is written in the official history of the world, but nothing is that simple. On September 3, 1948, Logan Emlet opened the valve on a toy steam engine connected to a water pipe running through the air-cooled X-10 graphite reactor at Oak Ridge, Tennessee. The reactor, running at one megawatt, was making steam in the pipe, and the little engine started puffing. The flywheel was belt-driving a classroom demonstration generator, which was connected to a 1/3-watt flashlight bulb. It wasn't much light, but it was definitely the first. Everything done on the X-10 was still secret, so this historical generation of electrical power by nuclear means was buried. Buried deeper was the first use of a reactor to heat a building, which occurred in England in the fall of 1948. The reactor was named "Bepo," meant to sound like an unknown Marx brother. I'm not making this up.

that if he pushed the temperature to beyond 852°F, then the effect would reverse, pulling the power down. The logical way to run the temperature up and test this idea was to stop the coolant pump and see what happened. The scram channel for core temperature that would normally shut the reactor down if the core was hot beyond a threshold was disconnected.

On November 29, 1955, the EBR-I was ready for the experiment. It was kept at a low power of 50 watts, so that things would not get out of hand. The coolant flow was stopped with the controls set for a period of 60 seconds, meaning that the power level was multiplying by about 2.7 every minute.[97] The pile immediately started acting crazy. In three seconds the power had climbed from 50 watts to one million watts, and the period was at 0.9 seconds and dropping. The thin, unsupported fuel rods had buckled under the high heat at the center of the reactor core, bending inward, getting closer together, and improving the reactivity of the non-moderated pile. As the heat increased, the fuel bent farther inward, becoming less like a metal-cooled reactor and more like a Godiva assembly, a touchy, dangerous ball of fissile metal. Harold Lichtenberger, the reactivity specialist on hand, yelled to the operator "Take it down!" The assistant operator at the controls, not knowing exactly how to respond to such a request, reached for the button to initiate a slow control drive-in. It took two seconds for this error to sink in, with the power rising faster and faster. Operating on instinct, the senior operator lunged for the scram button, and the power excursion ended immediately.

The crew resumed breathing. One strong advantage of a liquid-metal-cooled breeder reactor over a water-cooled reactor was immediately obvious. If they had been running a water reactor in that kind of runaway condition, the steam explosion would have wiped out the building. In this case, there was not a sound, not a whisper out of the pile as it went wild. The only way they could tell that anything was wrong was by the instruments on the control panel. Nobody had

97 In the popular literature, I have seen the initial rate of power-climb for EBR-I listed as "doubling every 60 seconds." Not quite true. The "period" of reactor power increase or decrease is expressed in "e-folding time," or the time required for the power to change by a factor of e, where e is the root of the natural logarithm, 2.718281828459045. . . . In breeder parlance, "doubling time" is actually the time required for the amount of bred plutonium to increase by a factor of two.

been exposed to any unusual radiation, and the reactor looked just as it had that morning before the startup. Fifteen minutes passed and they were feeling very good about their design concept when the building radiation alarms started going off.

It is an ear-damaging, hair follicle-killing sound that penetrates down to your center of mass. It sounds like a pterodactyl screaming into the side of your head, as if we knew what a pterodactyl sounds like. BRAK. You can almost feel his hot, nasty breath condensing on your temple and rolling down your cheek. BRAK. BRAK. Your nerves unravel at the urgency of the signal. BRAK. BRAK. BRAK. It means "make for the door," or, as they say in the accident reports, "all personnel evacuated the building." You unconsciously drop whatever you were holding and swivel to face the exit.

Careful analysis of the incident revealed that the core was completely trashed and the primary cooling loop was contaminated with fission products. The United States had experienced its first meltdown, with the top half of the core liquefied and running south into the bottom half. Nack in the center had vaporized and forced its way into the breeding blanket as the power reached 10 megawatts, or 7 times the pile's designed capacity if the coolant were running. No personnel were harmed, and the accident was undetectable outside the building. If the senior operator had not hit the scram button, the reactor would have shut itself down anyway. At the time of the scram, the melted fuel was foaming in the metallic coolant, diluting in the nack and beginning to lose its supercritical configuration.

The AEC kept the incident secret out of fear that news of a nuclear reactor accident would affect public opinion in a negative way, which it did a year later when an account was leaked into the journal *Nucleonics*. The editor was merciless. Keeping such incidents secret deprived the engineering community of experience, and it was incorrect to keep information concerning a system that was meant to be handed over to the public out of the public view, particularly if it indicated something bad. The AEC agreed in principle and resumed doing what it was doing. The public, watching movies about giant insects mutated by A-bomb tests and digging backyard bomb shelters, was considered incapable of handling routine nuclear news. They were too saturated with

Hiroshima, Nagasaki, deaths by radium, and Japanese fishing boats to think rationally.

The EBR-I core was redesigned and reinstalled in 1957. The new U-235 core had zirconium spacers between fuel rods, developed in Rickover's S1W program 10.47 miles northeast of the site, and this embellishment seemed to prevent further power excursions due to mechanical effects. In 1962 EBR-I finally got a plutonium core, and it ran with good performance until 1964 when the old pile was shut down permanently. It was time to move on to a bigger, more advanced power plant, the EBR-II. The improved plant ran without incident until it was finally shut down in 1994. In its lifetime it made over two billion kilowatt-hours of electricity and, with the production of direct space heating, kept the workers at Argonne West, which had become an official installation at NRTS, from freezing in the Idaho winters.

By some opinions, the most bizarre adventure in fission development in the 1950s was not the nuclear-powered tractor wheel hub, the plutonium-fueled coffee maker, or even the atomic land mine. No, the prize for the most money thrown at the least likely application probably belongs to the Aircraft Nuclear Propulsion program, or the ANP.

The ANP program began in 1947. There was talk in the Defense Department about a possible nuclear submarine project for the Navy, and the Air Force wanted its own counterweight, a strategic bomber that could stay in the air for months on nuclear jet engines. The working slogan was "Fly early!", and an unrealistic development period of five years was estimated. Some reactor physicists knew better and expressed doubt as to the practicality of the plan. Putting a light-weight reactor operating at 2,000°F in an airplane and running it without killing everyone within a few hundred feet was not as easy as it sounded. Airplanes had to be made of thin, lightweight materials, like aluminum or magnesium. Reactors were set in concrete, with a lot of lead and steel involved. The two design concepts, fission power and airplanes, were working at opposite ends of the materials spectrum. There were many unknowns for building such an engine, and a lot of piecemeal experimentation would be necessary before the concept could be called possible.

In 1948 a report was commissioned at MIT for studying the feasibility of developing a nuclear-powered airplane. The final document, called the "Lexington Report," contained some good news and some bad news. The good news was that building a nuclear-powered vehicle that could lift off the ground and fly on its own looked possible. The bad news was that it would take a billion dollars and 15 years to work out the details, and if the thing crashed somewhere, the debris field would be uninhabitable for thousands of years. The Air Force pounced on the good news. Together with the Joint Committee on Atomic Energy and a cadre of aircraft manufacturers, they overruled the nay-saying nuclear scientists, President Eisenhower, the Bureau of Budget, and the Secretary of Defense. The ANP got a green light to proceed in 1952.[98]

The project spread out all over the country, from Pratt & Whitney in Massachusetts to Lockheed in Georgia, from Convair in Texas to General Electric in Ohio. In July the largest chunk of the work went to the NRTS in Idaho, where the engines would be hot-tested out in the desert. There was jubilant celebration in every quarter.

A jet engine is basically a large metal tube, mounted with one open end pointing toward the front of the aircraft and the other end at the back. With the plane moving forward, air blows into the front of the tube. An axial compressor spinning at high speed at the front acts as a one-way door, encouraging air to come into the tube while preventing anything from escaping out. In the center of the tube is a continuous explosion of jet fuel mixed with the compressed incoming air. The mixture, burned and heated to the point of violence in the explosion,

98 Not to be outdone by the Air Force, the U.S. Navy in 1954 commissioned a study by E.W. Locke Jr. to find the feasibility of an atomic-powered dirigible. That's right, a rigid airship filled with two million cubic feet of helium with a feather-light nuclear reactor driving two T56 gas turbine engines. The design combined two dubious technologies: an enormous, lumbering military air vehicle blown around by the wind, and an unshielded nuclear power plant weighing less than 40,000 pounds. The Navy decided instead to build a nuclear-powered float plane (the Princess Project), but in 1959 Goodyear Tire and Rubber invested time in preliminary designs of a nuclear blimp. Francis Morse at the University of Boston proposed a 980-foot-long nuclear dirigible for the New York World's Fair in 1964. It would be a "flying hotel" carrying 400 passengers. He argued that the spread of radioactive debris in the event of a crash would be manageable, because dirigibles crash softly. Fortunately for the public impression of nuclear technology, none of these plans were implemented.

instead of blowing the airplane to pieces finds a clear path out through the back of the tube. The escaping explosion products create a reactive force, just as would be made by a rocket engine, pushing the engine and the vehicle to which it is attached forward. On its way out, the expanding gases spin a turbine, like a windmill, and it is connected forward to the spinning compressor wheel. The nuclear aircraft engine was to operate in this way, except that the continuously exploding jet fuel would be replaced by a nuclear reactor running perilously close to fiery destruction. General Electric got the contract for the engines, to be tested at the NRTS.[99]

A piece of desert about 30 miles north of the center of the NRTS was picked out and named Test Area North, or TAN. It was the farthest point from anyone else's reactor experiment. A large assembly building was erected, the control room was buried for safety reasons, and the test stand was built a mile and a half away, with a four-track railway connecting back to the main site. The engine was put together in the assembly building, and then it was rolled out to the test stand using a lead-shielded locomotive. The engine weighed a hefty eight tons.

The first test engine was called "HTRE-1," High Temperature Reactor Experiment One, or "Heater-One." The two jets were modified GE J-47s, and the reactor having enough power to superheat the intake air turned out to be too large to fit in the space normally occupied by the fuel burners. The reactor used enriched uranium clad in nickel chromium, with water as the moderator. The airstream was taken from the jet engine tube immediately after the compressor stage at the intake opening. Using a large conduit, this compressed air was fed through a honeycomb of passages in the reactor, where it was heated and expanded as it would have been in the fuel burner

99 There were two nuclear jet engine designs in the ANP program: the direct cycle at GE of Ohio, and an indirect cycle at Pratt & Whitney of Massachusetts. For the indirect cycle, an intermediate heat exchanger would transfer heat from a centrally located reactor to multiple jet engines, while it was assumed that in the direct cycle a small reactor would take the place of the fuel burners in the middle of the jet engine. The P&W design was complicated, and its development was at least a year behind the GE engine. The ideal circulating coolant for the indirect cycle engine turned out to be lead, which seemed a cruel material to be used on a high-performance jet.

in a normal engine. The air was then piped back into the engine in front of the turbine and out the exhaust nozzle. On November 4, 1955, the reactor was tested at criticality by itself, and on December 30, it was ready for a hot test in the fully assembled engine. The reactor was unrealistically large, meant to test the concept and not to be mounted in an airframe, and it heated the air for both tandem-mounted jet engines.

The assembly was rolled out to the test stand, bolted down to the concrete apron so it wouldn't fly away, and hooked up to a long, horizontal pipe used to direct the exhaust into a filter bank. This would prevent disintegrating fuel rods from being blown all over NRTS. The pipe ended in a 150-foot vertical smokestack staring right up into the big Idaho sky. The operation crew, hunkered down in the control room, spun up the two compressors using electric starter motors, lit the flames in the burners, and powered up the reactor. When both engines reached operating temperature, the jet fuel automatically shut off, and the jets spooled up to screaming speed on pure atomic power. It performed as predicted, but the gamma radiation was far greater than had been anticipated. Operational plans for the bomber would have to be modified, and perhaps more crew shielding would be needed.

Testing of Heater One continued, and work began on the world's first fully shielded bomber hangar. A special tracked vehicle, heavy with lead shielding, was built with robotic arms and a thick, lead-glass viewing port for the driver. It would be used by the mechanic to work on a radiologically dirty airplane, blazing with fission product contamination and neutron-activated metal parts. A 23,000-foot-long runway was surveyed, and test missions were scripted. An electric incinerator toilet was invented for use by the flight crew, and pre-packaged meals were planned. Money flowed.

There were problems with the extremely high temperature necessary in the reactor. Fuel and reactor internal components evolved into exotic ceramics. HTRE-1 was modified and renamed HTRE-2, changed into a high-temperature-materials test reactor by cutting a hexagonal, 11-inch hole in the middle of the reactor. Newly designed fuel elements were mounted in the hole and run up to 2,800°F. Progress was encouraging, and it was time for HTRE-3.

HTRE-3 was a complete redesign. One smaller, horizontally mounted reactor ran two tandem J-47 jet engines, mounted as they would be in the proposed airplane, with the reactor located at the center of balance of the airframe.[100] The reactor was sized realistically, such as would fit in the finished airplane, but it still dwarfed the big General Electric jets. The fuel pins and control rods took up a lot of space, but there still had to be enough air passageway to spin the turbines in two J-47s. The core diameter was 51 inches, and the length 43.5 inches. The moderator was a solid ceramic, zirconium hydride, and that also took up room in the reactor core. The whole thing was encased in a solid beryllium neutron reflector. The reactor would still be considered highly advanced 54 years later.

By November of 1958 Heater-Three was on the test stand and ready to show what it could do.[101] On the morning of November 18 the crew started the engine compressors and made the reactor supercritical by manual control. Power was increased slowly to 60 kilowatts and leveled off, just to "check for leaks." Everything seemed fine. The crew shut it down and went to lunch.[102] Feeling fed and frisky, the crew decided to proceed with the experiment

100 I first saw the HTRE-3 and HTRE-2 in 1980. They were near the security checkpoint for TAN. They were still considered to be radiologically unapproachable, and to examine them closely you had to use binoculars. A fellow who had worked on the project damned HTRE-3 with faint praise, saying "It was so powerful, it could practically lift its own weight off the ground." Today, the two engines are tourist traps. You can go up and take a picture of your kids pointing into the exhaust nozzles. They are in the parking lot of the EBR-I, which is a National Historic Landmark, opened for touring between Memorial Day and Labor Day.

101 There is a bit of confusion here. The NRTS records show that the HTRE-3 operated between September 1959 and December 1960, but this account is taken from *Summary Report of HTRE no. 3 Nuclear Excursion. APEX-509*, and it places the accident in 1958. This type of event was usually classified SECRET, and the operating schedule may have been distorted to hide it.

102 I'm not sure there was anywhere to eat near the test stand, so they probably brought lunch with them or had it trucked in. I once had lunch at what was possibly the only restaurant within 40 miles of the HTRE test stand on Highway 33, called the "Broken Wheel." As I stared out the window at the featureless, dun-colored desert, suddenly the ground-cover moved and a ripple went through it. An earthquake? No, as it turned out I had been looking at a vast herd of sheep, covered with dust and huddled close together against the extreme cold.

program and run it up to 120 kilowatts. The engines started smoothly, with power increasing by a factor of 2.7 every 20 seconds. When the power reached 12 kilowatts, they switched to automatic, released the control handles, and sat back to watch it happen.

The automatic control used an ion chamber to detect gamma rays originating in fission. The number of gamma rays detected per second was perfectly proportional to the power level of the reactor, and it was read out as kilowatts on a meter at the control panel. The same signal was fed to a pen-chart recorder, giving a permanent record of power history of the engines. The current from this same detector was also fed to an amplifier, and this enhanced signal controlled a set of electric motors connected to the reactor control rods, running them in or out of the reactor core to satisfy a pre-set level of gamma-ray production rate. The high-voltage line feeding electricity to the ion chamber had been modified as an improvement to the circuit. A filter had been installed to prevent clicks and hums originating in the electric motors from contaminating the gamma ray signal.

If the power got out of hand for some unforeseen reason, the reactor would scram automatically on the gamma-ray level signal at a pre-set point, below the level where any harm could come to the machinery. As a backup, a set of thermocouples in the core monitored the reactor temperature continuously, feeding another scram channel. There was no steam to explode, so the worst thing that could happen would be too much power making too much temperature and causing the thing to melt.

The power level passed the 60-kilowatt level. No problems. They were now in untested territory, but the engines were accelerating smoothly. 80 kilowatts. No problem. 90 kilowatts. Still increasing power. 96 kilowatts. The power level started to fall rapidly, as if the bottom had dropped out. The crew, watching the instruments, found this curious, and the reactor operators cocked forward in their seats. The automatic control, sensing that the power was dropping quickly, pulled out the controls, trying to bring it back up. The power seemed to keep going south, and the needle on the meter fell to the left. Twenty agonizing seconds passed and WHAP! The scram circuit, detecting that the thermocouples in the reactor core had all melted, shut her down automatically, throwing in all the controls at once. In those 20 seconds, every piece of fuel in the reactor had lapsed into the liquid state. The fact that it overheated had not disassembled the reactor, as it was made of some very rugged ceramic materials, and the reactivity had actually increased as the nickel-chromium-uranium-oxide fuel turned to fluid. When the indication had been that the power was falling, it was actually increasing very quickly. Although the fuel was designed to be very tolerant of extreme temperature, the power level had spiked beyond the capacity of the two jet engine air-intakes to keep the reactor core from melting. Only a few of the zirconium-hydride moderator sections were damaged.

A slight increase in background radioactivity was detected down-wind of the smokestack as some few fission products, evaporated off the white-hot fuel, made it through the filter bank. Aside from that and the pen-chart recording, there was no outside indication that anything had gone wrong, and no humans were harmed.

The steel annulus surrounding the reactor was pumped full of mercury from a holding tank. Mercury is a high-density metal, liquid at room temperature, and it makes an excellent gamma-ray shield. With it in place, the crew could approach the Heater-Three to discon-nect it without danger from the decaying fission products built up in the damaged fuel.[103] The engine was then dragged back to the hot-lab in the assembly building, where the core was rebuilt and the source of the problem was analyzed.

That filter in the high-voltage cable had unfortunately limited the number of electrons per second that could travel through the wire. That was fine, as long as not too many electrons were needed to reg-ister the number of gamma rays per second that were traversing the ion chamber. At the higher power level, at which the equipment had never been tested, the gamma flux overwhelmed the ability of the power supply to keep up. The current demand from the chamber was so high, the voltage dropped, and the detector stopped detecting. The automatic system interpreted this as a power loss, and it tried to com-pensate for it by pulling the controls. The power climb accelerated until

103 Hold on to your seat. Three quarters of the free world's supply of mercury was used to shield the HTRE-3. At the end of the project it had to be released slowly back into the market to prevent an economic crash in the liquid metals trade.

the engine was, as we say, outside its operating envelope. This incident went down as the first time in history a reactor was melted because of an instrumentation error. Human error was a fault only indirectly.

In December the AEC told the Air Force that the nuclear bomber could not be flown over the United States. The only way it could be flown was out over the Pacific Ocean, presumably taking off from the beach. The 8-million-dollar shielded hangar, recently finished, could no longer be used to house a nuclear bomber, and grading the runway would not be productive. On January 20, 1961, John F. Kennedy was sworn in as President of the United States. On March 28, he signed a paper canceling the ANP project, and that was that. The much disappointed staff at the NRTS knew that they were very close to a working atomic aircraft engine, but for the good fortune of nuclear power we will never know if it would have flown.

A cluster of three reactor accidents involved the explosive conversion of water into steam by nuclear means. Two of the incidents were accidents only by procedural technicalities, and the other one was a complete surprise. We know exactly how it happened, down to the millisecond, but we have no answer as to why.

Among the gifted reactor physicists at Zinn's Argonne lab was Samuel Untermyer II, grandchild of Samuel Untermyer I. Sam the First was a Jewish-American born in Lynchburg, Virginia, who became the most famous New York lawyer of all time. He was responsible for, among other things, maintenance of the 5-cent subway fare. His grandson, a graduate from MIT in 1934, did not believe that the revered pressurized water reactor, in which the coolant is kept at such a high pressure it cannot boil, was the only way to build a water-moderated pile. In those early days of reactor development in the late 1940s, the general opinion was that if the water in the reactor vessel were allowed to boil, then the neutron production would become erratic and unpredictable. The coolant voids caused by steam bubble formation were predicted to cause fuel melts, chemical explosions, unchecked power excursions, and probably boils on the reactor operators and tension headaches.

Untermyer proposed a contrary prediction. If the coolant, which is also the moderator, in the reactor vessel were allowed to boil into steam, it would simplify the power production mechanism. There would be no need for a complicated, expensive, failure-prone steam generator and a second cooling loop. The reactor vessel would become a boiler, like the boiler in a coal-fired power plant but simpler, without any boiler tubes. No pressurizer to maintain the lack of boiling would be necessary, and several pumps and valves could be eliminated. Instead of running wild and unpredictable, the neutron flux would be controlled. A great deal of boiling would result in moderator voids, which would degrade the fission process and lower the power level. If the boiling were quenched, the density of the moderator would increase, and the power level would rise to a threshold cutoff, floating between too much power and too little power. Such a boiling-water pile would control itself and follow the load demand. There would be a lesser need for an electronic feedback-control system, such as would cause the HTRE-3 to melt in '58. In the event of an uncontrolled upward power excursion, the moderator would boil away, and this would automatically turn off the fission process without the need for an electronically controlled scram system. These seemed like desirable qualities in a power reactor, but it would all have to be proven with physical experiments.

Untermyer convinced his boss, Walter Zinn, of the importance of the boiling-water-reactor concept, and together they petitioned the AEC for a contract to prove the principles. In 1952 he was given enough money to make a modest stab at it and a spot of desert at the NRTS. The test reactor would be laid out on the ground, without so much as a tin roof over it, and the control room was a small trailer parked half a mile away. Television cameras gave the experimenters views of the reactor, including one using a large mirror to show the top of the core. The reactor vessel was a steel tank, half an inch thick, 13 feet high and 4 feet in diameter, halfway sunk in the ground. The 28 aluminum-clad enriched-uranium fuel elements were surplus from the big-budget Materials Test Reactor being erected elsewhere on the desert. The total fuel loading was 4.16 kilograms of uranium. The control rods were inserted through the open top of the vessel, adjusted in and out of the core with electric motors and wired back to the control trailer. Pipes, cables, and tanks were all over the ground. The steam made by the reactor heat was simply vented into the air. Dirt was mounded around the exposed portion of the reactor to give it some gamma-ray shielding. Untermyer, expecting this to be the first in a glorious series of experimental reactors, named it BORAX-I.[104]

By May 1953 BORAX-I was loaded with fuel, filled with water, and ready to go. Over the next 14 months, Untermyer and his operating crew would afflict this frail collection of scrounged parts with every torture imaginable. Over 200 experiments were performed, subjecting the system to all the operator errors and component failures they could think of. Tourists riding by on Highway 20/26 would report a geyser to the authorities. That would be Untermyer seeing how far he could throw a plume of water if the thing went uncontrollably supercritical. The record, I think, was 150 feet straight up. Under all conditions, the system performed exactly as predicted, shutting itself down automatically while allowing no harm to the machinery. The self-regulating feature worked right on the money.

By July 21, 1954, they had done everything to the reactor that their fertile minds could conjure, and it was time to shut her down and vacate the site. As one last experiment, Untermyer wanted to blow out the control with compressed air and see how the thing would react in a fast transient, with the reactivity taken from stable criticality to prompt supercriticality as quickly as possible.[105] Although there was no conceivable physical situation in which this would normally occur, it would give them an upper boundary of reactor mayhem to work with. It would also be a fine spectacle to watch, a steam explosion in keeping with the Bill Crush trainwreck spectacle in Waco, 1896, only without the public being able to watch. They would, however, film it in slow motion, so that it could be watched and studied, over and over. Extra fuel was loaded, and the central control rod was beefed up.

The wind was blowing in the wrong direction. Thinking that it could waft fission products over someone else's reactor experiment,

104 Although I and everyone else call it BORAX-I, it may not have been known as a numbered unit until BORAX-II was built. Contemporary reports from the field simply call it "The Borax Experiment." Local newsmen called it "The Runaway," and this name stuck to it like cobalt-60 contamination.

105 I could not at this writing be certain how the central control was blown out of the core. Some accounts say compressed air, but others say it was an explosive charge or a cocked steel spring. The pneumatic cylinder seemed most likely, but other mechanisms are possible.

the site meteorologist canceled the test. The crowd of important guests, expecting a train-wreck, went away disappointed. The next day at 7:50 A.M., the smoke bombs ringing the site were sending fumes straight up. It was time. The drivers cranked up the emergency evacuation buses, and the State Patrol stood prepared to close down the highway. The visitors, having returned with renewed anticipation of a good show, were standing in rapt attention at the official Observation Post, growing quiet, spitting out gum, and polishing the dust off eyeglasses.

The crew was worried. Would the explosion be big enough to excite the crowd? Would the control rod hang in the guides and spoil the transient?[106] Harold Lichtenberger, the physicist sitting at the controls, promised to give it his all.

Lichtenberger hit the EJECT CONTROLS button. KABOOM! Up she went with the force equivalent of 70 pounds of high explosive in the reactor vessel. A total energy release of 80 megajoules had been expected. They got 135 megajoules instead, and this inaccurate prediction instantly qualified the test as a nuclear accident. The little reactor that usually shed energy at the rate of 1 watt ramped up to 19 billion watts with a minimum period of 2.6 milliseconds. In all other tests of explosive steaming, the thing would send up a geyser of water droplets, sparkling in the western sun. This time the core melted instantly and homogenized into a vertical column of black smoke. A shock wave rippled through the floor of the control trailer.

Walter Zinn, standing in the control trailer, shouted "Harold, you'd better put the rods back in!"

"I don't think it will do any good," Lichtenberger shouted back. "There's one flying through the air!"

In fact, the entire control mechanism, bolted to the top of the reactor vessel and weighing 2,200 pounds, was thrown 30 feet up in the air. Zinn watched as a sheet of plywood flew spinning across the desert like a Frisbee. BORAX-I was totally destroyed. Pieces were found 200 feet away, and all that was left of the reactor was the bottom plate of

106 There had been serious talk of burying a case of dynamite under the reactor vessel to ensure a visually satisfying event in case the steam explosion fizzled. I find no record of this ploy being carried out.

the vessel with some twisted fuel remnants lying on top of it. The slow-motion movies were fogged by unexpectedly high radiation, and the power to the cameras failed as the wiring was carried away in the blast, but there was enough footage to show the plume of debris as it shot upward. Pieces of fuel were shown catching fire in the air as they tumbled upward. A control rod could be seen rocketing away at the upper left and eventually the control chassis re-entered the frame from above. In the first few seconds of the film they watched something that always seemed a feature of a prompt fission runaway: the blue flash. It lasted only two milliseconds, but it lit up the top of the reactor assembly as if it had been struck with a bolt of lightning.

You could not have asked for a better show, and a fine time was had by all. Untermyer had pushed a boiling-water reactor as far over the brink of disaster as was possible, and there were no radiation exposures or physical injuries. With this one staged spectacle, Untermyer had brought the AEC, the Argonne Laboratory, the military services, and the NRTS face to face with their collective fear, that a nuclear power reactor out of control could wreak havoc with a nineteenth-century steam explosion. Instead of frightening everybody, it calmed them down to have seen it. The maximum disaster seemed finite and manageable, and it was clear that this accident could never happen again. No normal power reactor control panel would ever be built to pull out the rods at explosive speed, and nobody was crazy enough to jerk out the central control by hand.

The debris was buried on the spot, and plans for BORAX-II were implemented on the same site. In the next few years, BORAX-III, BORAX-IV, and finally BORAX-V, a full fledged commercial-grade power reactor, were built and tested, with further experimentation justifying Untermyer's dream of an inherently safe, self-controlling nuclear reactor. The good feelings, however, would barely make it into the next decade.

The United States Department of Defense was forming strategies for winning the Cold War, the global stand-off with the Soviet Union in which each side threatened to blow the other to kingdom come. These preparations were forward-looking yet practical, and they seemed somewhat less like science-fiction than the Air Force's ANP program.

Among the plans were the Distant Early Warning System, or "DEW Line," and Camp Century. The DEW Line would be a set of remote, long-distance radar stations along an arc inside the Arctic Circle, in Alaska and Canada, intended to give a heads-up of several minutes if Soviet nuclear missiles were bearing down on the United States. Camp Century would be an under-ice city in Greenland supporting a hidden array of mobile, medium-range guided missiles, capable of pounding Moscow down to bedrock with thermonuclear warheads.[107] The mission was code-named "Project Iceworm."

Each of these proposed measures would require electricity for lighting in the months of total darkness and operation of the radars plus space-heating to keep the soldiers from freezing to death. These types of installation, such as outposts in the Antarctic, were usually powered with diesel generators running 24 hours a day, but to give life to a base as large as the proposed Camp Century would take a million barrels of diesel fuel per year, which seemed impractical. Camp Century supported 200 soldiers, one dog, and—brace yourself—two Boy Scouts: Kent Goering of Neldesha, Kansas, and Soren Gregerson of Korsor, Denmark, all existing on a sheet of glacial ice 6,000 feet thick. The base commander, Captain Andre G. Brouma, is credited with the motto, "Another day in which to excel!"

The Army Corps of Engineers was assigned, on April 9, 1954, to develop a range of small reactors to provide electrical power and space heating for remote conditions in the Arctic and the Antarctic. The U.S. Army Engineer Reactors Group, headquartered in Ft. Belvoir, Virginia, was in charge, and their observers had been very impressed with Untermyer's BORAX-I experiments.

The list of specifications for the army reactors was interesting, and the boiling-water reactors being developed at NRTS were closer to meeting the requirements than were Rickover's submarine plant or Zinn's breeder. The reactors had to be small, relatively inexpensive, and as simple as possible, with a minimum number of moving parts. Nuclear reactors at this young stage were notorious for needing nuclear

107 Although the plans for Camp Century were profiled in the *Saturday Evening Post* in 1960, its military purpose was not revealed to the public until January 1997, when the Danish Foreign Policy Institute spilled the beans in a report to the Danish Parliament. Greenland, the site of the missile base, is owned by Denmark.

physicists, highly specialized engineers, and all manner of Ph.D. scientists on hand for normal operation. These reactors would need no such specialists at the plant site. Operators would be given a few months of training and a couple of weeks experience on a training reactor at Ft. Belvoir, but keeping a staff of thoughtful academics alive at a remote army base was out of the question. Moreover, the entire plant would have to be able to be pulled on a sled over ice, lowered into place, and hooked up with a minimum of effort. It could be broken down into sections, but they would have to screw together easily under hard conditions. Nothing could be welded together on site, and certainly no concrete, the basis of all nuclear construction, could be poured in temperatures far below freezing. There would be no calling the factory for replacement parts. They would have to be foolproof, such that an inexperienced operator or one deprived of sleep for three days would not be capable of pushing the wrong button and destroying anything.

The reactor design shop at Argonne took the challenge eagerly, and they came up with the Argonne Low Power Reactor. The name was changed to SL-1, meaning Stationary Low-power reactor, version one, and the small demonstration plant was built in a remote spot at the NRTS in Idaho. Construction and management of the project was transferred to Combustion Engineering in Stamford, Connecticut.

It was everything that the Army had asked for. It was a small boiling-water reactor, making three megawatts of heat. Of this, 200 kilowatts would be converted into electricity using a combined steam turbine-generator, 400 kilowatts would be hot air to be piped into living and working quarters, and the rest would be exhausted into the environment using an air-cooled heat exchanger. All the machinery was housed in a simple, cylindrical building made of quarter-inch steel, 38 feet in diameter and 48 feet high. In the ceiling was the "fan room," containing the heat exchanger and electrically powered blowers. The lower section was filled in with a combination of loose gravel and steel punchings, with the reactor buried in the middle and flush with the floor. The punchings, left over from making rivet holes in bridge girders, and gravel acted as a bio-shield to absorb all radiation made during and after fission in the reactor vessel. The mid-section of the building was an open space containing the turbogenerator, transformer, tanks, valves, and equipment that could be accessed by workers.

The reactor vessel was 14 feet 6 inches tall, filled with water to the 9-foot level, leaving 5 feet 6 inches of clear space at the top where the steam could collect.

A covered walkway led to the sheet-metal control building sitting on the ground next to the reactor building. Security was provided by a high chain-link fence surrounding the site, with a guard stationed during the day shift. It was functional and as simple as it could be, and the entire facility was built for about $1.5 million. As an Arctic power source, it would cost less to buy and run than the shipping charges alone for the fuel that it was to replace.

The SL-1 plant was started up on August 11, 1958. Its Argonne design had been slightly modified to extend the operating time between fuelings to five years. The fuel assemblies had been lengthened by a few inches so that more highly enriched uranium could be included. This increased the reactivity of the core considerably, and to bring it down to a controllable level some metallic boron was added to the lower ends of the flat, aluminum-covered fuel strips. The boron would soak up neutrons and negate the extended reactivity, but as the fuel in the upper core fissioned away, the boron's ability to absorb neutrons would also wane. Every time a boron-11 nucleus soaked up a neutron and took it out of the fission process, it would decay in multiple steps to carbon-14, which had no affinity for neutrons. Gradually, over five years, the active region of the reactor core would migrate down while maintaining the designed power level. There were better ways to do this, but for the SL-1 thin sheets of boron were riveted to the ends of the fuel assemblies.

The controls, like everything else in the plant, were designed to be as simple and foolproof as it was possible to make them. There was one main control, made as crossed cadmium-filled blades that would fit between the fuel assemblies, inserted down through the middle of the core. It was moved up and down in the reactor to control the neutron population by a rack-and-pinion mechanism, driven by a DC electric motor and gearbox bolted to the floor off to the side. The control rack penetrated the top of the reactor through a watertight seal called the "shield plug."

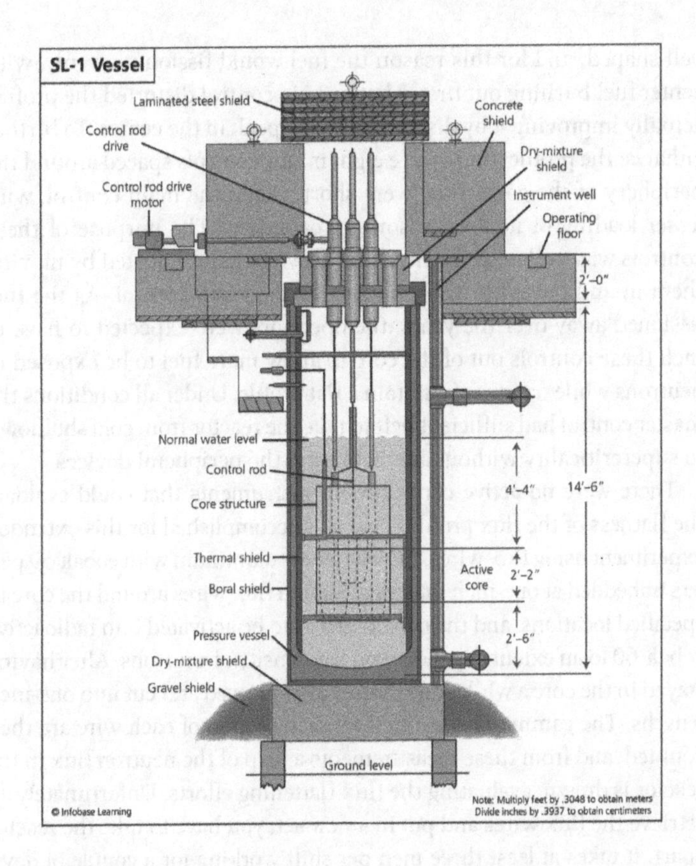

In its operating configuration, the SL-1 reactor was radiation-shielded on top by large concrete blocks and a stack of steel plates. All this shielding had to be removed for servicing the inside of the reactor vessel. The control "rods" were actually cross-shaped, and only the large control in the center of the core was used to regulate the fission. The smaller rods were used to smooth out the neutron flux so that the fuel would be consumed evenly across the diameter of the reactor core.

Having one big, master control was fine, but for the fuel to last for more than a few years the flux profile would have to be flattened. In theory, the number of fission neutrons in the reactor at a given time would tend to peak at the center of the core. A plot of neutron population across the diameter of the reactor, or the "flux profile," would be bell-shaped, and for this reason the fuel would fission unevenly, with center fuel burning out first. That one big control distorted the profile, actually improving it by discouraging the peak in the center. To further enhance the profile, there were eight minor controls spaced around the periphery of the core. They were shorter than the main control, with lesser loading of neutron-absorbing cadmium. The purpose of these controls was to flatten the flux, and they could be adjusted by moving them in and out with motors, just like the main control. As the fuel fissioned away over the years, the operators were expected to have to inch these controls out of the core to allow more fuel to be exposed to neutrons while trying to maintain a flat profile. Under all conditions the master control had sufficient heft to take the reactor from cold shutdown to supercriticality without any help from the peripheral devices.

There were no active control-room instruments that could evaluate the flatness of the flux profile. That was accomplished for this extended experiment using flux wires. Flux wires are aluminum with cobalt-59 pellets imbedded at one-inch intervals. Station flux wires around the core in specified locations, and the cobalt-59 was to be activated into radioactive cobalt-60 to an extent dependent on the density of neutrons. After having stayed in the core a while, these wires are retrieved and cut into one-inch lengths. The gamma rays emitted by each section of each wire are then counted, and from these measurements a map of the neutron flux in the reactor is drawn, evaluating the flux-flattening efforts. Unfortunately, to retrieve the flux wires and put in a new set, you have to take the reactor apart. It takes at least three men per shift working for a couple of days, and the reactor, of course, has to be completely shut down.

By January 3, 1961, the SL-1 had been producing power for over two years, and the experiment was going well. This was more than just a test to see if a small boiling-water reactor would produce power. There was not really anything to learn on this point. The purpose of the test was to see if such a plant could be run and managed by minimally trained soldiers without the advantage of expert scientists looking over their shoulders. Could the inhabitants of a remote station, even more cut off from civilization than the NRTS, maintain a power reactor no matter what happened?

There had been a few problems. Water seals and gaskets were leaking, but it was nothing fatal. Those boron strips tacked onto the fuel turned out to be a bad idea. The boron, when it was changed into carbon-14, would curl up and crack. Pieces of it were falling into the bottom of the vessel, and what was left on the fuel was binding up the controls. Under any normal reactor operating conditions, the words "binding up the controls" would result in a quick shutdown, a study, a redesign, assignment of blame, papers written, operating rules changed, and so on. In this case the problem was treated as an interesting perturbation thrown into the exercise. It was just the sort of unexpected problem that could show up on the glacier in Greenland. The Army said "Let's see how it plays out," while the AEC and Combustion Engineering kept a detached, mildly interested stance. Any time the reactor was down and apart for maintenance, the peripheral rods had to be jacked up and down by hand to clear out the bent boron at the bottom of the core, or the motors would not be strong enough to move them against the resistance. The main rod, weighing a hefty 100 pounds, was big enough to take care of itself, crashing its way through the twisted metal and having no particular trouble moving with the motor.

It was a clear night and bitterly cold. The graveyard shift consisted of Senior Reactor Operator Jack Byrnes, an Army private; Assistant Operator Dick Legg, a Navy Seabee; and trainee Richard McKinley, Air Force. Byrnes was going through a marriage crisis, he wasn't making enough money, and he had problems with being managed. Legg was sensitive to comments about his stature and enjoyed playing tricks on his colleagues to pass the time. McKinley was there to learn how to run a reactor plant.

It was time to change out the flux wires, and the previous shifts had done all the heavy lifting. The reactor was basically put back together, with the vessel filled to 9 feet with water and the head screwed down. All the night shift had to do was reconnect the controls and put the big concrete blocks that shielded the top back into place. It was a three-man job. Byrnes and Legg would reconnect the rack to the main control while McKinley would stand off and act like a health physicist, pointing a "cutie pie" ionization chamber at Byrnes and Legg while they worked. The day shift would actually have an "HP" on the staff, monitoring all the activities in the reactor building to constantly check for abnormal radiation, but the night shift was a minimum crew.

The rack that engaged the pinion for up-down motion was screwed to the top of the control, which was at its lowest position in the core, keeping the reactor at cold shutdown status. A C-clamp had been tightened on the rack to hold it in a slightly raised condition, just above the top of the shield plug. This position allowed a three-foot metal rod to be temporarily screwed into the top of the rack so that a man could handle it with the motor disconnected.

The instructions were clearly mimeographed. Byrnes would take hold of the handling rod with all ten digits and lift the 100-pound center control by one inch. With the load off the C-clamp, Legg, crouching over the shield plug, would unscrew it and lay it aside, and then Byrnes would gently lower the control until it rested on the bottom of the core structure. McKinley was standing off the reactor top, in front of one of the man-sized concrete shield blocks, idly watching the show and pointing his radiation detector.

All that we know for sure is that at 9:01 P.M., Byrnes, against the written directions and everything that the instructors had drilled into his head, with one massive heave jerked the master control clean out of the core as fast as he could. If it were lifted four inches, the reactor would go critical, blasting the three workers with unshielded fission radiation. Byrnes managed a full 23-inch pull, and the reactor went prompt critical with a 2-millisecond period, producing a steam explosion in the reactor such as has never been seen before or since. The water covering the reactor core instantly became superheated steam.[108] The four-foot slug of still water over the core, not becoming steam, was pushed with incredible speed to the top of the reactor vessel, through the 4 feet 6 inches of clear space, until it hit the screwed-down top like a very big hammer. The force of the hammer-hit picked up the 13-ton steel vessel and shot it nine feet out of the floor, shearing away its feed-water and steam pipes. The nine shield plugs on top of the vessel shot off like cannon shells, burying control-rod fragments in the ceiling.

Byrnes and Legg were killed instantly, not by the intense radiation surge, but by the explosive shock of two billion billion fissions, 15

108 Some accounts say that the rod was pulled 20 inches, but the revised figure of 23 inches is based on careful reevaluation of the data. The rod was pushed out an additional 7 inches by the upward force of the steam blast. The weight of the master control was originally recorded as 85 pounds, but this did not take into account the weight of the rack plus the handling rod.

megawatt seconds of energy, and an air pressure wave of 500 pounds per square inch. McKinley died two hours later of a massive head wound, inflicted by the concrete shield block as he was thrown backwards. All three men had fission products, built up by the reactor running at full power for two years and turned to nascent vapor in the sudden heat of prompt fission, buried deep in their bodies. There was no way to simply wash away the contamination. It seemed embedded in every tissue.

Legg was pinned to the ceiling with a piece of the master control. The reactor internals were an unrecognizable tangle of twisted parts, and water, gravel, and steel punchings were scattered all over the reactor building floor. The crude steel building, which was meant only to house the equipment and keep the rain off, managed to prevent a scattering of radioactive debris, and outside it was hard to tell that anything had happened. The plant was a total loss, and everything would have to be carefully disposed of, leaving not a trace of radioactive contamination or subjecting any worker to an abnormal dose. The three bodies were so deeply and severely contaminated, they would have to be treated as high-level radioactive waste. Autopsies were performed quickly, behind lead shields with instruments on 10-foot poles.

A commendable job was done analyzing the accident and cleansing the site of any trace of it. It required a great deal of skill and planning to decontaminate the site, as the inside of the reactor building was too radioactive for normal work. A full-scale mock-up of the building was constructed for decontamination practice and to figure out the actions of Byrnes and Legg before and during the accident. Television cameras were inserted into the reactor core using remote manipulators, and a Minox subminiature camera was used on the end of a pole to take photographs through small openings. A seldom-mentioned technique called the "gamma camera" was used to spot where highly radioactive fragments of the reactor had landed in the ceiling and on the floor of the building.

The gamma camera is a variation of the old "pinhole" camera. It is possible to make a picture on a piece of photographic film by mounting it on one inside face of a light-proof, square box. On the opposite face of the box is a tiny pinhole. Light enters the box through the hole, very dimly, and it forms an image on the film without the use of a lens. Light can only travel in a straight line, so individual rays of light from the scene outside the box are organized into an image simply because they all have to come through at the same point, the pinhole. If the box is made of lead, through which radiation cannot pass, then the pinhole on the front face of the box will image gamma rays onto the film the same way it uses visible light. Gamma rays are photons, just like visible light but at a much higher oscillatory frequency and energy. To use the camera, the investigators first made a light image of the ceiling in the reactor building using the open pinhole, then covered the pinhole with a light-tight but gamma-transparent shutter. The gamma ray image was allowed to expose the film for 24 hours through the same pinhole, superimposing a gamma-ray image atop the light image. When the film was developed, it identified radioactive objects in the picture as shining brightly, like points of light.

The reactor building and its contents were buried in a trench 1,600 feet away from the original SL-1 building site. The cleanup took 13 months and $2.5 million. Much was learned about managing power reactor disasters, and the question of what had happened was answered in great detail, down to the millisecond.

What was never answered by detailed forensic analysis was: Why did Byrnes pull that control out? He knew good and well what it would do if it came up to four inches. Why jerk it out all the way? There are, to this day, a few conflicting opinions, ranging from it was a murder-suicide to he was exercising the rod to clear the bent boron. In my opinion, Byrnes was showing off for McKinley, the new guy from the almighty Air Force. The Air Force was running the dangerous, super high-tech HTRE experiments down the road at Test Area North, and the Army was stuck with this cheap, low-power rig that was just sitting here making a slight turbine-hum. Byrnes wanted to give McKinley a thrilling blip on the cutie-pie radiation detector he was holding by bouncing the main rod. He knew that if he could bring it up to supercritical for just a split second, the power would drop again quickly as the control went back down. No harm done, but he bet himself that he could make Air Force lose control of his bladder. The thing was heavier than it looked. He wiped his sweaty palms on his pants, braced, and put both arms into it. Up she came. They never knew what hit them. Their nervous systems were destroyed before the senses had time to register the violent event.

It was a tragic loss of life, and the general attitude toward small, simple, cheap nuclear power was affected. The Army went on to deploy the PM-2A portable medium-power reactor in Greenland, the PM-3A at McMurdo Station in Antarctica, the MH-1A mobile high-power reactor at the Panama Canal, the SM-1A stationary medium-power reactor at Ft. Greely, Alaska, and the PM-1 at Sundance, Wyoming. They even developed the ML-1 at the NRTS, the first nuclear power plant that would fit on the back of a truck. It all ground to a stop in 1977 when the need for remote power and the money to support it both drifted away. It was not at all a bad idea, but commercial nuclear power production moved off in the opposite direction, toward bigger, more complicated, more expensive installations, and the SL-1 incident was a reminder of the dangers of making it too simple. The accident had proven that there was no such thing as foolproof.

The snowcap over Greenland into which Project Iceworm was dug turned out to be one big, slow-moving glacier. The rooms and tunnels so carefully carved out of the ice deformed and shrank with time. Every month, 120 tons of ice had to be shaved off the inside walls, and by the summer of 1962 the ceiling in the reactor room had come down five feet. In July 1963 it was clear that this bold step in missile positioning was not going to work, and the PM-2A reactor was shut down and shipped back home. By 1966, Camp Century was unlivable, and it was given back to nature. When last visited in 1969, it was completely wrecked and buried under a great deal of new snow.

The deep snow cores taken at Camp Century are still in use today. These long cylinders of snow cut from the glacier are a record of the climate and atmospheric conditions for the last 100,000 years, and they have been used to map the carbon dioxide history on Earth since the emergence of mankind.

The last steam explosion in Idaho was the final in a series of experiments with the SPERT-I, or the Special Power Excursion

Reactor Test One, on November 5, 1962. One would think that enough was learned in the BORAX-I and the SL-1 blow-ups to pretty much give us what there was to know about water boiling too fast in a reactor tank, but this odd experiment was funded by the AEC. It, like BORAX, gave a bit more of a show than was anticipated, and it went down in history as another accidental power excursion at the NRTS. The incident was all recorded as a movie in slow-motion at 650 frames per second.[109]

The reason for the SPERT experiments had to do with the expanding need for nuclear specialists in the 1950s. The AEC was aware of a projected shortage of nuclear engineering and health physics graduates in American universities, and technical campuses needed small research reactors in place to encourage these studies and excite some interest in nuclear topics. The Aerojet General Corporation had already introduced an inexpensive "swimming pool" reactor for use in schools. It was simply a concrete-lined pool of water, sunk in the floor, with uranium fuel assemblies and control rods clustered in the middle. "Is it safe?" asked the AEC. "What would happen if a control rod came loose and dropped out the center of the core?"

It would explode, of course, with the water shooting out the open top of the reactor and hitting the roof really hard, but that was not a sufficient answer. Details of everything that could possibly happen to an open-topped, water-moderated, low-power reactor were demanded. The SPERT contract was given to the Phillips Petroleum Company, and a site was chosen about 15 miles west of the eastern boundary of the NRTS. The first of four SPERT reactors was started up in June 1955, and it was run through a series of torturous accident scenarios.

The setup was very much similar to BORAX-I, but it was enhanced with a metal shed covering the reactor. Controls were inserted through the bottom of the reactor core leaving the top completely open, with one master control in the center. Two periscopes looked down into the core from above, and a tilted mirror showed the open reactor top from the side. These optical features were used to make motion pictures of every experiment simulating wrongful operation by

109 See it on YouTube. http://www.youtube.com/watch?v=0FIhafVX_6I&feature=related

undergraduates horsing around with the controls.[110] Nuclear instruments and recording procedures had improved since BORAX, and better data collection was anticipated. Experiments with fast power transients blew the water out the top of the reactor, just as seen in the BORAX excursions. The fuel was the same as was used in most of the test reactors at NRTS, thin plates of aluminum-uranium alloy clad with pure aluminum. It was not meant as a high-temperature fuel, and some plate warpage and slight melting were observed in the core.

November was the end of the open experiment season in Idaho, as the temperature began to drop to Greenland levels, and the team was out of tortures for the SPERT-I. SPERT-II was designed and ready to build. As one last experiment for 1962, the team wanted to simulate AEC's worst fear, that the main control rod would fall out. The control-rod drive was modified to break free and fall with gravity, and the metal roof was removed so that the driven water would have nothing to blow away. A 3.2-millisecond period was predicted, with some fuel melting this time. The test was not cut out to be such an event as the BORAX-I excursion. SL-1 had taken the thrill out of seeing water reactors explode.

It was a sunny day in the desert, and the wind was calm. Three . . . two . . . one . . . RELEASE. The main control rod went into freefall. The little reactor suddenly lit up with a blue flash in the mirror as 30.7 megajoules of energy came alive and started heating the fuel. All 270 fuel plates melted, and the fission process shut down completely, as predicted. Everything was going according to plan for

110 On May 3, 1958, Frederic de Hoffman announced the introduction of the Test, Research, and Isotope reactor of General Atomic, or the TRIGA. It was a brilliant design, started as an exercise for young nuclear engineers by Dr. Edward Teller, co-inventor of the H-bomb and the inspiration for the movie "Dr. Strangelove, Or How I Learned to Stop Worrying and Love the Bomb." Dr. Freeman Dyson, also a nuclear genius, contributed to the concept, and the design was influenced by findings at the NRTS in the BORAX and SPERT programs. The TRIGA pool reactor was "safe even in the hands of a young graduate student," which is saying a lot. The unique fuel formula was a mixture of uranium and zirconium hydride. Being co-located with the uranium, the hydrogen moderator would heat instantly in the event of a fission runaway, and the hot hydrogen would lose its ability to moderate the neutrons down to the advantageous low speed. The result was a reactor that absolutely could not be made to explode or melt down. The TRIGA is still available from General Atomics of San Diego, California. Three new ones were sold recently overseas.

about 15 milliseconds, and then all hell broke loose as the unexpected transpired.

The melting fuel sagged and changed shape. Heat transfer between the fuel and the water was suddenly improved, and the resulting steam explosion was more energetic than expected. It completely destroyed the reactor. Contents of the reactor vessel were pounded out of shape and thrown skyward. There was no roof to carry away, but a flying periscope hit a steel roof-beam and bent it outward. Bits and pieces of reactor were scattered all over the place. There were no injuries except to a few egos, and the large bank of instruments could find no harmful release of radioactive gasses into the atmosphere. I will not say that the fine engineers at the NRTS were slow to learn, but this sort of behavior in a suddenly uncontrolled water reactor was hardly a new finding.

This was not the last reactor blown up at NRTS, but all the others were predictable, controlled, and not considered accidents. NASA planned to send up a nuclear reactor into space in the System for Nuclear Auxiliary Power program, and the SNAP-10A nuclear power generating station was scheduled to be launched into orbit in 1965.[111] There were, of course, concerns that the booster could fail before achieving orbit and a power reactor could come down in the ocean. What would be the hazards if this worst launch failure happened?

At NRTS the recently vacated ANP test facilities at Test Area North were converted to test the SNAP-10A in a simulated high-speed crash into the Pacific Ocean. It was correctly predicted that a severe nuclear power transient would result from slamming into the salt water. The reactor was naked of any shielding, and it was moderated by zirconium hydride mixed with the uranium fuel. Hitting the surface would suddenly introduce extra moderating material favorable to fission (water) and the reactor would go prompt supercritical.

A huge water tank was mounted on one of the ANP double-wide rail cars that carried a nuclear jet engine with a SNAP reactor mounted in the middle. The reactor was protected from the water with a Plexiglas shield until an explosive charge threw it out of the way and let the

111 There were actually seven SNAP-10A's built. Three were blown up at the NRTS, one was a spare, two were used for flight system ground tests, and only one was orbited.

water crash in. The resulting fireball on April 1, 1964, threw reactor fragments far and wide.[112] To keep thermocouples and radiation instruments from being melted and lost in the explosion, the measurements were done remotely, using infrared pyrometers looking into a mirror behind a lead wall. Just to be sure, the test was run three times in experiments SNAPTRAN-1, 2, and 3, and three perfectly good SNAP reactors were destroyed.[113] They took movies.

There was also the LOFT (Loss Of Fluid Test) reactor and Semiscale, which was not really a reactor. Both experiment series were used to find what would happen if a water-cooled reactor breaks a major steam pipe. The LOFT reactor was a half-scale, fully operational pressurized water reactor, while Semiscale was a slightly safer experiment using an electrical heat source instead of fission to make the steam in a simulated power plant. These programs and the accidental experiments dating back to EBR-I were all valuable in finding how to build a safe power reactor and how not to build one. This knowledge was put to use in designing the Generation II nuclear reactors that now produce 20 percent of the electrical power in the United States. There would be commercial reactor accidents in America, but never a steam explosion, and the three men who stood on top of the SL-1 were the last people ever to die in a power reactor accident in this country.

Today, there are no ongoing reactor safety experiments, anywhere in the world.

Torturing nuclear reactors to make them give up the secrets to safe power production was not the only activity at the NRTS. Among the first building sites at the new reservation in 1948 was the Idaho Chemical Processing Plant, known to all as the "Chem Plant." If the

112 What burned, causing a "fireball?" Ask yourself, why is fire visible? Why does fire glow bright yellow? The answer is sodium contamination. Sodium atoms glow yellow when heated by an otherwise invisible flame, and the slightest trace of sodium in a fire is what makes it look like fire. In the case of SNAP-10A, it used nack, or a sodium-potassium alloy, as the coolant. There was plenty of sodium on hand to light up the event when the reactor violently overheated. The sodium and potassium reacting explosively with the water topped it off.

113 I have been unable to find a report that describes the destruction of SNAPTRAN-1. I'm not sure whether it was too insignificant to warrant a report, or too significant to be declassified.

techniques for building civilian power plants were to be sorted out in the Idaho desert, then it made perfect sense to also work on fuel reprocessing and waste handling. A model plant was built to recycle the fuel used in reactor experiments and to develop practical methods for extracting the various components made in the fission process.

Commercial reactor fuel is uranium oxide, with two out of every three atoms in the solid material being oxygen. The uranium content in fuel is usually enriched to 3.5% fissile uranium-235. The rest is uranium-238. After about 4.4% of the uranium-235 has fissioned, the fuel can no longer support the self-sustained chain reaction, and it must be replaced. Approximately 20.5% of the waste product embedded in the spent fuel is plutonium resulting from neutron activation of the U-238. The rest of the waste, 79.5%, is fission products, 82.9% of which are stable. That leaves 1.1% of the spent U-235 as radioactive waste which must be either disposed of or put to use as industrial and medical isotopes. The idea of fuel reprocessing is to remove the remaining U-235 and Pu-239 from the spent fuel and return it to the energy production process. As you can see from the breakdown of the used fuel components, the waste to be buried is a tiny part of the original fuel. In 1948 when uranium was thought to be scarce, it made no sense to bury an entire used fuel-load without breaking it down and sorting the portions.

The Foster Wheeler Company of New York, experts at making oil refineries, designed the plant, the Bechtel Corporation out of San Francisco built it, and American Cyanamid ran it. The Army had wanted to operate it under military control, but the AEC wisely argued that if it was to model a commercial process, then civilians should learn how to do it. Construction took 31 months on 83 acres of flat-as-pool-table desert north of Big Southern Butte. The first shipment of fuel arrived to be processed in November 1951.

Spent fuel arrived in heavily shielded casks, strapped down to flatbed trucks. A truck would make it past the security checkpoint and roll into a special bay in the storage building. Remotely controlled mechanical hands would take off the top of the cask and gently place the spent fuel in stainless steel buckets suspended from the ceiling. Using motorized overhead tracks, the fuel rolled into the long crane building and was diverted into its appropriate place to be processed.

The fuel was dissolved and then broken down into components by specific chemical reactions. The fuel solution ran through stainless steel pipes and tanks, with the routings automatically controlled by pneumatic valve sequencers behind a shield wall in the building. Pressures, temperatures, flow rates, and tank levels were monitored and recorded. Every aspect of the facility was designed to shield human workers from any contact with radiation. There were radiation detectors and alarms everywhere, and to prevent possible power outages the plant had its own motor generator.

In general, the Chem Plant was well managed and well designed, and there was never a radiation injury or an accident that damaged the equipment. There were, however, three criticality accidents in which uranium dissolved in solution managed to find itself in a critical mass and go supercritical. Enriched uranium dissolved in water or an organic solvent will become an active nuclear reactor, increasing in power, if a specific "critical mass" is accumulated. The hydrogen in the water or the solvent acts as a moderator, slowing the fission neutrons to an advantageous speed, and even a fairly low U-235 enrichment level, like 3%, will overcome neutron losses by non-productive absorption in the moderator. This has been realized since the earliest days of reactor engineering, and those who work with uranium solutions are quite aware of the possibility.

The volume or mass of the potentially critical solution is not the only factor. Any shape that reduces the surface-area-to-volume ratio is a bad choice for a vessel that is to contain a uranium solution. In a worst-case configuration, such as a spherical tank, a minimum critical mass is needed to start a chain reaction, because the neutron leakage from the surface is minimized. Another example of a shape that is good for making an impromptu reactor is a cylindrical tank, especially one built with the "tomato can" height-to-diameter ratio. Tin cans holding food are purposely made to optimize the amount of material they can contain using the least amount of sheet metal, to lessen the manufacturing cost. A uranium solution shaped like a can is a disaster waiting to happen. If any concentration of uranium in solution is held in a tank in a reprocessing plant, then the tank has to have a seriously bad magnitude of surface area compared to its volume. It absolutely must promote neutron loss by surface leakage.

The tanks are therefore vertically mounted, thin tubes. They look like thickened pipes.

The Chem Plant at the NRTS was not the only one being built in the world. Eventually Canada, England, France, the Soviet Union, and Japan would have fuel processing and reprocessing plants serving their nuclear power plants and, in a few cases, bomb manufacturing. I would like to think that not one of these plants was built having a container inside the building that could hold a critical mass and configuration of uranium in solution, but I would be wrong. Incredibly, there has never been a Chem Plant made that could not support a critical mass in at least one tank. Given time, uranium will eventually find this tank, usually by inappropriate means, and a problematic supercriticality will result. A death in a power-reactor accident is exceedingly rare, but many people have died worldwide in criticality accidents involving nuclear fuel.

On October 16, 1959, the graveyard shift was working on 34 kilograms of uranium fuel enriched to 91% U-235 in the form of liquid uranyl nitrate diluted with water. The next step of the processing was to extract impurities by mixing the nitrate solution with sulfuric acid in three steps. The first step had been completed, and the mixture was held in two vertical "pencil tanks." These vessels were specifically designed to defeat criticality, being too small to hold the entire fuel load in one tank and being thin, measuring 3.050 meters tall and 0.125 meters in diameter. There was a tube leading from the drain end of the pencil tanks to a 5,000-gallon waste-receiving tank, looking like a tomato can. This tank was never, of course, supposed to have any uranium in it. It was there to hold the non-fissile waste products that were being extracted from the fuel. Just to make sure that it was impossible to siphon uranium solution out of the pencils or start a gravity drain, there was a loop in the connecting tube, located way above the top of the tanks. Only deliberate sabotage, it was thought, could cause any uranium to get into the receiving vessel.

An operator, reading his instructions, turned on two valves to spray some air into the two pencil tanks to stir it up and make sure that the acid and the uranyl nitrate were thoroughly mixed. Pencil tank number one had a pressure gauge on it so that the operator could make certain that he was applying the air at the correct sparge rate. Pencil tank number two did not have a pressure gauge, so the operator just opened the valve until he was sure the thing was good and sparged. Unfortunately, the excessive blast of air forced the liquid up through the anti-siphon loop, defeating it, and causing uranyl nitrate to siphon into the tomato can at 13 liters per minute.

The tank waited until it had 800 liters of 91% U-235 in solution before it went supercritical. Radiation alarms started going off all over the place. They were ignored. If you evacuated the building every time one of those hyper-active alarms sounded, nothing would ever get done in this place. The third time somebody hit the evacuation alarm, people started moving. Nobody took the clearly marked evacuation routes, only the well-worn paths that they took out of the building every working day. As a result, there was a log jam at the door. Fortunately it was a small staff for this shift, and everybody showed up at the guard shack alive and well.

In the initial blue flash, hidden by the stainless steel tank walls if anyone had been looking, a hundred million billion fissions occurred. There were several excursions, with the contents of the tank boiling furiously, which would dilute the moderator with steam and kill the reaction. The steam bubbles would then collapse, and the supercritical condition would start over and do it again. After about 20 minutes, half the water content of the solution had boiled away, and the unplanned reactor aborted itself.

The lessons learned from this accident were sobering. Those particular pencil tanks were seldom used, and nobody on shift that night had any experience with them, so they had to read the written instructions, which were out of date. Routinely used waste tanks had better anti-siphoning systems. This one did not. The evacuation procedure, which had never been used before, obviously needed work. It did not take a sabotage to put uranium where it was not supposed to go. It only took the turn of an unfamiliar valve.

On January 25, 1961, the Chem Plant had been down for a year of renovation and had been operating 24 hours a day for the past four days. At 9:50 in the morning, another batch of uranyl nitrate found an "unfavorable geometry" in the upper disengagement head of the H-110 product evaporator. This was thought impossible because of an overflow line in the head that was supposed to prevent it from holding enough uranium solution to be critical. Someone had cleared out a clogged line downstream using compressed air, and it managed to blow enough uranyl nitrate into the cylindrical vessel to cause another big boil-off. Nobody was hurt, and nothing was damaged, but again it called into question several philosophies and engineering practices. The general boredom of working a shift in a chemical processing facility may have had something to do with it.

Finally, on October 17, 1978, a slow leak in a water valve eventually let the uranyl nitrate concentration in the lower disengagement section of scrubbing column H-100 grow to a supercritical level. It was another non-damaging, zero-death, embarrassing reactor where there should not have been one.

The Chem plant is still there, but it is now called INTEC, the Idaho Nuclear Technology and Engineering Center. It is working on a liquid-waste-processing method as part of the Department of Energy's Idaho Cleanup Project. The exciting days of reactor experiments are long gone, and the purpose of this federally funded effort is to erase radiological traces of the old NRTS off the desert. Now, irrigated circles of land, planted with Idaho potatoes, encroach on the property once alight with the technology of the future.

Despite all the fine development work and all the logical reasons for reprocessing, such as greatly reducing the throw-away waste product, there is no commercial fuel reprocessing plant in the United States. Our fuel would be buried whole, if it were buried. There is currently no place to bury spent fuel in this country, so it just piles up at the power-plant sites. All other countries relying on nuclear power as a base-load source of electricity have routine reprocessing and waste burial, from Great Britain to Japan.

In spite of three criticality accidents, the Idaho Chemical Processing Plant was probably the safest fuel reprocessing facility in the entire world. We will visit this topic again in glorious detail, but now let's remain calm, go across The Pond, and see what the reserved and quiet Brits were up to.

Chapter 5

MAKING EVERYTHING ELSE SEEM INSIGNIFICANT IN THE UK

> "Will you please issue the following operating instructions to the operator engaged in controlling the Wigner Energy Release. If the highest Uranium or Graphite temperature reaches 300°C, then Mr Fair, Mr Gausden and Mr Robertson are to be informed at once, and the PCE alerted, to be ready to insert plugs and close the chimney base."[114]
>
> —D R R Fair, Manager, Pile

THE ATOMIC ENTHUSIASM OF THE 1950s was supposed to acclimatize the general public to all things nuclear and prepare us for a future that was in the advanced planning stage, but just the opposite seemed to

114 I am not sure what "PCE" means. It was possibly a typographical error in the original memo, meant to be "PEC," or Production Executive Committee.

be happening. Instead of getting used to it, people seemed to develop general fears and anxieties that in the previous generation had existed in railway stations. This was true even among the servicemen who would be tasked with deploying nuclear weapons systems and remote power plants and fighting in conflicts using this new paradigm of warfare. In the case of the civilian public, a problem was the overbearing secrecy of the entire nuclear business. There is nothing more frightening than the unknown. In the case of the armed forces, it was an excess of knowledge.

Servicemen had been subjected to the flash, the thunderclap, and the ground shock of an atomic-bomb detonation in multiple test exercises using real weapons and real soldiers. In the next war, the infantry would have to move forward into territory that had just been sterilized by the radiation pulse of a nuclear detonation, moving toward ground zero as the mushroom cloud formed in front of them, wearing only light protective gear. This test series was both a confirmation that this maneuver was possible and an effort to allay the soldiers' fears. They had to be convinced that they could advance into the freshly destroyed target area without dropping dead from some invisible, undetectable agent. The men had been prepared for the experience by sitting through several mass lectures, hearing about rems, rads, roentgens per hour, the importance of beating the dust off your clothing, symptoms of radiation poisoning, and how to aim a rifle while wearing a respirator. It was more than they wanted to know. The fact that the AEC observers were wearing full-body radiation suits and waving Geiger counters did not help.

As early as 1951, the Air Force had pushed for an atomic weapon that could be used against massed air attacks, and by January 1, 1957, Project Bell Boy had produced an air-to-air nuclear-tipped missile that could be launched from an F-89J pursuit plane. It looked short and fat, with a bulging nose cone and stubby fins. It had a solid-fuel rocket engine in the tail and no guidance system. The beauty of firing a nuclear warhead at another airplane was that you didn't need a guidance system or even a particularly good aim. Any airplane within a mile of it would be destroyed as the 2-kiloton warhead exploded and made a hole in the atmosphere.

The warhead designation was changed from EC 25 (Emergency Capability) to fully operational MK 25 Mod 0 in July 1957, just before its proof of function and weapons effectiveness trial at the Nevada Test Site. The test was scheduled for July 19 in Shot John of Operation Plumbbob. The rocket-propelled weapon was designated MB-1, or, affectionately, "Genie."

Genie was a well-designed system, and eventually 3,150 units would be manufactured and stored safely away, waiting for an appropriate war. There were two problems with Genie, mostly of a philosophical nature: there was nothing to use it against, and the troops were scared to death of it.

The concept of a massed air attack, in which a squadron of heavy bombers would blot out the sky and rain gravity bombs down on cities, was from World War II. It was a strategy used by the United States Army Air Force against enemy assets in Germany and Japan, and not the other way around. America had actually never seen a bomber group dropping explosives on it. We had therefore invented the perfect weapon for blowing up an entire sky full of multi-engine aircraft in one swat, so we could have defeated ourselves two wars back. If the UK had had a few Genies in 1940, the Battle of Britain could have been over in about 20 minutes, but the idea of attacking North America using a close formation of airplanes had never been seriously considered. In the Cold War era there was basically nothing to shoot down that an inexpensive Sidewinder air-to-air missile would not take care of.

Be that as it may, the problem of human squeamishness at having an A-bomb explode overhead could be addressed. Simply explaining to soldiers that a nuclear detonation at 10,000 feet was not the same as having it go off at 1,000 feet was true but insufficient. To the soldiers it was a matter of degree. At the high altitude there would be no ground disturbance. No radioactive dust kicked up into a mushroom cloud, no neutron activation of the ground, and negligible fallout. It was all a function of range. The fission neutrons could not travel that far in air before they decayed into hydrogen gas, and the gamma ray pulse would be short-lived and dissipated in a spherical wave-front with a diameter of four miles when it hit the ground.

Air Force Major Don Luttrell had a master's degree in nuclear engineering, and he did not require convincing, but he came up with an idea to calm the fears of those unwashed by the firehose of knowledge. He persuaded four other officers and a camera operator to stand with him directly under the detonation point of Shot John and prove on film that the weapon was perfectly safe unless you were flying in a bomber group.

The camera man was George Yoshitake. For protective gear he wore a baseball cap as he filmed five officers squinting up into the sky, standing by a crude sign saying "GROUND ZERO, POPULATION: 5," disregarding George as a participant.[115] The F-89J interceptor fired the Genie at a coordinate in the air directly over the men. It traveled the 20,000 feet from the plane in 4.5 seconds and exploded, causing a brilliant burst of light. In the bright desert sun the photon flash fell over the men and briefly washed out the picture. They felt the wave of heat, like opening the door on an oven. The shock wave followed in a few seconds, causing them to cringe from the thunderclap as they watched the bright spot turn into a red ball of fire, eventually disassembling into a weird atmospheric display. It was a cloud of white mist surrounded by an orange donut. There was no debris to fall, as the rocket and its payload had been reduced to individual atoms.[116]

The film was shown to military personnel likely to deal with nuclear weapons, but not to the public. The average citizen had no idea that such a weapon existed, regardless of President Dwight Eisenhower's Peaceful Atom proclamation that declassified much of the atomic bomb development activities. The engine of the atomic submarine *Nautilus*, which was destined to become the model for standard civilian nuclear power plants, was also classified secret.

Typical of the public's concept of a nuclear reactor was shown in 1953 on the very popular television program, *The Adventures of Superman*, starring George Reeves as the Man of Steel. It was episode 33, "Superman in Exile," in which Clark Kent detects an atomic pile running out of control 100 miles away. He rips off his glasses and

115 See the film at http://dvice.com/archives/2012/07/this-is-what-st.php.

116 Two of the men, Luttrell and Yoshitake, are still alive, although all six participants eventually suffered from cancer. It is doubtful that any disease was caused by standing under Shot John. It was only a short pulse of radiation that hit them, and not the chronic exposure that can cause cancer. They, as well as the majority of soldiers in the armed forces, smoked tobacco, and this dangerous habit typically overwhelms any signal of radiation-induced cancer in the thousands of nuclear test participants. The study of cancer in Cold War veterans continues.

ducks into a janitor's closet where he can change his clothing. Jumping out an open window, he flies to the reactor site and finds a scientist in a lab coat, obviously suffering badly from the radiation load, trying to push a control rod back into the face of the pile. Superman, who can withstand just about anything, pushes the rod home without breaking a sweat, and the world is saved from an impending reactor explosion.[117]

From a nuclear engineering perspective, there are several things wrong with this picture, the first being the manually operated control. A reactor control rod had not been moved by hand since 1942, when the original CP-1 was in operation in Chicago, and even that rudimentary pile would have automatically shut down if any part of its operation were out of normal bounds. The fact that the reactor was shown as a vertical wall painted white with a matrix of round holes in it was modeled after the X-10 reactor (the "Clinton Pile") at Oak Ridge, which everyone had seen in the newsreels and a few still pictures. It was the graphite-moderated, air-cooled, low-power job that had been banged together during the war to test fuel slugs for the plutonium production reactors being installed at Hanford, Washington, and it was probably the only reactor configuration that civilians had ever seen. The loading face of the pile, which was painted white, was a seven-foot-thick concrete wall, acting as a bio-shield for workers standing in front of it.

An impressive science fiction film in 1956, *Forbidden Planet*, starring Leslie Nielsen and Anne Francis, did greater harm with its depiction of a nuclear reactor. In the movie, which was loosely based on William Shakespeare's shot at sci-fi, *The Tempest*, Nielsen and his crew land the flying saucer C57-D on the planet Altair-IV and find two survivors of an exploration party that had landed 20 years ago,

---

117 Superman finds that his few seconds in front of the reactor face have infused him with radiation, and he must exile himself to prevent harm to anyone around him. It has been over 55 years since I saw that episode, but I also remember the radiation making noise and electrical sparks, all of which is wrong. However, episode 69, "Peril," has Perry White, the editor of the *Daily Planet*, exercising his inner scientist by engaging in astonishingly prescient nuclear research. Perry is working on a process that extracts uranium from seawater, named "formula U183." A large percentage of the Earth's available uranium is dissolved in the ocean, and Japanese scientists are currently working on a system to extract it for use in power reactors. I have been advised of their secret uranium-extraction filter material, but there is not room in this footnote to discuss it.

living in an abandoned extra-terrestrial civilization. Nielsen decides to contact Earth for instructions on dealing with this odd situation, and in a refreshing twist of drama such a message is not easy to send. Earth is 16 light-years away.

To accomplish this feat it is necessary to dig a shielding trench and remove the ship's power reactor to drive the transmitter, which must be assembled *in situ*. At this point any remaining credibility is blown away, as the men are shown hauling the graphite reactor out of the engine room. Nuclear engineers watching this movie cringe as they see an unshielded core, which had just been used to hurl a ship across interstellar space, pulled casually through the crowd of crewmen with a motorized crane. The naked core would have painted them all with a withering blast of fission-product radiation. This is what the public saw of reactor design and practice: a technology full of danger that was safe as long as it was turned off, and neatly blocked piles of graphite with cross-holes drilled in them and stuffed with uranium.

It is an exaggeration to say that the British Atomic Energy Authority after the end of World War II had about the same picture of how a nuclear reactor looks as the set designers of *The Adventures of Superman*, but not by much. The only reactor that the British scientists and engineers had ever seen was the X-10 pile at Oak Ridge, the first and almost the last megawatt-capable air-cooled reactor ever built in the United States.[118]

The British had contributed greatly to the war-time atomic bomb project with a special delegation working at the Los Alamos lab in New Mexico, providing us with personnel ranging from William Penney, a brilliant mathematician, to Klaus Fuchs, the man who spilled every secret he could get his hands on to the Soviet Union. The Brits were useful and trusted, and they knew everything about atomic bomb theory and construction techniques. Everything, that is, except for the reactors. They were not allowed anywhere near the plutonium

118 An updated replica of the X-10 was built at the Brookhaven National Laboratory in 1948, and it was in constant use for research until final shutdown in 1968. The Brookhaven Graphite Research Reactor was not a military secret, and newsreels showing workers pushing fuel rods or experiment samples into the holes on its front can be difficult to distinguish from similar films made at X-10. I think that the power-scram point on the X-10 was set for 1.8 megawatts.

production reactors or fuel-processing facilities at the Hanford Site. The bombs were top secret, but the water-cooled graphite reactors, being extremely valuable industrial assets, were even more so. Just about any country having some physicists and engineers could figure out how to build an A-bomb, but producing the materials for this bomb was where the secrets lay. After the war was over, the British were very disappointed to learn that all the camaraderie and warm feelings of brotherhood were blown away by the United States Congress Atomic Energy Act of August 1946, forbidding any sharing of atomic secrets with anyone, even close allies. The Brits were left to fend for themselves in the twisted new world of Cold War and Mutually Assured Destruction. All they had in the world was Canada, an ambitious prime minister, a tour of the X-10 pile, and a willing spirit.[119]

William Penney, fresh from working for the Americans on blast effects studies at the Bikini Atoll tests in 1946, was put in charge of the British atomic bomb development. Bill Penney was born in British-owned Gibraltar in 1909 and reared in Sheerness, Kent. Early interest in science and primary education in technical schools led eventually to a Ph.D. in mathematics from London University in 1932. After that he did time at the University of Wisconsin-Madison as a foreign research associate. This work persuaded him to change his career from math to physics, and he made it back to England, applied to the University of Cambridge, and earned a D.Sc. in Mathematical Physics in 1935.

In front of newsreel cameras, Penney always had the silliest grin, speaking with a back-country drawl and answering questions with the intellectual presence of a slow-witted kindergartner, but underneath it he was a razor-sharp scientist with experience and an ability to make people work together in harmony. As the guiding light for the steadfast British nuclear weapons program, he was awarded every honor, culminating in Queen Elizabeth II granting him the title The Lord William George Penney, Baron Penney.

Putting things in the order of priority, the first thing would be to figure out how to build an atomic pile. A heap of graphite called

---

119 All was not lost. Some leeway was granted quietly in 1948 by an agreement called the "Modus Vivendi." It allowed American scientists to tell their British colleagues that they were perhaps going in the wrong direction in their bomb-production development, but not exactly how.

GLEEP (Graphite Low Energy Experimental Pile) was assembled at the new research center at Harwell, an abandoned World War II airfield in Oxfordshire, in an airplane hangar. It was first started up on August 15, 1947, and was used to get the hang of how a nuclear reactor works.[120] Time was of the essence. A month later, before there was any significant experimental work using GLEEP, construction of the first plutonium production pile began at a spot called Windscale.

Seascale was a vacation spot on the northwestern coast of England, or it had been during the Victorian age. Now it was a depressed area. A couple of hundred yards inland was an abandoned royal ordnance factory named Windscale, and here the British nuclear industry would begin. It was in the middle of dairy-farming country, and putting open-loop, air-cooled reactors there made as much sense as installing a fireworks stand in the middle of a high school auditorium. So be it.

The two production reactors would be like the X-10, only larger, with more capacity for making plutonium. X-10 was basically a short cylinder of solid neutron-moderation material, not perfectly circular but octagonal, laid on its side with holes bored clean through it to hold uranium fuel. A long platform in front moved up and down and would allow men, working like window washers, to poke little fuel cartridges into the holes on the vertical face. Each fuel element was metallic, natural uranium, having the small U-235 content allowed in nature, sealed up in an aluminum can, about the size of a roll of quarters. Using metal poles, workers could push in new fuel on the face of the reactor, and used-up fuel, still hot from having fissioned to exhaustion, would fall out the back, landing in a deep pool of cooling water sunk into the floor. It would be a terribly clumsy way to make a power reactor, but as a plutonium production pile the fast turnaround of the inefficient fuel and the enormous size necessary to keep a chain

120 GLEEP was initially thought to be the first working reactor in the Eastern Hemisphere, but the Soviets had beaten them to it. The F-1, which was an excellent copy of the Hanford 305 plutonium production reactor, first started up at 6:00 P.M. local time on Christmas Day, 1946. The Soviet espionage network, which was second to none, obtained the plans for this top secret system at the Hanford Works, and this was a short-cut to production without a lot of rediscovery and experimental development as was necessary in the UK.

reaction going were perfect. Instead of wasting a lot of effort trying to make electricity, all exertion would be to convert the otherwise worthless uranium-238 in the fuel into plutonium-239, and the power was thrown overboard. A short time in the neutron-rich environment for the fuel meant that the probability of plutonium-239, the ideal bomb material, being up-converted into the undesirable plutonium-240 was minimized.

The two Windscale reactors were huge. The core, built like X-10, was 50 feet high and 25 feet deep. The fuel-loading face was covered with a four-foot-thick concrete bio-shield, backed with six inches of steel and drilled with the holes for inserting fuel cartridges into corresponding channels cut in the core and running clean through, horizontally. There were 3,440 fuel channels. Each square array of four fuel channels was serviced from one hole in the loading face, normally sealed with a round plug. Every fifth horizontal row of access holes had the positions numbered from left to right. When running, the core would be loaded with 21 fuel cartridges in each channel, or about 70,000 slugs of metallic uranium. The fissions would be controlled using 24 moving rods made of boron steel, engineered to absorb neutrons and discourage fission, trundling in and out using electric motors wired to the control room, located in front of the building. A scram of an emergency shutdown was handled separately, using 16 additional control rods inserted into vertical channels. Under normal operation, these rods were withheld using electromagnets. If anything went wrong, electricity to the magnets was cut and the heavy rods would fall by gravity into the core, shutting it completely down.

The first set of plutonium production reactors at Hanford were water-cooled, but even the Americans admitted that this scheme, while elegant and impressively complicated, invited a disastrous steam explosion. Let a water pump fail, and the coolant stalled in the reactor would flash into steam, sending the works skyward. Besides all that, unlike the Yanks, the Brits had no mighty river like the Columbia to waste as an open-loop cooling system. Using seawater was a possibility, but there would be destructive corrosion. Penney's design team decided to do it the easy way, just like X-10, blowing air through the loose-fitting fuel channels and up a smokestack, sending the generated heat and whatever else managed to break loose up into the sky and out over the Irish Sea. Each reactor would be equipped with eight very large blowers, two auxiliary fans, and four fans dedicated to shutdown cooling.[121] Each main blower was a monster, driven by a 2,300-horsepower electric motor running on 11,000 volts. The exhaust stacks were designed to be an imposing 410 feet high, made of steel-reinforced concrete.

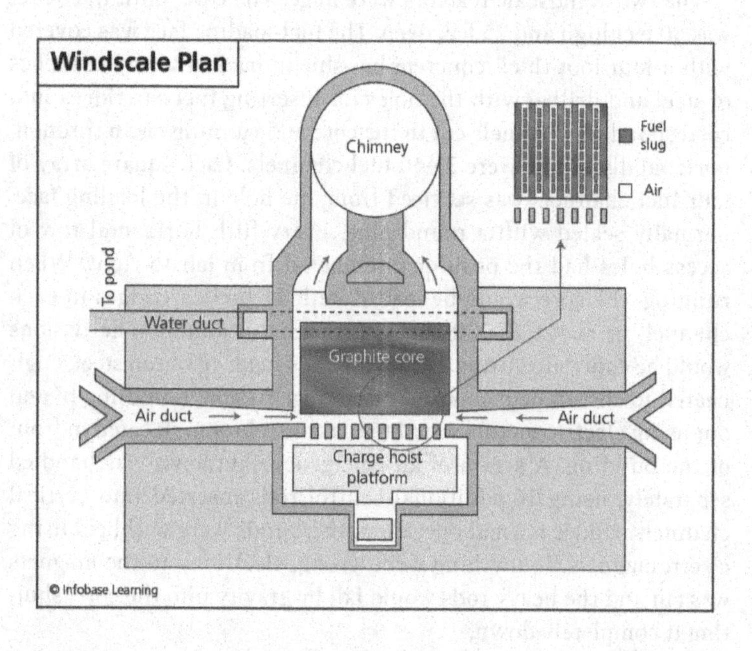

The Windscale Unit 1 plutonium production reactor, shown from the top looking down. Not shown are the two auxiliary buildings, left and right, containing air-blowers the sizes of double-decker buses. The water duct was used to cool off the used fuel as it was poked out the back of the graphite reactor core and moved out of the reactor building for plutonium extraction.

121 No published plan diagram of the Windscale reactors shows the blowers. That is because the blowers were located in two separate buildings per reactor, one left and one right, connected by large concrete tunnels. The Windscale reactors are still there and can be seen on Google Earth. Windscale Unit 1 is at latitude 54.423796°, longitude -3.496658°. The stack on Unit 1 has been torn down and the base is filled with concrete. Unit 2, to the left of Unit 1, looks complete, but the west-side blower building has been torn down and made into a parking lot.

Such a reactor could be made using heavy water as the neutron moderator, as the Canadians were demonstrating, but it was cheaper and easier to use graphite, just like the X-10. It would take 2,000 tons of precision-machined graphite blocks to make one reactor.

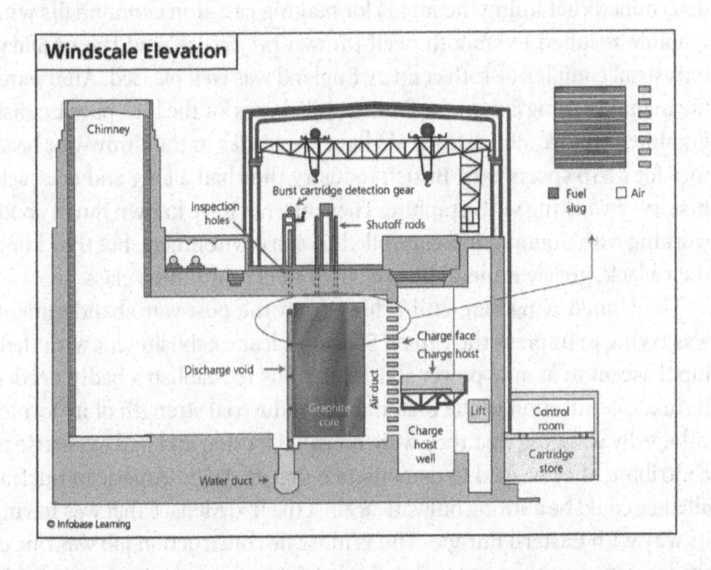

The Windscale Unit 1 reactor, shown from the left side. Only a small portion of the chimney is shown so that the diagram would fit on the page. The charge hoist was a movable platform used to position workers on the front of the reactor, located behind a thick concrete shield and the space for cooling air to enter the fuel channels at the front.

Graphite is a strange and wonderful material. It is a crystalline form of carbon. So is diamond, and a diamond is the hardest material on Earth, hard enough to saw through a sapphire. Graphite is completely different, being soft enough to cut with a butter knife and be used to lubricate automobile window channels. In recorded history, an enormous deposit of natural graphite was first discovered on the approach to Grey Knotts from the town of Seathwaite in Borrowdale parish, Cumbria, England, about 13 miles northeast of Windscale. Locals found it exceedingly useful for marking sheep, and news of this proprietary livestock branding method leaked out in 1565 after decades of use. Expanding on this theme, entrepreneurs in Cumbria proceeded to invent the lead pencil, and from this wonderful application the mineral acquired its official name from the Greek word *grapho*, meaning to write or draw.

As happens to many such discoveries, a military use was soon found and the English government clamped down on the source. It had been determined that lining the molds for making cast-iron cannonballs with graphite resulted in smooth, well-thrown projectiles, and the military-industrial complex of 16th century England was well pleased. Afterward, the manufacturing enterprise found myriad uses for the fine, pure English graphite from Cumbria, from arc-light electrodes to the throw-out bearings for MGB sports cars. British industry thus had a long and complete history of working with graphite. They may not have known much about working with uranium in a controlled-fission environment, but they knew their black, greasy mineral like the bottom of a Guinness glass.

The United Kingdom, still aching from the post-war abandonment, was trying to impress the United States nuclear establishment with their rapid ascent to atomic power status and thus reestablish a badly needed intimate connection to the material and industrial strength of its former colony. By showing that they were on equal footing and had expertise to contribute, they wished to demonstrate that an Anglo-American nuclear alliance could be a strong bulwark against the Red Menace that was having its way with Eastern Europe. The Windscale construction job was one of the largest, most time-pressed in English history, employing nearly 5,000 workers, including more than 300 surveyors, architects, and building engineers. Everything was in short supply except determination.

The Americans, irritatingly miles ahead of the Brits in these matters, were hard to impress, but they were concerned with what they had heard about the British plutonium production initiative. In 1948, a team from Los Alamos slipped into the headquarters at Harwell to give some advice. There were a couple of days of classified seminars with Bill Penney's top scientists. All the talk came down to this: *Whatever you think you know about graphite is wrong.* The conversation between the American scientific delegation and their British counterparts might have gone like this:

"For example," offered the Yank, "there's the *Wigner Growth*."

The Brit lowered his teacup slowly, seeming to twitch slightly. "Say what?"

"The Wigner Growth. If graphite is exposed to fast neutron flux, as it will be in your production pile design, the recoil action of neutrons colliding with carbon atoms will displace these atoms in the graphite crystal. Over time, changing the distance between atoms changes the physical size of your graphite block. It seems to grow."

"But how do you . . . ," Nigel sputtered.

"Control it? You don't. You have to allow for it in the design. The only thing you can control is the direction in which it grows. If you make your blocks by extruding them, then the blocks will lengthen only at right angles to the extrusion axis."

"Well, dash it all."

It was best to learn of this wrinkle early on in the construction. Heroic redesign of the core allowed each block to expand horizontally, at right angles to the axis of extrusion, while being held in place with graphite slats. By March 1949, word had arrived from hastily convened experiments at Harwell, trying to replicate the effect claimed by the Americans. British graphite, so wonderful for making pencils, was completely different from the synthetic graphite they used in the U.S.A. It expanded in all directions. That was bad. The prediction was that after running just 2.5 years, the pile would be warped and unusable. The fuel channels would close up and trap the aluminum canisters in the core. The Canadians came through with a modified prediction. Yes, British graphite would expand on all axes, but only at one fifth the length of the American product. That was good. Under that condition, the pile might operate for as long as 35 years, if the internal stresses did not crumble it like a tea biscuit. Other potential problems were revealed too late to be designed out so easily.

Later in 1948, Cockcroft, head of the entire British nuclear enterprise, came back from a fact-finding tour of the X-10 at Oak Ridge with some more disturbing news. They would have to filter the pile-cooling air before releasing it to the general atmosphere. Of all the 70,000 fuel canisters in one pile, if the delicate aluminum case around just one of them were to break open, then its highly radioactive contents would go straight up the chimney and rain down on the dairy farms of Cumbria. That could be a problem. When building their improved version of the X-10 up on Long Island, New York, the Americans had put filters between the atomic pile and the air exhaust stack, just to prevent a fission product spread if the fuel broke apart or caught fire. "We shall do the same," pronounced Cockcroft.

Impossible. The base of the Unit 1 chimney had already been poured, and 70 laboriously laid feet of vertical brickwork was already in place. If filtering were necessary, then it should have been designed in to begin with, in the gallery between the core and the chimney base. There was no way to tear the thing down and start over, and, at this late stage, the only place to put filters would be atop the chimneys. Cockcroft was adamant, and he was in command. The filter assemblies, aptly named "Cockcroft's Follies," were built as specified, using 200 tons of structural steel, concrete, and bricks hauled up 400 feet to the tops of the air stacks.

Several filter packs were tried. In early 1953, a temporary lash-up was settled on. The filters would be fiberglass tissue, sprayed with mineral oil to trap dust made radioactive by going through the reactor core and any fuel that might disintegrate and escape. There was much grumbling, not only among the men who had to retrofit this ghastly mess, but from Penney's bomb makers. To make plutonium at the necessary rate, a ton of cooling air had to go up each of the two stacks every second. That meant that the hot air had to be moving at 2,000 feet per minute as it exited at the top, but at that point it would hit the drag-inducing filter pack like a brick wall. Delicate fiberglass tissue would quickly come apart, and the oil would be blown off the fibers. The air speed would have to be reduced, and this would mean less plutonium per month. The Canadians would have to pick up the slack using their heavy-water reactor at Chalk River.[122]

With further improvements and changes in the filter packs, including substituting silicon oil for the mineral oil, Cockcroft's follies were preventing the occasional radioactive dust particle from ejecting into the atmosphere. An exception was the volatile fission

---

122 The Canadians did indeed supply plutonium for the time-critical atomic bomb project, but unlike the Windscale reactors, the NRX reactor at Chalk River was not designed specifically to make Pu-239. The Windscale pile operations would purposefully run the fuel through very quickly, whereas NRX would use the fuel to useful depletion. As a result, the fissile material for the bomb contained a lot of Pu-240, which fissions spontaneously and threatens to melt the bomb core before it can explode. The Brits solved this problem by using a "levitated pit," in which the bomb core was suspended in the middle of a hollow sphere of high explosive, removing it from close contact with any neutron-reflecting material. It seemed a brilliant innovation, but the Americans had been using this design feature since 1948 in the MK-4 A-bomb, tested in Operation Sandstone.

product iodine-131. As a gas, it was largely unhindered by the filters. There was nothing to worry about. When one was striving to build an arsenal of weapons, one of which can obliterate a large city, it seemed almost inappropriate to fret over probabilistic health effects on individuals. At this point, production speed outweighed all but the most basic safety concerns.

On they labored. Engineering problems were knocked down one at a time. Of particular concern were the fuel canisters. The metallic uranium, which would catch fire spontaneously if exposed to air, had to be cast into small cylinders a few inches long and sealed up in aluminum cans with heliarc-welded seams. The hot uranium would react chemically with the aluminum, so the inside of each can was coated with insulating graphite. The uranium slug would expand as it heated and burst the can, so it had to be made to fit loosely. Each can was filled with pressurized helium to ensure heat conduction to the aluminum, and the assembly was covered with cooling fins. In early summer 1950, Cockcroft threw another serious monkey wrench into the business. New calculations seemed to indicate that the critical mass of Pile. No. 1 would have to be increased by as much as 250 percent, or it would not work.

Reeling briefly with this news, the design team proceeded to inventory every material in the reactor that would parasitically absorb neutrons, affecting the core reactivity. There was one item that could be reduced enough to compensate for the lack of critical mass: the fins on the fuel cartridges. In three weeks the team managed to trim one sixteenth of an inch off each of the one million cooling fins, thus solving the problem.

In October 1950, craftsmanship triumphed over knowledge, and the Windscale Pile No. 1 achieved self-sustaining fission. By January 1951, used fuel cartridges had been chemically processed, and Tom Tuohy, the Works General Manager, held the first lump of British plutonium in his hands. In June 1951, Pile No. 2 became operational, and plutonium production was proceeding at full capacity. The people of the United Kingdom under hard postwar circumstances had put their backs into it and prevailed.

There were problems. On May 7, 1952, there was an unexplained temperature rise in Pile No. 2, only in the upper portion of the core.

The operators were able to bring the temperature down with the fans, but there was not a clue as to why this had happened. In September the same thing happened to Pile No. 1, but this time there was smoke coming out of the stack. Was the graphite burning? They had been warned by the Americans: *Whatever you do, do not let the graphite catch fire*. Once it gets going, water will not extinguish the fire. It will only make it burn hotter, and the graphite sucks the oxygen out of the water and leaves you with explosive hydrogen. Analysis showed that the smoke was caused by lubricating oil leaking from a fan bearing, but the high temperature remained inexplicable.

In May 1952, Pile No. 2 was taken down for maintenance. The workers found that 2,140 cartridges had migrated out of the fuel channels. Some were hanging precariously out the back of the core, and some had flown over the pool of water below and into the wall behind it. The air flowing from front to back had literally blown them out of the reactor. Each fuel cartridge had a little graphite boat under it, so that the cooling fins would not make it stick in the channel, but this would have to be changed. A worse finding was that the Burst Cartridge Detection Gear (BCDG) had punched a hole in the end of a cartridge, putting fission products in the air stream.

There were eight BCDGs, called "Christmas trees," in each reactor. A Christmas tree was a rectangular matrix of 32 vacuum nozzles, arranged to fit over any 4-by-8 array of exit holes in the back of the reactor core. It could be positioned remotely from the control room to sniff the holes, looking for escaping radioactivity that would indicate a broken fuel can, but the operation was completely blind. From the control room or even from the face of the pile where fuel was pushed in, you could not see the BCDGs in action. During normal operation the eight of them would crawl over the back end of the pile, like very large insects, looking for trouble. It was a good idea to have equipment that could tell exactly which hole was leaking, but it could be too strident and cause what it was trying to detect, punching the vacuum nozzle through the aluminum. Some fuel cartridges were even jammed into the frames of the Christmas trees, never making it to the recovery pool below for plutonium extraction.

The American delegation made another informative visit, this time bringing the big guns, Dr. Edward Teller, the brilliant Hungarian theoretician and Mother of the H-bomb, and Dr. Gioacchino Failla from the Columbia University Radiological Research Laboratory. They could explain the mysterious temperature transients in Piles No. 1 and 2.[123] Technically, it was illegal to give such information to a foreign power, but the Brits had to be informed before their entire atomic bomb program went up in smoke.

There is another Wigner effect, called *Wigner Energy*. The same phenomenon that would bend the crystalline structure of graphite would store a great deal of potential energy in the material. If you allowed this energy to accumulate over a long period of time, it could reach a break-down point, at which it would all release at once, igniting the graphite with extremely high temperature and sending your plutonium production machinery up the chimney. The way to keep this from happening is to stage a periodic annealing, in which the core is heated to an abnormally high temperature, about 250°C, using nuclear fission with reduced cooling. At this point a "Wigner release" should occur, with the graphite providing its own heating without the need for fission. Shut down the reactor, and adjust the fans to slowly bring down the temperature.

It was not quite that simple, but the Atomic Energy Authority appreciated the heads-up, and thanks for not waiting until the Windscale facility was wiped out to let us know what had happened. The Windscale piles were very large, and Wigner Energy could collect in various regions or pockets in the graphite stack, meaning that the pile would have to be selectively heated and annealed by skillful use of the controls. There was never a written procedure for annealing. It was more of an art than a science. The first anneal on Pile No. 1 was in August 1953, and they were scheduled for once every 30,000 megawatt days of operation.[124] There were eight anneals from the first until July 1957, and during the first operation three scientists hovered over the

---

123 Inside nuclear engineering, Dr. Teller is considered to be the mother of the H-bomb and not the father because he "took it to full term."

124 The operating power of a Windscale reactor has never been published, but near as I can tell it was 90 days between anneal no. 8 and anneal no. 9 on Pile No. 1. The plutonium/tritium production was so pressed at this time, it was decided to go 40,000 megawatt days between anneals for no. 9. Assuming that Pile No. 1 ran 24/7, a power rating of over 400 megawatts can be calculated. That is a lot of power for an air-cooled reactor.

operators, giving advice and asking questions. After that, it seemed like a routine task, and no science was needed.

The two Windscale piles, with some help from the Canadian NRX, produced enough plutonium for an official British atomic bomb test on October 3, 1952. The blast, occurring under water a few hundred yards from Trimouille Island, Australia, released a respectable 25 kilotons of raw energy, digging a crater 20 feet deep and 980 feet across. William Penney's bomb design crew and the efforts of thousands working at Windscale had paid off in the most visible way. This would prove to the United States that Britain had achieved parity in the arms race, and an immediate summit meeting was expected.

Unfortunately for this motive, a month later the United States set off "Ivy Mike," the world's first thermonuclear bomb on Elugelab Island at Enewetak Atoll, releasing the energy equivalent of 11 million tons of TNT high explosive. Elugelab Island vanished, replaced with a crater 64 feet deep and 6,240 feet across. The British arms industry would have to go a bit farther to come up even with that.

Penney's team had the imposing task of coming up with a thermonuclear weapon design quickly, starting from scratch. They had not so much as an encouraging word from the Americans, and making hydrogen isotopes fuse explosively was not an easy problem. It was only obvious that a source of tritium, the heaviest isotope of hydrogen and one most likely to fuse, would be necessary.

There is no natural source of tritium, and, like plutonium, it must be manufactured using fission reactors. To accomplish this, a rod of lithium-magnesium alloy, about half an inch in diameter, is encased in a sealed aluminum can and pushed into one of many special "isotope channels" in the reactor. Neutrons from the fission process are captured by the lithium-6 nuclide component of natural lithium, and the atom subsequently breaks down into one helium atom and one tritium atom. The helium pressurizes the can, and the tritium combines chemically with the magnesium component of the alloy, becoming magnesium-tritide. After being in the reactor for about a week, the rods in several cans are harvested and chemically processed to remove the pure tritium. So far, so good.

The tritium production cartridges were given the code name "AM," and the Mark I cartridges were built using standard fuel cans with a lead weight added so that they would not fly away. The Mark II cartridges were an improvement, having the lithium-magnesium rod diameter increased to 0.63 inches. Under the extreme needs of the hydrogen bomb program, the even more improved Mark III cartridges were built using thinner cans, no lead weight, and alloy rods an inch in diameter. The Mark III's were admittedly dangerous. The lithium-magnesium rods were pyrophoric, meaning they would burst into hot flame and burn like gasoline if they were exposed to air. The cans were as thin as possible, and the helium pressure could blow them open if the temperature was as high as 440°C. In addition, there were now so many AM cartridges in the Windscale cores, they dragged the fission process down. Absorbing neutrons without adding any to the fission process made the reactor subcritical. To work at all, the reactor fuel had to be beefed up with a slight uranium-235 enrichment, coming from the new gaseous diffusion plant at Capenhurst.[125]

On January 9, 1957, the British Prime Minister, Anthony Eden, having just presided over the political disaster of the Suez Crisis, resigned from office after being accused of misleading the parliament. The British Army had done a splendid job of tearing up Port Said, Egypt, with French assistance, but internationally it was seen as the wrong force applied at the wrong time as a reaction to Egypt having nationalized the Suez Canal. Eden was succeeded by Harold Macmillan, who desperately wanted to recapture the benevolence and camaraderie of the United States, and the hydrogen bomb initiative was the point of the spear. It would proceed at an ever-accelerated pace.

By May 1957, Penney's group had put together the first British thermonuclear test device, code named Short Granite, to be exploded in Operation Grapple I off the shore of Malden Island in the middle of the Pacific Ocean. Time was becoming crucial, as a nuclear test-ban treaty was in the works, and the end of unrestricted nuclear weapon experimentation was in sight. The expected yield was north of a

125 The piles were also loaded with LM cartridges. These special canisters were used to produce polonium-210 from neutron capture in bismuth oxide for use in A-bomb triggers. The LM cans were not considered flammable.

megaton. On May 15 Short Granite was dropped from a Vickers Valiant bomber, and it was an embarrassing wipeout, giving an energy release of only 300 kilotons. The first attempt at a British hydrogen bomb had failed.[126] They would get it right the next time, but a lot more tritium was required.

After Grapple I, the Windscale piles were operating in emergency mode at the highest possible power level using flammable metallic uranium fuel and flammable lithium-magnesium in piles of flammable graphite with gale-force air blowing through them. Fire was prevented by coverings of thin aluminum on the fuel and AM cartridges sitting in graphite bricks that could become superheated on their own at an unpredictable time for reasons that were not entirely understood. The annealing interval was increased to once every 40,000 megawatt days for the sake of more plutonium and tritium production. It is not unreasonable to predict an eventual problem under these conditions.

Sunday night, October 6, 1957, was in the middle of a local influenza epidemic, and Windscale Pile No. 1 was well overdue for an anneal. The front lower part of the graphite had not really released any energy at the last anneal in July 1957, and it probably had 80,000 megawatt days of Wigner energy buildup ready to let go at any time. Production was halted, the pile was cooled down, and the ninth anneal would begin Monday morning.

The temperature of the pile was monitored using 66 thermocouples placed at three depths in the reactor face in selected fuel channels. Considering that the core occupied 62,000 cubic feet, and a detailed map of the temperature throughout the graphite was necessary to monitor an anneal, having only 66 thermocouples was pitifully inadequate. There were also 13 uranium-temperature monitors in the control room

126 I am told that Short Granite was a full-scale two-stage hydrogen bomb weighing 4,550 kilograms, and the active fusion component had to be the light solid, lithium deuteride. So what was all the furious tritium production for if not a liquid tritium-deuterium bomb? A tritium-based two-stage thermonuclear weapon, such as the American EC-16, would weigh about 32,000 kilograms, which would have nailed the Vickers Valiant bomber to the ground. As it turns out, the tritium production in the Windscale piles was not for the Short Granite bomb. It was used in the electrically driven neutron generator, or "initiator," used in the A-bomb "Orange Herald," tested over Malden Island on May 31, 1957. Short Granite used a beryllium-polonium solid-state initiator, made using the polonium-210 produced in the Windscale piles in the LM cartridges. The accelerated production at Windscale was to make bomb-trigger components.

and another seven on top of the pile in the crane room. The operations staff had performed these measurements before, and there was no concern about this weakness in the instrumentation. On October 7, the main blowers were turned off at 11:45 A.M., and an extremely slow and cautious approach to criticality in the lower part of the core was begun. Stopping twice to look at the temperature readings, it took the operators seven and a half hours to withdraw the controls and reach criticality. There was a holdup to work on some faulty thermocouple connectors. Once the pile was maintaining a low power level, the crew wrangled with some control rod positions to concentrate the fissions in the lower front of the core, where annealing was obviously needed. The goal was to bring this section in the core up to 250°C.

The facts that Windscale Pile No. 1 was able to achieve criticality and maintain it could give one with a religious bent the belief that God was on the side of the United Kingdom. Or, perhaps the Devil was. Driving it was like trying to steer the RMS *Titanic* around an iceberg. The core was so big, the peak of the neutron flux curve could be put just about anywhere in the massive block of graphite by artistic positioning of the controls, but it was ponderously slow to respond to control movements, and the instrumentation was a bit numb. The pile would not do anything quickly, and the dangers of a sudden runaway or an explosion were nonexistent.

An hour after the beginning of Tuesday, October 8, the pile seemed to be running cool, reading between 50° and 80°C in the graphite. One thermocouple indicated 210°C, and this was obviously where the annealing process had begun. An anneal would always start somewhere and then spread out slowly, hopefully infecting the entire pile. The operators ran in the controls to shut it down, letting it cook, and sat back to watch it happen. For some odd reason, two uranium thermocouples read 250°C. Uranium does not anneal; only graphite. Ian Robertson, the pile physicist on duty, had the flu, and he went home to collapse in bed at 2:00 A.M. The anneal seemed well in hand by the operating crew.

Robertson came back on duty at 9:00 the next morning, sick as a dog, to find that the temperature was not spreading the way it was supposed to, and the operators had decided to reheat the reactor by bringing it back to criticality. In a couple of hours they had it back running at low power, but something was not right. One uranium thermocouple showed a sudden temperature rise to 380°C. What's that all about? This type of reactor was run by the graphite temperature and not necessarily by the neutron counts, and it was meandering all over the place, from 328° to 336°C. Pile No. 1 had its own quirks, but meandering was not one of them. Something was amiss, but the annealing had to proceed. The fission process was shut down again at 7:25 P.M. Robertson was starting to get woozy, and the operating crew insisted that he go home.

By Wednesday morning, things seemed to have calmed down. The crew shut the inspection ports atop the core and turned on the shutdown fans to give it some air and bring the wayward temperatures down. At about midnight, the temperature readings off the graphite thermocouples started to go wild.

It was now early Thursday morning. One thermocouple in the middle of hole 20-53, 20 rows up from the bottom and 53 across, was reading extremely high at 400°C. Air dampers were opened. In 15 minutes the temperature rose to 412°C. It did not make sense. At 5:10 A.M. the dampers were opened again, and this time the smokestack radiation monitor indicated a bit of radioactivity in the filter. At about the same time, the stack on Windscale Pile No. 2, which was running plutonium production at balls-to-the-wall power, indicated increased radioactivity. Obviously a fuel cartridge had burst in No. 2. They really should do something about that.

Attempts to bring the annealing under control continued. Air dampers were opened and closed again, with no expected results. By 1:30 P.M., the stack radiation was unusually high, and it was clear that there was a problem. The uranium temperature was at 420°C and rising quickly. The Pile Manager, Ron Gausden, was called in. He gave instructions to open the dampers, turn on the shutdown fans, and try to bring down this temperature. Obviously, a fuel cartridge had broken open and fission products were collecting in the stack filter. The Christmas trees, which were parked off to the side at the back of the core, would not work with the pile at this elevated temperature, so they could not identify the broken cartridge. The temperatures were not going down, and at 2:30 P.M. Gausden told the operators to turn on the main blowers. He picked up the phone to call Tom Hughes and share the excitement.

The Windscale operation was critically understaffed, particularly for this running emergency to produce bomb material. Of the 784 professional posts, 52 were vacant. Hughes had worked at Windscale since 1951 as the Assistant Works Manager in charge of the chemical group. He was also Acting Works Manager for the entire site and had just been assigned a third post as Works Manager for the pile group. His response to the call from Gausden would mark his first visit to an actual pile.

Huw Howells, the Health Physics Manager, was already there to inquire about radiation coming out of the stack. With a short rundown of the situation in the control room, Hughes and Howells decided to call H. Gethin Davey, the Works General Manager with bad news about Pile No. 1. A crew from the control room rode the lift to the refueling platform and pressed the UP button to take them to row 20 on the loading face and have a peek behind plug 53.[127]

The loading face was a bio-shield, built to protect workers who could stand there and push fuel into the graphite pile behind the shield with long poles. When not in use, each hole was plugged with a metal cylinder. A worker pulled out the plug in 20-53 and looked in the hole. Normally, it would take a flashlight to illumine the black-as-carbon fuel channels to see what was inside, but in this case a bright red glare greeted them. All five holes behind the plug were glowing.

The pile was on fire, and it had been burning since Tuesday. Windscale Pile No. 1, running too fast in the dark, had crashed into its iceberg.

Gausden ordered the crew to make a firebreak by removing all the fuel around the burning channels. Easier said than done. Even some fuel cartridges that were not on fire were so swollen from the high temperature, they would not budge. The men dressed out in rubber suits, full-mask respirators, gloves, and dosimeters to monitor their radiation exposures. It was going to be a long night. They started dumping fuel into the cooling pool in back of the core, pushing it out with the heavy rods. It was hot, difficult work, but the main blowers put positive air pressure behind the loading face, and clean, cool air

127 An interesting bit of trivia from B. J. Marsden of the Nuclear Graphite Research Group at the University of Manchester: the refueling platforms at the Windscale reactors, called "charging hoists," were airplane elevators salvaged from WWII aircraft carriers.

blew out the loading holes and onto the men. The airflow was all that made the work possible.

The graphite on the front of the pile below where the men were ejecting fuel was now a mass of flames, and the highest temperature reading had passed 1,200°C. At 5:00 P.M. on Thursday, Davey picked up his phone and rang his deputy, Tom Tuohy. "Come at once," he said. "Pile No. 1 is on fire."

Tuohy was at home caring for his wife and children, who were all down with the flu, and when he got off the phone he instructed them to please keep all the windows closed and not to leave the house.

Tom Tuohy, excitable, bubbly, and auburn-haired, was born at the eastern end of Hadrian's Wall in northern England in a town named Wallsend in 1917. He studied chemistry at Reading University, worked for the Royal Ordnance Corps during the war, and had joined the nuclear effort in 1946. He arrived at the stricken pile in minutes and went straight to the loading face. He saw men struggling to punch the fuel out the back of the pile and got a grim impression of the situation. They were pulling back rods that were glowing yellow on the ends and dripping molten uranium. Back in Davey's office he found a knot of scientists arguing over using carbon dioxide or argon to quench the fire. By 7:00 P.M. he decided to get a better look. He took the stairs to the crane room and lifted an inspection plate so he could see the front of the reactor. It was glowing. He went back at 8:00 P.M. and found flames shooting from the fuel channels, colored yellow by the sodium in the construction workers' sweaty handprints on the graphite blocks. At 11:30 he pried up the viewing port, which was becoming quite warm, and saw blue flames hitting the concrete wall many feet behind the back of the pile, and a new fear gripped him. The flames were hot enough to ionize the nitrogen in the air. Not only could the hot flames burn through the concrete wall, but he was not certain about the floor under him staying put much longer.

Early on Friday, October 11, Davey was sent home sick and in pain with the flu. Between 4:00 and 5:00 A.M., the largest tank of carbon dioxide that could be found was expended into the fire, with zero effect. There was nothing to do but start pumping water into the fuel channels, a desperate move. The concept of flooding a graphite reactor with water was so remote, there were no fire-hose taps in the building.

The works fire brigade was called in with pumper trucks, and nozzles were jury-rigged to force water through the holes in the loading face and deep into the fuel channels, two feet above the highest point of the fire. The water was turned on just before 9:00 A.M. K. B. Ross, the Director of Operations, dispatched the following message to the Authority Chairman, Sir Edwin Plowden, in London:

> Windscale Pile No. 1 found to be on fire in the middle of lattice at 4:30 pm yesterday during Wigner release. Position been held all night but fire still fierce. Emission has not been very serious and hope continue to hold this. Are now injecting water above fire and are watching results. Do not require help at present.

Plowden dashed off a note to the Minister of Agriculture, wisely anticipating trouble with radiation contamination of the dairy farms in the area.

Tuohy was relieved when there was no explosion from the water interacting with the graphite. Watching through the viewing port at the back end of the pile, he could see the water gushing out of the fuel channels and cascading down the pile into the cooling pool at the bottom. The water looked good tumbling down, but still the fire raged out of control. The back of the core remained a wall of flames. He watched for an hour, starting to feel that the situation was hopeless, when he was struck by a brilliant idea. All the men had evacuated the loading face. There was no longer any need for the blowers. He decided to turn them off.

Almost instantly, the flames died away.

This time when Tuohy went up to the crane room to look, the viewing port seemed firmly cemented to the floor. The fire in its dying gasp was sucking oxygen out of the remaining air and causing a vacuum in the reactor core. With no more air forced through the fuel channels, there was no more fire. By noon Tuohy was able to report to Davey that the fire was out and the situation was now under control. Just to make certain of his pronouncement, the water was kept flowing for another 30 hours.

The core of Windscale Pile No. 1 was a total loss. Over 10 tons of uranium were melted and five tons were burned. Very little of the uranium or even the fission products went up the chimney. A brittle oxide crust had formed in the extreme heat of the fire, and the heavy oxides were bogged down in it before they made it to the air outlet. The immediate concern was the volatile fission product, iodine-131. The filter packs, which were no longer called "Cockcroft's follies," were not expected to capture any of the 70,000 curies of iodine-131 that were present in the fuel when the fire began, but there was a fortuitous happening. The LM cartridges, containing bismuth oxide meant to activate into polonium-210, burned up, and the light bismuth oxide dust went up the stack and caught in the filters. This and some vaporized lead from the weights in the AM cartridges reacted chemically with the free iodine and captured some of it in the filters. Although 20,000 curies of iodine-131 blew out the top of the stack and over the dairy farms, about 30,000 curies of it were caught in the filters. It was better than no capture at all.[128]

The immediate bad thing about loose iodine-131 is that it falls on the grass, cows eat it, it winds up concentrated in the milk, people drink it, it goes straight to the thyroid gland, and the tightly contained radiation load has a good chance of causing thyroid cancer. The good thing is that it has a short half-life of only eight days. On Sunday night, October 6, the reactor was shut down, iodine-131 production by fission stopped, and the countdown for the decay of the iodine-131 began. By eight days later, half of it was gone. On Saturday it was estimated that 20,000 curies of it were contaminating 200 square miles of territory north of the reactor, but in about 80 days that contamination would drop to 1/1,000 of that level.

However, 20,000 curies divided by 1,000 is 20 curies, and that is still a great deal of radiation. It is 7.4×10^{11} becquerels, or 7.4×10^{11} iodine-131 disintegrations per second, emitting first a 606 kev beta ray and then a 364 kev gamma ray, becoming a stable xenon isotope. The dairy industry in Cumbria was thus in serious trouble. On Saturday

---

128 There was also a great deal of radioactive xenon-133 gas released at Windscale, 16 times more than the iodine-131. This sounds terrible to the general public, but xenon-133 has zero body burden and it cannot react chemically with anything, unlike iodine-131. Animal body chemistry has no use for xenon. The xenon-133 floats away and decays to stable cadmium with a half-life of 5.24 days. There is no xenon filter, and radioactive xenon escapes regularly from every nuclear power plant in the world.

morning there was enough active iodine-131 on the grass, 132,000 disintegrations per second per square foot, to rattle a Geiger counter in a most alarming way.

This was new, untrod territory in the nuclear business, in which a private industry would have to be shut down because of an unfortunate situation in the government-owned, secret, dangerously handled fissile-isotope-production endeavor. There were no safety standards; no maximum limit for the amount of radiation to which a toddler should be subjected in its otherwise wholesome cow's milk. The government did the right thing and bought every pint of contaminated milk. It was poured away into ditches, and the countryside smelled of sour milk for quite a while.[129] After a year, not a trace of iodine-131 could be found in the farmland. One good thing about radiation contamination is that it improves with age.

Neither Windscale reactor would ever be started up again, and there would be no more open-looped, air-cooled reactors built anywhere in the world. Pile No. 2, which was in perfect operating condition, was shut down permanently five days after the fire was stopped, and as much remaining fuel as could be moved was carefully taken out of the wrecked Pile No. 1 and processed to remove the plutonium. Unusable radioactive debris was dumped offshore into the Irish Sea. The cores of both reactors were sealed off with concrete, and the buildings were used for several decades as offices, labs, workshops, and for bulk storage. The 6,700 damaged fuel cartridges and 1,700 AM and LM cartridges in Pile No. 1 will be removed by remote-controlled robots by 2037. The core is still a bit warm. The name Windscale, carrying bad karma, was buried and the site was renamed Sellafield. It is now owned by the Nuclear Decommissioning Authority (NDA).

The people who worked at Windscale in 1957 and who participated in the cleanup operations have been studied, looking for radiation-caused diseases. Statistical analysis indicates that the widespread radiation release should have caused 240 thyroid cancers, but no

---

129 There are also reports that the milk was diluted with water and then dumped into the Irish Sea. It seems a bit odd to dilute it with water before pouring it into the ocean, and the aroma of decomposing milk was mentioned in local newspapers. It's possible that farmers who were not found to have contaminated milk disposed of it anyway on their own initiatives, but more likely they took advantage of the inflated milk prices and sold it.

correlation has been found. The last accounting in 2010 could find no evidence of health effects from the radiation release back in 1957. Tom Tuohy, who stood alone in the most hazardous conditions of radiation exposure at the height of the Windscale fire, died in his sleep on March 12, 2008, at the advanced age of 90 years.[130]

Bill Penney, the man whose relentless demands for more tritium, plutonium, and polonium-210 had caused Pile No. 1 to be overworked to the point that it burned down, was assigned as the chairman of the inquiry committee. On October 26, 1957, 16 days after the fire was put out, he submitted his report to the Chairman of the United Kingdom Atomic Energy Authority. His committee found the operating staff at fault for causing the fire. During the annealing process, they had reapplied nuclear heating to the reactor too soon after the first try. He did not point out that the fire had started in hole 20-53 when the overheated lithium in a Mark III AM cartridge diffused through the aluminum and set fire to the adjoining fuel, and he assigned no responsibility to himself or to the questionable engineering of the isotope production cartridges.[131] In retrospect, the Windscale piles were a disaster waiting to happen. Panic-containment photos were released showing contented cows grazing in the grass with the Windscale chimneys looming in the background.

The secrecy covering the Windscale site and its mission did not help the publicity crisis that arose when it hit the newspapers in London and Manchester. Simply cautioning people about drinking milk or leaving windows open was enough to cause some concern. At one leak point a scientist at Windscale, Dr. Frank Leslie, wrote a letter to the *Manchester*

---

130 See an interesting British documentary on the Windscale fire from 2007 at http://www.youtube.com/watch?v=ElotW9oKv1s&feature=relmfu. Some important technical details are garbled, and the recreated scenes of men working at the loading face are seriously wrong.

131 The actual cause of the Windscale fire remains a controversial and unresolved topic. It is probable that it was started in an isotope channel by one of the many AM cartridges, which were the most temperature-sensitive components in the entire pile. There was one isotope channel for every four fuel channels, and when you opened an access plug, the isotope channel was in the center, with four fuel channels surrounding it. The first man to open plug 20-53 reported that "it was glowing bright red, and so were the four channels around it." It is often assumed that the center hole was a fuel channel, but clearly he was referring to the isotope channel behind 20-53.

Guardian on October 15, expressing disappointment that the government had given no warning to the public until the matter was resolved. Harold Macmillan, the excitable Prime Minister, had to be restrained from seeking out Dr. Leslie and strangling him manually. The Penney report was finally declassified and released to the public in January 1988.

On November 8, 1957, about a month after the Windscale fire, the British bomb program achieved what it had so dearly sought. Above the southern end of Christmas Island in the Pacific Ocean, the two-stage thermonuclear device Grapple X went off with a bang, yielding 1.8 megatons. Penney's development crew had done it almost correctly, and a bomb that was built to give one megaton unexpectedly wiped out the military installations on the other end of the island with 80% over yield. A triumphant British delegation of scientists was invited to Washington to talk about thermonuclear topics with some fellows from the Lawrence Livermore Lab, and the shutout was broken. From that point on, the United Kingdom was granted technical parity with the United States in matters of nuclear weapons and use of the Nevada Test Site.

The replacement plutonium production reactors for the Windscale units were already up and running, as of August 27, 1956, at the Windscale site. Four large graphite reactors using closed-loop cooling by carbon dioxide, named Calder Hall Piles 1 through 4, produced 60 megawatts of electricity each, and were considered to be the world's first nuclear reactors making significant commercial power. In reality, the purpose of these reactors was to convert uranium-238 into plutonium-239 and to manufacture tritium. Calder Hall Pile No. 1 ran for nearly 47 years before it was finally shut down in 2003.

A sister plant named Chapelcross was built on an abandoned airfield in southwest Scotland near the town of Annan. The plant, comprising four 180-megawatt Calder Hall reactors, was completed in 1959. Its purpose was to produce plutonium and tritium, with 184 megawatts of electricity on the grid being a secondary byproduct. The Calder Hall and Chapelcross reactors were safer and more complex than the antique Windscale units. The carbon dioxide cooling gas worked just like air being blown through, but it was contained in a closed loop, pumped back into the bottom of the core after the heated gas had been used to make steam in heat exchangers. The steam, which was used to turn the turbo-generators, was also in a closed loop, and the ultimate heat sink for disposing of the waste energy was one large cooling tower per reactor, dumping heat into the atmosphere. There was not much worry of fire in a graphite pile cooled with fire-suppressing carbon dioxide. Even an unlikely steam explosion would not release fission products into Scotland.

For efficient plutonium production, the fuel was still metallic uranium, this time enclosed in cans made of magnesium oxide. It was codenamed MAGNOX, and this became the type designation for a series of 26 power reactors built in the UK. Magnesium oxide is perfectly fine as a canning material, as long as it does not get too hot or too wet. If left too long in a cooling pond, the magnesium oxide corrodes and leaks, and the fuel inside can melt easily if deprived of coolant while fissioning.

Given the Cold War pressures of producing Poseidon and Trident missile warheads and the risky fuel design, these improved graphite reactors were remarkably safe. There were no meltdowns or fires of the Windscale severity, but there were some interesting incidents.

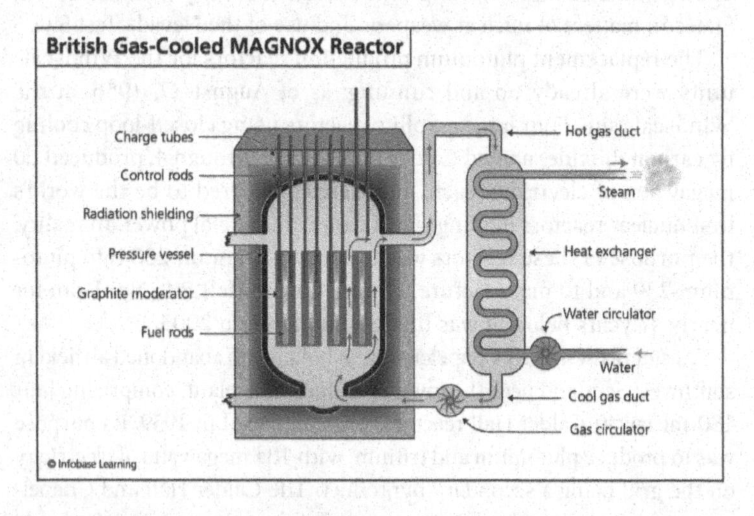

The British MAGNOX reactors used graphite neutron moderators and a fast-flowing gas for cooling. This design was different from the Windscale reactors in that the primary cooling system was closed, using the same gas over and over as it was recycled back into the reactor after it had given up its heat to a steam generator. In this design, the gas was carbon dioxide or helium, and with the closed system the graphite would never be exposed to oxygen and would be unlikely to burst into flame.

The Chapelcross piles were just like the Calder Hall piles, only different. They were built in series, starting with Pile No. 1, and improvements were added as the construction progressed. Pile No. 1 required annealing, similar to the Windscale piles, but, starting with Pile No. 2, the graphite fuel channels were lined with graphite sleeves. The sleeves were fragile and were something else that could break, but they were designed to let the graphite run so hot it would never require annealing. In Chapelcross Pile No. 2 on May 19, 1967, after seven years of running at full power, the graphite sleeve inside one of 1,696 fuel channels crumbled and blocked the flow of carbon dioxide coolant over the fuel canisters. At least one fuel canister overheated, the uranium melted and caught fire, breaking open the can and sending radiation alarms into a tizzy.[132] Considering that a MAGNOX reactor is loaded with more than 9,000 fuel canisters at one time, this would seem a small statistical problem, but full-blown disasters can start small and memories of the Windscale fire were still fresh. The reactor was scrammed immediately, and the accident investigation began.

Unlike the Windscale piles, the more advanced gas-cooled reactors were built with the fuel channels cut vertically, and fuel handling was semi-automated to prevent workers from having to get anywhere near the reactor cores. The entire core was encased both in a steel pseudo-sphere as protection from fission product leakage and a pressure vessel, containing the cooling gas and keeping broken fuel cartridges off the landscape. This was an excellent plan, but peeking into a fuel channel was not as easy as it was in the old days. A special television camera was built, mounted on the end of a pole that could be positioned over the burned fuel channel to look in and see what had happened. This device confirmed that the fuel channel was blocked with graphite debris and that it would have to be cleaned of this and the remains of the burned fuel, but unfortunately the camera broke off and fell into the reactor. It did everything but scream as it tumbled off into the abyss, with the monitor screen image spinning and then going black. This would complicate the repair.

132 Quoting the OSHA *Occupational Safety and Health Guidelines for Uranium and Insoluble Compounds*, under *Reactivity*: "Contact of uranium with carbon dioxide causes fires and explosions." The carbon dioxide coolant probably protected against graphite fires, but it was dangerous in contact with metallic uranium. Ironically, the suggested method for putting out a uranium fire is to smother it with graphite chips.

It took two years, but the men at Pile No. 2 were able to design a workable plan for restoring the plant to full operation. Three men were outfitted in fully sealed radiation suits with closed respirators. Each man was given only three minutes to work, during which he would receive his entire year's allowed radiation dose working at the top end of the core in a concentrated radiation field. One at a time, they moved quickly through the forest of control rods to the clogged fuel passage and ran a scrubber on a long pole down the hole, cleaning the passage and restoring its roundness.

The cleanout was successful, and Chapelcross Pile No. 2 resumed operation. All was good until 2001, when the Chapelcross power plants seemed to trip over a streak of bad luck.

Fuel has to be changed-out at the MAGNOX reactors at least twice a year. First, the burned fuel, blisteringly radioactive with fission products, must be pulled out of the fuel channels and lowered down a chute and into a lead coffin. This is accomplished using the discharge machine, which runs on rails on the floor of the crane room, addressing holes that are opened one at a time, giving access to the reactor below. An operator rides in a seat on the back of the machine, which weighs a hefty 60 tons. Half of that weight is lead shielding to protect the operator from what he is pulling out of the reactor, and if the machine is operating properly, the workers never see or touch the fuel cartridge.

The operator positions the discharge machine over a hole in the floor, grabs a fuel cartridge down in the reactor, and pulls it up into the shield. The machine then rolls on rails to the end of the room, where the well hoist grabs the cartridge away from the discharge machine and lowers it on a cable down five floors, through the concrete-shielded discharge well. It drops into a lead barrel, which then rolls out on rails to the transit hoist for a trip to Windscale for fuel reprocessing. After you have discharged a few hundred thousand fuel cartridges, it seems routine and foolproof; but one day early in 2001, one fuel cartridge out of 10,176 had a problem.

The operator sent the cartridge down the discharge well and into the lead barrel. For some reason it did not release, and the cable came back up, ready to snag another cartridge after it had been snaked out of the core. Instead of an empty grabber, up came a live cartridge, broadcasting gamma radiation all over the refueling floor. Alarms jangled everyone's nerves as they leaped from the chairs on their discharge machines and hot-footed for the door. Nobody was injured by the radiation, but it was considered a serious accident, and inadequate design, improper operation of the discharge machine, and statistical probability were blamed. All refueling of the early MAGNOX reactors at Chapelcross and Calder Hall was halted while the incident was investigated.

In July 2001, the workers at Chapelcross discovered that steel drums filled with depleted uranium trioxide tend to rust and develop holes when they are left out in the rain for several years. It was decided to substitute stainless steel barrels.

Later in July, corrections to the problem experienced earlier that year, when a fuel cartridge grabber would not let go, seemed to cause the opposite problem. After hearing something heavy crash into the door at the bottom of the discharge well, investigators sent a remote-controlled TV camera into the area and had a look. Twelve fuel cartridges had let go of the grabber and fallen over 80 feet into the transport barrel, splattering the irradiated uranium over the discharge bay. A careful cleanup was accomplished.

In August it was time for the annual maintenance and cleanup at Chapelcross, starting with Pile No. 1. There was a disturbing finding. It was known that graphite could "grow" in just about any direction under neutron bombardment at temperatures lower than 300°C, but in Pile No. 1 the graphite had shrunken. The pile was now not quite as tall as it had been, and the steel charging pans, designed to guide new cartridges into the core when refueling, were now hanging in space, supported perilously by the nozzles on the burst-can-detection gear. The other piles were found to have similar defects, but not as bad as Pile No. 1, which was built differently.

Only Piles 2 and 3 were in decent enough shape to be restarted. In June 2004, the entire power station was shut down for good. The sister plant, Calder Hall at Sellafield, had been shut down permanently in 2003. The British graphite reactors had made a good run of it, but they were now obsolete and it was time to call it quits. By 2004, there was no longer a need to manufacture plutonium, and better ways to build a power reactor had been developed. On May 20, 2007, at 9:00

A.M., the four 300-foot hyperboloid cooling towers, visible on a clear day from a distance of 50 miles, were brought down in 10 seconds by demolition explosives, and that was that.

The fuel-reprocessing industry in the UK has reported only one criticality accident in which fissile material managed to come together accidentally in a supercritical configuration. Just about every other country that has tried to separate plutonium from uranium in spent reactor fuel has experienced at least one such excursion, and this was Britain's. This incident at the Windscale Works on August 24, 1970, is described by the criticality review committee at Los Alamos as "one of the most interesting and complex because of the intricate configurations involved." An entire book could be written to describe this accident, but I will be brief.

It is the goal of every chemical engineer working on a reprocessing plant design to allow no vessel large enough or of an optimizing shape, such as a sphere or tomato can, to exist in the complicated pipe-works of a reprocessing plant. Occasionally, however, when there is no conceivable way that fissile material, such as plutonium, can collect there, a round tank finds its way into the layout. At Windscale, plutonium was recovered from spent MAGNOX fuel, and near the end of the process it was refined in 300-gram batches. The plutonium was dissolved in a mixture of tributyl phosphate and kerosene and fed to a conditioner vessel, where the amount of plutonium in solution was adjusted to between 6 and 7 grams per liter. This was safely less than the concentration required for criticality.

From there, the small batch was lifted through a pipe to a transfer vessel by vacuum, where it would be fed through a U-shaped trap to a refining operation called the pulsed column. The transfer vessel was a short, cylindrical tank having hemispherical top and bottom. It was a near-perfect size and shape to house an impromptu reactor, but surely no such concentration of fissile material could be fed to it from the conditioner in one gulp, and the trap prevented any backflow into it. It was a mistake to think this.

It turned out that after every transfer of a subcritical amount of plutonium to the transfer vessel, a small amount of plutonium was stripped out of the solution by water sitting in the bottom of the tank. Ordinarily, that would be no problem, because the concentration of plutonium in the tank would never be greater than 7 grams per liter of solvent, but the amount of plutonium dissolved in the water grew gradually. This went on for two years, until the bottom of the tank held a whopping 2.15 kilograms of fissile plutonium-239.

On the last batch through the system, 30 grams of plutonium were dumped in on top of the water, mixing the solvent and water together and causing a supercritical nuclear reactor to suddenly exist in the plumbing for 10 seconds.

It was not a fatal accident. Nobody was injured, and the intricate piping in the plant looked exactly the same before and after the supercriticality, but the fact that it happened in such a carefully designed process was unnerving. It goes down in history as proof of how difficult it is to predict what will happen in a maze of pipes, valves, tanks, and traps carrying a fluid for which its volume and shape are important. The concept of impossibility becomes murky.

Windscale became Sellafield, and the THORP, or the Thermal Oxide Reprocessing Plant, became operational in 1997. It took 19 years to build the facility at the place formerly known as Windscale. Its mission is to take used reactor fuel from Britain, Germany, and Japan and separate it into 96% uranium, 1% plutonium, and 3% radioactive waste, using a modified PUREX process.

In July 2004, a pipe broke and started filling the basement with highly radioactive, pre-processed fuel dissolved in nitric acid. The loss of inventory after nine months had climbed to 18,250 gallons. This went unnoticed until staff reported the discrepancy between solution going out of one tank and not arriving in another tank.

As it turned out, the liquid flow was monitored by weighing it periodically in an "accountancy tank." This tank had to be free to move up and down as it accumulated the heavy uranium-plutonium waste in dissolved form, and the extent of sag as it filled was translated into weight. As the tank was being installed, it was decided to leave off the restraints that would keep it from wagging side to side as it accumulated liquid. This move, thought to protect it in case of earthquake, proved to be too much for the pipe connecting it to another process downstream, and it broke under the floor where it could not be seen.

On April 19, 2005, a remote TV camera was sent down to see what was going on, and it confirmed that the radioactive soup had formed a lake in the secondary containment under the Feed Clarification Cell. The spill was nicely contained in the stainless steel liner of the containment structure, but it contained 44,092 pounds of uranium, 353 pounds of plutonium, and 1,378 pounds of fission products. It was the fission products that made it a problem, as no human being would ever be able to enter the room and repair the broken pipe, even when the solution was removed using existing steam ejectors. The only possibility is to repair the break using a robot, but building a mechanical plumber is proving to be difficult. THORP will possibly complete its existing reprocessing contracts, bypassing the accountancy tank, and close down in 2018.

The history of atomic accidents in Britain is thus a story of ambition, impatience, originality, and accomplishment under hard circumstances, marked by a spectacular incident the year that David Lean filmed *The Bridge on the River Kwai*. If you find yourself cornered by a force of Brits who rib you mercilessly about the SL-1 explosion, it is correct to remind them of the British Blue Peacock nuclear weapon, in which the batteries were kept warm by two chickens living in the electronics module aft of the warhead. It is an effective diversion, and it is almost true.[133]

133 Although the use of live chickens, installed with a one-week supply of food and water, as a heat source was a serious proposal for the Blue Peacock (originally Blue Bunny) in October 1954, accounts of its implementation are difficult to find. Unfortunately, this specification was declassified on April 1, 2004, and it was immediately assumed to be an April fool's joke. Tom O'Leary, head of Education and Interpretation at the National Archives, responded with: "The civil service does not do jokes."

Chapter 6

IN NUCLEAR RESEARCH, EVEN THE GOOF-UPS ARE FASCINATING

"If the oceans were filled with liquid sodium, then some crazy scientist would want to build a water-cooled reactor."

—Hyman Rickover, grousing about the sodium-cooled USS *Seawolf*

ADMIRAL HYMAN RICKOVER PUSHED HIS passion for a nuclear-powered submarine as hard as he could without being formally charged with criminal intent, and he was rewarded with one of the most successful projects in the history of engineering. His finished prototype submarine, the USS *Nautilus*, was all that he had hoped. First put to sea at 11:00 A.M. on January 17, 1955, she broke every existing record of submersible boat performance, made all anti-submarine tactics obsolete, and never endangered a crew member.

At the end of World War II the Navy, whether it knew it or not, was ready for the improved submarine power plant of a nuclear reactor. All the groundwork was in place after the Army's startling success in the development of atomic bombs and the infrastructure for their manufacture. As an engineering exercise, the task of reducing the size of a power reactor from that of a two-story townhouse to something that would fit in a large sewer pipe was seen as possible, but there were a couple of serious considerations.

First, there was the nagging dread of a steam explosion. It was one thing to lose a boiler or two on a battleship, but in the tightly confined spaces of a submarine everyone would be killed and the boat lost if a steam line broke. If a reactor were to lapse into runaway mode, the water coolant could flash into high-pressure steam and tear the sub in half under water. Experience was in short supply and it was hard to convince mechanical engineers that this was a very unlikely occurrence in those early years of nuclear power. The use of water in the primary coolant loop was considered too dangerous to pursue.

Second, in the early 1950s there was concern over the availability of uranium as reactor fuel. To build exactly one uranium bomb in World War II the United States depended on all the uranium ore that could be mined in the Belgian Congo and Canada. There was no guarantee that enough uranium would ever be available to power a fleet of submarines, even from multiple foreign sources working at maximum efficiency. Plutonium, on the other hand, was a perfectly good power reactor fuel and we had plenty of it. It was manufactured at the Hanford Works in Washington State. It was therefore obvious that a nuclear submarine should be fueled with plutonium, and to fission plutonium fast neutrons were optimal. It was best to have no neutron moderation in a plutonium reactor, and this meant that there could be no water coolant in the primary loop. The coolant would have to be something heavy and liquid, such as melted metallic sodium. It would run hot and thermally efficient at atmospheric pressure, putting no expansive stress on pipes and associated hardware.

Rickover strongly disagreed with this "fast reactor" philosophy. Unlike many engineers assigned to the nuclear submarine project, Rickover had paid his dues riding around in the ocean in submarines S-9 and S-48, which were feeble death-traps built in the 1920s.[134] He knew from miserable experience and alert observation that there is no such thing in a submarine as a pipe, tank, flange, or valve that will never leak. In fact, with the combined stresses of being crushed, twisted, hammered, vibrated, abused, and built by the lowest bidder, a submarine's bilge ditch would slosh with a sickening mixture of sea water, diesel fuel, sweat, lubricating oil, salad dressing, vomit, battery acid, head overflow, and coffee. Anything in the boat that conducted or contained compressed air or any fluid was capable of leaking, regardless of how well it was welded together or tightened with threaded fasteners. The captain in an S-boat had to wear a raincoat when operating the periscope, and dampness covered every surface inside the craft. Running in cold water meant that standing in the control room you could peer down the centerline and lose sight of the end of the compartment in the fog.

The concept of cooling an engine with liquid sodium thus seemed wrongheaded to this experienced submariner. The slightest sodium leakage would react with water or with water vapor in the air, burning vigorously and leaving a highly corrosive, flesh-eating sodium hydroxide ash. Sailors could perish just from breathing the hot vapor. Furthermore, Rickover had confidence that when there is an attractive price put on a mineral, such as uranium ore, people will dig up the earth to find it. In time he would be proven correct on both issues, the difficulties of using sodium and the abundance of uranium. With the safety of the crew being his primary design factor, his uranium-fueled pressurized-water reactor became the standard

134 The USS S-48, built in 1921, was possibly the only submarine to find peril in a heavy snowstorm. She was returning home to Portsmouth, Connecticut, on the night of January 29, 1925. At about 6:30 P.M. off the coast of New Hampshire, the wind picked up and heavy snow started falling. With visibility zeroed, she ran onto the rocks off Jeffery Point, rolled off, then grounded again in Little Harbor. Early next morning, still stuck and listing badly, the battery compartment started taking water, and the ensuing electrolysis action broke down the salt water and released chlorine gas inside the sub. The Coast Guard picked up the crew an hour and a half later, and they were transferred to the hospital at Fort Stark for treatment.

for submarine propulsion and for most of the civilian nuclear power industry. With a well-trained and disciplined reactor operations crew, there was no safer way to generate power in a confined space.

There were several experimental liquid-metal cooled and breeder reactors built in the United States, beginning in 1945 with Clementine, and unforeseen problems with these exotic designs were experienced. Never was a serious sodium leakage encountered.[135]

The United States Atomic Energy Commission (AEC), starting work on January 1, 1947, had among its tasks the job of persuading private enterprise to build nuclear power plants. It was a noble goal, born of some very long-term projections. It was seen, even back in the late 1940s, that civilization would require more and more electrical power, and we could not generate it by burning coal forever. Coal and oil were seen as limited resources that were not being regenerated, and there were a finite number of rivers left to be dammed. Wind, used to make power since the days of the Roman Empire, was seen as too feeble and unreliable to meet the growing power demand, and solar power was limited to hot-water heaters in Florida. A plausible power supply for the future was nuclear fission, and it was not too early to begin an experimental phase of development.

A stellar committee met to come up with five nuclear reactor concepts to be prototyped and tested as candidates for the standard civilian power reactor. Uranium, the fuel of choice for all military reactors, was not seen as plentiful. Even if a hundred mines were dug, it was just another commodity that we would run out of eventually. Therefore, the ideal civilian nuclear power reactor would be a breeder, which paradoxically produces more fuel than it burns. Furthermore, the danger of a steam explosion should be minimized by vigorous application of the engineering art.

There are two types of breeder reactor, the fast breeder and the thermal breeder. The fast breeder runs on plutonium-239. The thermal breeder runs on uranium-233. The first of the five prototype reactor programs

135 At least no leakage problem while running. The USS *Seawolf* submarine with a liquid-sodium-cooled fast reactor did experience an unscheduled sodium expulsion, but the boat was in dock at the time. The Superphoenix fast breeder reactor in Creys-Malville, France, has suffered from sodium leaking and corrosion in the cooling system, and it has been out of service since September 1998.

to be funded by the AEC in June 1954 was a sodium-cooled thermal breeder named the Sodium Reactor Experiment, or the SRE. Winner of the contract for the project was the North American Aviation's Santa Susana Field Laboratory, located in a rocky wilderness about 35 miles northwest of Los Angeles, California, called Simi Hills.

The 2,850-acre site was divided into four sections. Three of the sections were devoted to testing high-performance rocket engines and explosives, and the fourth was populated with exotic nuclear-reactor experiments. The lab was thus blessed with all the fun stuff of the era except above-ground nuclear weapon and aircraft ejection-seat tests, and it should have been located in Idaho instead of in a place that would experience a population explosion in the next few decades. As the years passed, the valley below would jam tight with Californians, some of whom would find fault with the Santa Susana lab for no good reason other than what it was doing.

Admittedly, over the decades Santa Susana saw its share of accidents. The hot lab facility at the lab was the largest in the United States at the time. Workers could take apart highly radioactive reactor fuel assemblies using robotics behind windows three feet thick, offering complete radiation protection. Every now and then a fire would break out behind the hot-cell glass, causing massive internal contamination. The cores in four experimental reactors on site strayed outside the operating envelope and melted. Highly toxic waste disposal was handled by shooting at the barrels at a safe distance with a rifle until they exploded, sending the contents high into the air and wafting away into the valley below. In July 1994 three workers were trying to test some rocket fuel catalyst and it exploded unexpectedly, killing two, seriously injuring the third, destroying a steel rocket fuel test stand, and setting a 15-acre brush fire.[136] In

136 Otto K. Helney, a 53-year-old engineer, Larry A. Pugh, a 51-year-old physicist, and Lee Wells, a 62-year-old assistant, seemed old enough to know better when they mixed 10 pounds of gycidal azide polymer and nitrocellulose (gun cotton) together in an aluminum pan, hoping to measure the shock wave when the two chemicals ignited. Helney and Pugh died instantly, and Wells was blown against vertical terrain and burned over 20% of his body. Six other men died at Santa Susana in various incidents in the early 1960s. It was an exciting place to work.

2005 a wildfire swept through Simi Hills and burned everything flammable in its path.

By April 25, 1957, the SRE was up and running and soon providing 6.5 megawatts of electricity to the Moorpark community, using a generator courtesy of Southern California Edison. It was the first civilian nuclear power consumed in the United States. In November, Edward R. Murrow featured the SRE power plant on his *See It Now* program on CBS television.

By current standards, the SRE was an odd power-reactor design. It was to run on fissions caused by neutrons slowed down to thermal speed, and yet the coolant was to operate at atmospheric pressure. The moderator material was graphite, a stable, solid-state material. Preventing any chance of a graphite fire, the moderator was formed into columns with a hexagonal cross-section and covered with gas-tight, pure zirconium.[137] The coolant was liquid sodium. Liquid sodium, being a dense, heat-conductive metal, is a very efficient coolant, and under foreseeable operating conditions, it never boils or produces dangerous gas pressure, as could water. It does, however, absorb an occasional neutron in a non-productive way, to much the same extent as ordinary water. Although the ultimate purpose of the SRE was to begin development of a civilian thermal breeder, its first fuel loading would be metallic uranium, slightly enriched to make up for the neutron losses in the coolant. As the basic configuration was proven by experiments, thorium-232 breeding material and uranium-233 fuel would be introduced later.[138]

---

137 Zirconium was an excellent choice. It is transparent to neutrons and able to withstand very high temperatures, but it was expensive. For graphite moderator units out on the periphery of the reactor core where the temperature would never be very high, stainless steel was substituted for the zirconium.

138 From whose stockpile would this U-233 come? U-233 does not occur in nature, and it must be made by production reactor from Th-232. Experimental U-233 reactor fuel probably came from a stockpile at Los Alamos. In 1955 a composite Pu-239/U-233 atomic bomb named MET was built and tested in Operation Teapot. The U-233 was made using the B Reactor at the Hanford site, and this material was used in several thermal breeder experiments in the late 1950s and early 1960s. The stockpile wound up at Oak Ridge, and a $511M contract from the DOE to dispose of this material is currently under way.

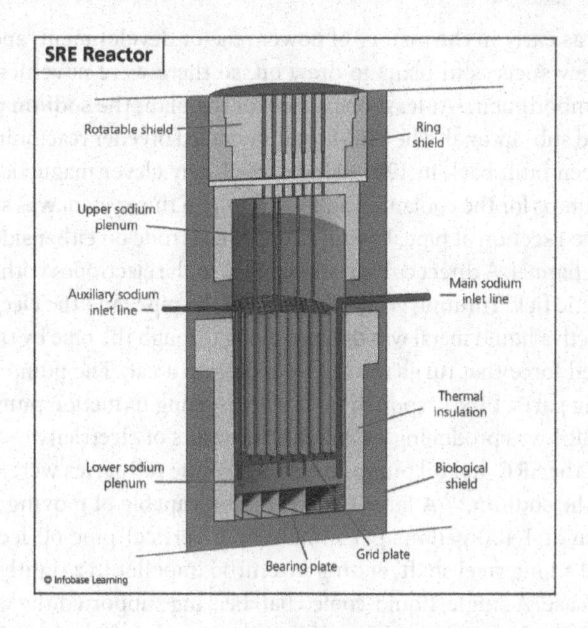

The Sodium Reactor Experiment ran with a graphite moderator at very high temperature, giving an efficient means to make high-pressure steam. The coolant was liquid sodium, which would never boil away, could operate at atmospheric pressure, and would never react chemically with the graphite. It was a sound concept, but the problems of dealing with a flowing liquid metal that reacts explosively with water have plagued reactors with sodium-based coolants.

There was another good reason to use sodium instead of water as the coolant. In water-cooled graphite reactors, such as the plutonium production reactors at Hanford, the unintended loss of coolant in the reactor always improves the neutron population, and the reactivity of the pile increases. The power starts going up without human intention. Graphite is a near-perfect moderator material, and any action that reduces the amount of non-graphite in a reactor, including the formation of steam bubbles, is fission-favorable. There was no such worry if sodium, with a boiling point of 1,621°F, were used instead of water. With an intended coolant outlet temperature of 650°F at full power, bubble formation in the reactor was hardly a concern, and the reactor could operate at an ideal temperature for external steam production while running under no pressure at all.

It was early in the history of power reactor development, and there were few successful plans to draw on, so there were novelties in the SRE embodiment. At least one aspect of the plan, the sodium pumps, seemed sub-optimal. The EBR-I sodium-cooled breeder reactor in Idaho had been built back in 1951 using exceedingly clever magnetic induction pumps for the coolant. A sodium pump in this system was simply a modified section of pipe, having a copper electrode on either side of the inner channel. A direct current was applied to the electrodes with a static magnetic field running vertically through the pipe, and the electrically conductive liquid metal was dragged along through the pipe by the same induced force that turns the starter motor on a car. The pump had no moving parts. EBR-I reported no problems using induction pumps, but the EBR-I was producing a scant 200 kilowatts of electricity.

For the SRE, hot oil pumps used in gasoline refineries were used to push the sodium.[139] A large electric motor, capable of moving molten sodium at 1,480 gallons per minute up a vertical pipe 60 feet high, turned a long steel shaft, ending in a turbo impeller in a tightly sealed metal case. A single, liquid-cooled ball bearing supported the working end of the shaft. The problem of keeping liquid sodium from leaking past the impeller and into the bearing was solved by modifying the end of the pump. The shaft was sealed with a ring of sodium, frozen solid in place by a separate cooling system. The coolant to be pumped into the seal could not, of course, be water, which would react enthusiastically with the sodium. It had to be a liquid that had zero trouble being next to sodium. Tetralin was chosen.

1,2,3,4-tetradhydronapthalene, or "tetralin," is a solvent, similar to paint thinner you would buy at the hardware store, first synthesized by Auguste George Darzens in 1926. Its molecule is ten carbon atoms and a dozen hydrogens, looking like two benzene rings stuck together. It has no particular problem with sodium, and it evaporates at about 403°F. The tetralin was circulated through the sodium seal, keeping

139 Electromagnetic sodium pumps were not unknown to the SRE design group. In fact, one was used to load the cold trap, designed to remove oxidized sodium and any other contaminants from the primary cooling loop. This pump failed about 60 days into operation at power in late June 1957. Similar electromagnetic devices, called "eddy current brakes," were used on the primary and secondary coolant loops to stop the coolant from moving after a reactor scram.

it solidified, in a continuous loop using two parallel evaporation coolers to shed the heat from the seal. Electrically driven pumps kept it moving, with a gasoline engine for a backup in case of an electrical failure. It was a complicated sub-system in a complicated power plant, requiring pipes, valves, pumps, wiring, instrumentation, tanks, and coolers, just to keep the sodium off the pump bearings. The fact that it had to be backed up was ominous. It is always a better system when if everything fails, the wreckage reduces to an inert, safe condition.

The cooling system used three closed loops. The primary loop was liquid sodium running through the reactor. The natural sodium was constantly being activated into radioactive sodium-24 by contact with the neutrons in the reactor. To eliminate the potential accident of radioactive sodium leaking into the steam system, a second sodium loop took the heat from the first loop and took it outside the reactor building, where it was used to generate steam for the turbo-generator in a water loop. For low-power experiments in which electrical power was not generated, the water loop was diverted into an air-blast heat exchanger, dumping all the power into the atmosphere. There was no danger of broken or melted fuel leaking radioactive fission products into the surroundings, as the second sodium loop was a well-designed buffer. No expensive containment structure was needed for the reactor, because there was no chance of a radiation-scattering steam explosion in the building. The steam was generated out in the yard, and it was not connected directly to the reactor.

Radioactive gases produced by fission, such as vaporized iodine-131 and xenon-135, were controlled in the stainless steel reactor tank by a bellows structure at the top, giving the tightly sealed system room to expand. Helium was kept over the reactor core to prevent air from leaking in and reacting with the sodium. The fission gases were piped off and compressed into holding tanks for controlled release into the environment after having been held long enough for the radioactivity to have decayed away. There were outer space reactors, rocket engines, and military systems being developed at Santa Susana, and all had their considerations of performance, weight, and reliability, but this SRE was to be a prototype civilian power plant. As such, the prevention of harm to the public was a primary and noble design consideration.

All newly designed sub-systems were actively tested at Santa Susana before they were integrated into this new type of power reactor. It was an experimental setup stacked with many unknowns, and there was a lot to learn about graphite-moderated, sodium-cooled reactors. The operation log shows that almost immediately there was trouble with the sodium pumps. Hours after the first power startup on April 25, 1957, the shaft seal on the main primary pump failed. A few weeks later, tetralin was leaking from an auxiliary secondary pump. A week later, the main secondary pump was replaced. In August, the sodium was found to be contaminated with tetralin. In November, the cold trap was clogged with sodium oxide. Air was getting in somewhere. In January 1958, sodium smoke filled the high bay, and men had to go in with oxygen masks to find the leaking bellows valves. In May the shaft seal on an auxiliary primary pump failed, and out of a main primary pump they could smell the strong odor of tetralin. Two months later all the pumps were taken apart and the ball bearings were replaced. The entire main primary pump was replaced with a spare. A month later, the electromagnetic pump clogged with sodium oxide again, and in September the main secondary pump was leaking tetralin. By April 1959 the main primary pump was leaking tetralin from the sodium seal, and the entire unit had to be replaced. A month later, a tetralin fire was extinguished without causing reportable damage.[140]

So, pumping sodium around in a nuclear reactor was not as easy as it seemed on paper, and the SRE was being taken apart and worked on all the time. That is the life of an experimental reactor, and Rickover's insistence on no sodium in his precious *Nautilus* seemed to make sense if you were outside looking in. Every time a component from the reactor tank or the primary coolant loop was removed for repair or examination, the sodium frozen to it had to be cleaned off. For this purpose, the pump or the fuel assembly was moved to the wash cell, a special setup behind an atmospherically sealed hot-cell window. Using remote arms, a technician would hose down the sodium coating with warm water. It would instantly turn into hot sodium hydroxide and wash down the drain at the bottom of the stainless steel hot-cell chamber, leaving the once-contaminated piece bright and sparkling clean.

140 I can't tell exactly where the fire occurred. It was in the "XX vault," where X indicates an unreadable character in the SRE operations log summary. There was an NAS vault, a main pipe vault, an aux pipe vault, and a primary sodium fill tank vault. There was no tetralin in any of these locations.

Real trouble did not start until RUN 13, from May 27, 1959 to June 3, 1959. The crew was supposed to run the coolant outlet temperature up to 1,000°F to see if the system could stand to work at higher power. They hoped to log 150 megawatt days. All was well until two days after startup at 11:24 A.M., when the reactor scrammed due to an abnormal sodium flow rate. Not hesitating to contemplate why the flow rate was wrong, the operating crew restarted the reactor immediately and ran it back up to power. There it stayed until 9:00 the next morning, May 30. At that point, the reactor system went squirrelly.

First, the reactor inlet temperature began to rise slowly over three days. On June 1, the temperature difference across the heat exchanger rose sharply, indicating that something wasn't working. The thermocouple in one fuel assembly, number 67, showed a temperature increase from a normal 860°F to 945°F. The temperature in the graphite abruptly jumped by 30°F, also on May 30, and the thermocouple in fuel assembly number 16 showed a similar increase in temperature. They did not notice it at the time, but the automatic control-rod positioner was compensating for a slow increase in reactivity in the reactor. Obviously, something had occurred that was impairing the coolant flow, and by June 2 the main primary pump casing was reeking of leaking tetralin. The reactor was shut down on June 3 to examine the fuel and repair the coolant pump.

As the pump was being torn down, fuel assembly number 56 was removed using the impressive automatic fuel-removal machine and transferred to the wash cell for examination. To quote the accident report exactly, "During the washing operation a pressure excursion occurred of sufficient magnitude to sever the fuel hanger rod and lift the shield plug out of the wash cell." Translation: The damned thing exploded and put the wash cell out of commission for a year. Nobody was killed, thanks to the three-foot-thick window and aggressive ventilation.[141]

Retrospective analysis would find that the vent at the bottom of the fuel assembly, where coolant was supposed to flow in and past the hot fuel rods, had been blocked by a substance technically referred to as

141 The window was three feet thick, but not the glass. Hot cell windows were made of two very thick panes of glass, mounted in parallel and spaced about three feet apart in a sealed tank, very much like an aquarium. The space between the glass windows is filled with zinc bromide dissolved in water. Objects in the hot cell are given a yellow cast by looking through the zinc solution, which is an excellent high-energy gamma-ray shield.

"black stuff," and this left a large remainder of sodium in the bottom of the assembly. When the technician aimed the hose into the top of the assembly, the big sodium wad went off like a hand grenade. This was not noted at the time, and number 56 was put back in the reactor. Damage to the wash cell diverted attention from further identification of the black stuff, but a working explanation was that it was residue from tetralin decomposed in the hot coolant. There was not supposed to be any tetralin in the coolant, but three pints of it were found in the cold trap plus a couple of quarts of naphthalene crystals, or tetralin with the extra hydrogens stripped off. No connection in particular was seen between these contaminations and the strange behavior at the end of RUN 13. The troublesome tetralin-cooled seal on one pump, the main primary, was replaced with a nack-cooled sodium seal, and the reactor was ready for RUN 14.[142]

RUN 14 was started on July 12, 1959. The experimenters expected some trouble with the fuel-channel outlet temperature, but they were not sure why. Perhaps if they could intensify the effect, the cause would snap into focus. The reactor was brought smoothly to criticality at 6:50 A.M. At 8:35 they increased the power level to a modest 500 kilowatts, and the graphite temperature started to flop around wildly, running up and down by about 10°F, and various fuel-channel temperatures started to diverge by 200°F. Not seeing this as a problem, they kept going until 11:42 when the reactor, hinting that something was amiss, scrammed due to a loss of sodium flow in the primary loop.

At this point I must find fault with the way they were operating the SRE. Something about the reactor was not working, and yet they kept restarting it without knowing why. Today, this disregard of trouble signs would be unheard of, I hope. You would never restart a reactor without knowing exactly why it scrammed. There would be inquiries, hearings, lost licenses, and firings, but apparently not in 1959, when

142 The "nack" cooled seal used a mixture of sodium and potassium, or NaK, as the cooling fluid into the sodium seal to keep it frozen solid. Nack is liquid at room temperature, and it flows as freely as water. It was used in the fuel pins to fill the space between the solid uranium fuel pellets and the stainless steel tubes in which they were encased, giving a thermal coupling that would ensure heat flow to the coolant, which flowed around the tubes. The tetralin was a low-viscosity solvent, and it was hard to keep it from finding a way past the sodium seal and into the coolant. The nack would leak less, and if it did, so what? In all the other motorized sodium pumps in the system, tetralin was retained as a coolant.

the screws of bureaucracy weren't tightened as they seem now. Start her up again and let's see if that was just a fluke. By 12:15 they had SRE back up to power and were increasing the power and the outlet temperature. At 1.5 megawatts the temperature was fluctuating inexplicably by 30°F.

At 3:30 P.M., both air-radiation monitors in the reactor building indicated a sharp rise in activity. By 5:00 P.M., radioactive air was going up the exhaust stack and into the atmosphere, and the radiation level over coolant channel seven was extremely high, at 25 roentgens per hour. Something had obviously broken, but for some reason it did not occur to the experimenters that radioactive products that would cause such an indication are produced in the fuel, and the fuel is welded up tight in stainless steel tubes. If all the fuel is intact, then nothing radioactive should be leaking into a coolant channel, and certainly not through the gas-tight reactor vessel and into the air in the room. By 8:57 P.M. they had shut down the reactor. They fixed the problem of leaking radioisotopes by replacing the sodium-level indicator over channel seven with a shield plug, and they restarted the reactor at 4:40 A.M. the next day, July 13.

By 1:30 that afternoon, they noticed that the graphite temperature was not going down when they increased the coolant flow. They knew not why. At 5:28 P.M., they were running at 1.6 megawatts and commenced a controlled power increase. The power seemed to increase faster than one would expect up to 4.2 megawatts, but then the reactor went suddenly subcritical and the power was dropping away. They started pulling controls to bring it back, and by 6:21 P.M. they had managed to coax it to run at 3.0 megawatts.

Up to this point, the reactor had been recalcitrant and unusual, but now it went rogue. Power started to run away. Control-rod motion was quickly turned around, sending them back into the reactor to soak up neutrons and stop the power increase. Instead, the power rise speeded up. By 6:25 P.M., the power was on a 7.5-second period, or increasing by a factor of 2.7 every 7.5 seconds. The fission process was out of control, and a glance at the power meter showed 24 megawatts and climbing.

Deducing that if this trend continued, then in a few minutes the reactor would be a gurgling puddle in the floor, the operator palmed the scram button, throwing in all the controls in the reactor at once and bringing the errant fissions to a stop.

As the reactor cooled down, only one question came up: Why was there not an automatic scram when the reactor period, spiraling down out of control, passed 10 seconds? The period recorder, which leaves a blue-line-on-paper graph of period versus time for posterity, had a switch that was supposed to be tripped when the pen hit 10 seconds on the horizontal recorder scale. This switch was supposed to trigger an automatic scram.[143] Testing found that the switch would have worked, but only if the period had been falling slowly. The trip-cam was modified so that the switch would operate even when a scram was most needed, with the period falling rapidly.

Seeing nothing else to fix, the operating staff brought the reactor back to criticality at 7:55 P.M. and proceeded to increase the power. By 7:00 A.M. the next morning, July 14, they were running hot, straight, and normal at 4.0 megawatts. Two hours later, the radioactivity in the reactor building was reading at 14,000 counts per minute on the air monitors. Technicians put duct tape over places where fission products were found escaping. There was only one other automatic scram for the whole rest of the day, when workers setting up a test of the main primary sodium pump accidentally short-circuited something.

On July 15, it was seen as pointless to be trying to run the electrical generator while trying to test at high temperature, so the staff drained out the secondary coolant loop and switched to the air-blast heat exchanger. The next day, at 7:04 A.M., the SRE was made critical once more. On July 18, the motor-generator set, which was supposed to prevent power surges into the control room instruments, failed. The operators switched power to unstabilized house current and continued operation. At 2:10 A.M. on July 21, the reactor scrammed suddenly, having picked up another fast power rise. The scram was attributed to the unstabilized power, and the reactor was restarted 15 minutes later. One more scram at 9:45 A.M.

The next day, July 22, channel 55 was giving trouble. This assembly contained various experimental fuels, and the temperature was fluctuating in the 1,100 to 1,200°F range. There was only one automatic scram on July 23, probably just a fluke, but by 1:00 P.M. the temperature

143 At least I assume it was a scram actuation. In the SRE plans, this switch is listed as a "setback " actuator. Does this indicate that it was meant only to limit the power rise rate and not stop it completely? If so, it would not have worked as planned. With power increasing exponentially, a scram is quite appropriate.

in channel 55 was up to an eyebrow-raising 1,465°F. Operation continued. In the early morning on July 24, eight hours were spent trying to dislodge some apparent debris stuck in the fuel channels by jiggling the assemblies. It was noted in passing that four of the fuel modules seemed jammed and stuck firmly in place. There were two annoying automatic scrams later that day.

At 11:20 A.M., the Sodium Reactor Experiment RUN 14 was terminated, and the long-suffering machinery was allowed to rest quietly while the experimenters poked around in the core with a television camera. To their surprise, they found that the core of a nuclear reactor that had been acting oddly for six weeks, subjected to overheating and temperature fluctuations, many automatic scrams, pump seal coolant failures, oxidizing sodium, radiation leakage, and a power runaway, was wrecked. Of the 43 sealed, stainless-steel fuel rods in the core, 13 had fallen apart, scattering loose fuel into the bottom of the reactor vessel. How the thing had managed to run at all under this condition was an amazement in itself. Attempts to remove the fuel rods came to an end when the contents of channel 12 became firmly jammed in the fuel handling cask. An investigation of the damage and its cause was started immediately.

The interim report, "SRE Fuel Element Damage," was issued on November 15, 1959. It was found that tetralin had been leaking into the sodium coolant through the frozen sodium seals in the pumps. In the high-temperature environment of the active reactor core, the solvent had decomposed into a hard, black substance, which would tend to stick in the lower inlet nozzles of the graphite/fuel modules and prevent coolant from flowing. In the blocked coolant channels, the sodium vaporized, which had been thought unlikely, and denied coolant to the fuel. The stainless steel covering the cylindrical fuel slugs melted, and structural integrity of the fuel assemblies was lost. Naked uranium fuel, having fissioned for hundreds of megawatt-hours, was able to mix with the coolant. Gaseous fission products presumably escaped the sealed reactor vessel, probably through the same leakpoint that allowed the sodium to oxidize, and other fission waste dissolved in the coolant, making the primary loop radioactive.

The most puzzling part of these findings was: Why did the stainless steel melt instead of the metallic uranium? The 304 stainless used for fuel cladding melts at 2,642°F, and the temperature in the reactor was nowhere close to that. Uranium melts at 2,060°F, and yet the stainless steel melted away, leaving the uranium unclad and unsupported. The fuel assemblies were, however, operating at temperatures for which the reactor was not designed. Where fuel slugs within the tubes were leaning against the inner walls, uranium diffused into the stainless steel, making a new alloy, a stainless steel/uranium eutectic. This uranium/steel mixture melts at 1,340°F, so with the reactor core overheating, the fuel assemblies fell apart. This problem would have been hard to foresee.

Another puzzle was harder to figure out. Why did the reactor run away, increasing power on a short period? The power excursion was simulated using the AIREK generalized reactor-kinetics code running on an IBM 704 mainframe. If the parameters were tweaked hard enough, the simulation could even be forced to agree with the recorded data, but even a well-tuned digital simulation could not indicate why the transient had occurred. It was easier to explain the subcritical plunge than the short-period lift-off. Further calculations and physical experiments proved that it was the sodium void in the blocked coolant channels that caused the fission process to run wild. It was the same phenomenon that caused the fuel assemblies to melt. Graphite is a better neutron moderator than just about anything, and the fission process improves when there is no other substance, such as coolant, in the way.

Those inert, radioactive gases produced in the accident that had not leaked out and made it up the exhaust stack, xenon-135 and krypton-85, were kept in holding tanks for a few weeks for the radiation to decay away and then slowly released up the stack and into the environment. The iodine isotopes had apparently reacted with something in the building and were not found in the released gas. On August 29, 1959, a news release was issued to the Associated Press, United Press International, *The Wall Street Journal*, and seven local newspapers, informing the public of the incident. The wording tended to downplay it to the point that one would wonder what made it news, beginning with "During inspection of the fuel elements on July 26 . . . a parted fuel element was observed." So? What's a "parted fuel element"? No big deal was made.

The SRE fuel melt was a unique accident, and yet it was typical. It was unique in that so many obvious trouble clues were ignored for so long. I am not aware of another incident quite like this. It was typical of reactor accidents in that nobody was hurt, and the only way you could tell from looking at it that something had happened was to take it apart. As in many cases of reactor accidents, the fuel was damaged by a lack of coolant. There was no steam explosion. Lessons were learned, extensive modifications were made to the system to improve its reliability. For the newly funded SRE Power Expansion Program, the hot-oil pumps were junked and Hallum-type primary and secondary sodium pumps were installed, and SRE was restarted on September 7, 1960.[144] It ran well, generating 37 gigawatt-hours of electricity for the Moore Park community. It last ran on February 15, 1964, and decommissioning of the nuclear components was started in 1976. In 1999, the last remainder of the Sodium Reactor Experiment was cleaned off Simi Hills. The idea of a graphite-sodium reactor died.

Real trouble did not begin until February 2004, when locals filed a class action lawsuit against the Santa Susana Field Laboratory's current owner, The Boeing Company, for causing harm to people with the Sodium Reactor Experiment. They had been stirred to action by a new analysis of the 1959 incident by Dr. Arjun Makhijani, an electrical engineer. In a way, the suit was like building a house in the glide path of an airport and then suing because airplanes were found to be flying low overhead. When Santa Susana was built, Simi Valley was a dry, desert-like landscape. Makhijani estimated that the accident released 260 times more iodine-131 than the Three Mile Island core melt in Pennsylvania in 1979, which speaks well of Three Mile Island. His estimate was speculative, because no iodine-131 contamination could be detected at the time. Over 99% of the volatile isotope was captured as it bubbled up out of the naked fuel into the coolant, becoming solid sodium iodide, which collected in the cold

144 The Hallum Nuclear Generating Station is a little-known, second graphite-sodium reactor built near Lincoln, Nebraska, starting in 1958. This experiment was also funded by the AEC. Unlike the SRE, the vertical-shaft centrifugal sodium pumps for this reactor were designed from scratch and were considered far superior to the ill-fated SRE pumps. Unfortunately, the modified Hallum-type pumps used in the rebuilt SRE had undersized overflow loops which produced a serious gas-entrapment problem. The Hallum reactor operated only briefly, from 1962 to 1964. The graphite moderator cans were clad in stainless steel, and stress cracking and corrosion caused irreducible problems. By 1969, evidence of the Hallum reactor was erased from the prairie, but the Hallum-type pump remains as a credible means of moving liquid sodium.

trap in the primary cooling loop. Any pollutant that might have made its way up the stack was diluted in the air, and 80 days later there could be no detectable trace, as if there were any to begin with. There were no milk cows living in Simi Valley and no edible grass to contaminate. There was no detectable thyroid cancer epidemic. Nobody was hurt. Boeing settled with a large payout to nearby residents.

In those early decades of nuclear power, it was an unwritten rule in the AEC that the public was not to be burdened with radiation release figures or the mention of minor contamination. It was true that the general population had no training in nuclear physics and radiation effects, and if given numbers with error bars and a map of an airborne radiation plume, imaginations could take control in nonproductive ways. Nobody wanted to cause a panic or unwarranted anguish or to undermine the public's fragile confidence in government-sponsored research. The results of such a policy are worse than what it is trying to forestall, as the government is commonly accused of purposefully withholding information, and misinformation rushes in to fill the vacuum. Conspiracy theories thrive. This fundamental problem of nuclear work has yet to be turned around.

Our next adventure in sodium takes us to Lagoona Beach, which sounds like a secluded spot somewhere in Hawaii, but it's not. It is in Frenchtown Charter Township, Michigan, 27.8 miles from downtown Detroit, looking out onto Lake Erie.

Walker Lee Cisler was born on October 8, 1897, in Marietta, Ohio. An exceptionally bright student, Cisler sealed his fate by receiving an engineering degree at Cornell University in 1922. With that credential, he was hired at the Public Service Electric and Gas Company in New Jersey, was named chief of the Equipment Production branch of the U.S. War Production Board in 1941, and in 1943 was tapped as the chief engineer for Detroit Edison.

In 1944 he joined the Supreme Headquarters, Allied Expeditionary Force (SHAEF) in Europe, assigned the task of rebuilding the electrical power systems on the continent as the German army retreated. By 1945, he had the power system in France generating more electricity than it had before the war. Impressed, the early embodiment of the AEC named him secretary of the AEC Industrial Advisory group in 1947. Resuming work at Detroit Edison, he became president in 1951 and CEO in 1954.

As a visionary, pushing for a greater and everlasting energy supply in his native land, Cisler became an early advocate of the breeder reactor concept, and by October 1952 he established the Nuclear Power Development Department at Detroit Edison. His dream of a civilian-owned, commercial breeder played right into the AEC plans, and it was the second breeder concept in their set of demonstration power reactors to be built. The first, the thermal U-233 breeder, would be built at Santa Susanna. The second, the fast plutonium breeder, would be Cisler's baby. A kick-off meeting was held with the AEC at Detroit Edison on November 10, 1954. Present at the meeting was Walter Zinn, the scientist who headed the slightly melted EBR-I project at National Reactor Testing Station in Idaho. The plant would be named Fermi 1, in honor of the man whose name, along with Leó Szilard's, was on the patent for the original nuclear reactor. Ground was broken at Lagoona Beach in 1956, after Cisler had secured $5 million in equipment and design work from the AEC and a $50 million commitment from Detroit Edison.

It would be a long crawl to implementation of Cisler's plan, fraught with ballooning costs, many engineering novelties, and strong opposition to the project by Walter Reuther. Reuther was an interesting fellow. A card-carrying Socialist Party member, anti-Stalinist, and a fine tool & die machinist, he became a United Auto Workers organizer/hell-raiser and was attacked by a phalanx of Ford Motor Company security personnel in the "Battle of the Overpass" in 1937. This, at the very least, made him a well-known figure in Detroit.[145] Reuther, the UAW, and eventually the AFL-CIO filed suit after suit opposed to the building permit for the plant

145 Walter worked as a "wage slave" at the Ford Motor Company starting in 1927. Henry Ford sent him to Nizhny, Novgorod, Soviet Union to help build a tractor factory, but he became overly interested in the proletarian industrial democracy, and Ford fired him in 1932. After working for a few years at an auto plant in Gorky, Reuther returned to the U.S. and became a very active member of the UAW. On May 26, 1937, at 2:00 P.M., he and Richard Frankensteen were in the middle of a leaflet campaign ("Unionism, Not Fordism") and they were asked by a news photographer to pose on the pedestrian overpass in front of the Ford sign. Ford's modest army of about 40 security specialists walked into the frame from the left and proceeded to discipline the uninvited visitors. The Dearborn police stood out of range and shouted advice while the union men were beaten, kicked, dragged by the feet, slammed down on the concrete, and thrown down two flights of steps. Reuther, always a champion of the underpaid and the underappreciated worker with too little money, died on May 9, 1970, when his privately chartered LearJet smacked the runway at the Pellston, Michigan, airstrip in rain and heavy fog. He was on his way to the UAW recreational facility at Black Lake.

and later the operating license, based on multiple safety concerns and the fact that it was not an automobile. The suits ate up a vast amount of time and money, and court decisions finding against Fermi 1 were taken all the way to the Supreme Court. In the summer of 1961, the court decided seven to two in favor of Cisler, the AEC, and Detroit Edison. Construction could proceed, and the projected cost had risen to $70 million.

Because Fermi 1 was designed as a plutonium breeder, liquid sodium was chosen as the coolant. A fast breeder core was much smaller than a thermal reactor of similar power, so the coolant had to be more efficient than water, and to hit the activation cross section resonance in uranium-238, the breeding material, the fast neutrons from fission had to endure a minimum loss of speed. Sodium was the logical choice, and two sodium loops in series were used, similar to the SRE design. It doubled the complexity of the cooling system, but it ensured that no radioactive coolant could contaminate the turbo-generator.

The use of sodium made things difficult and complicated. Refueling, for example, was comparatively simple in a water-cooled reactor. To refuel a water-cooled reactor you shut it down, and after it cools a while, unbolt the vessel head, crane it off, and lay it aside. Flood the floor with water, which acts as a coolant and a radiation shield. You pick up the worn-out uranium in the core, one bundle at a time using the overhead crane, and trolley it over to the spent-fuel pool. Carefully lower new fuel bundles into the core, drain the water off the floor, and replace the steel dome atop the reactor vessel. You're done.

There is no way to do this with a sodium-cooled system. For one thing, the sodium gets radioactive running in the neutron-rich environment of fission, and it would take a week of down-time just to let it decay to a level of marginal safety. You cannot let it cool down, because in this state the sodium is solid metal, and you could not pull the fuel out of it. It cannot be exposed to air, and everything done must be in an inert-gas atmosphere. The entire building would have to be pumped down and back-filled with argon. Also, sodium, liquid or solid, is absolutely opaque. You cannot see through it, so, unlike looking down from the refueling crane through the water to see what you are doing, you would have to be able to refuel the reactor blindfolded. If anything weird happened in a water-cooled reactor core, you could conceivably stick a periscope down in it and have a look. Not so with sodium coolant. Fermi 1 was to be an electrical power-production reactor, running at 200 megawatts. Although in later terms this would be a very small reactor, in 1961 it was a clear challenge to design a machine of this size running in sodium.

The problems of dealing with refueling a sodium-cooled fast reactor were solved by enclosing the works in a big, gas-tight stainless steel cylinder, about two stories tall. In the bottom of the cylinder, off to the side, was an open-topped stainless tank containing the U-235 reactor core, surrounded by the U-238 breeding blanket.[146] Liquid sodium in the primary cooling loop was pumped into the bottom of the reactor tank through a 14-inch pipe. The sodium flowed up, through small passages between fuel-rods, taking the heat away from the fissioning fuel. The thick, hot fluid was removed by a 30-inch pipe at the top and sent to a steam generator for running the electrical turbo-generator. A "hold-down plate" pressed down on the top of the fuel to keep it from being blown out of the can by the blast of liquid sodium coming up from the bottom. Under normal operation, the liquid sodium in the structure was 34 feet deep. Any open space left was filled with inert argon gas.

An ingenious, complex mechanism was used to refuel the reactor and extract and replenish the irradiated U-238 rods in the breeding blanket. To one side of the reactor tank and inside the stainless vessel was another round can, this one containing a revolving turntable loaded with fuel and blanket rod-assemblies. It worked like the cylinder in a revolving pistol or the carousel in a CD changer, indexing around and stopping automatically with a rod assembly in position to be extracted or inserted. An exit tube led vertically to an airlock on the top floor of the reactor building, where a refueling car ran on rails and was able to reach down into the reactor vessel, through the air lock, and remove or replace rod assemblies in the rotor. The car, gamma-ray shielded by 17,500 pounds of depleted uranium, was driven back and forth on the rails by a human operator, under absolutely no danger from radiation in the reactor, the coolant, or in fuel being inserted or removed.

146 A breeder reactor is supposed to run off fuel that it produces, which in this case was plutonium-239. Until it has actually made fuel and it has been extracted from the breeding blanket, the breeder usually starts off with a load of uranium-235. Unlike a water-cooled reactor, a fast breeder runs off pure fissile isotope, so the starter fuel must be highly enriched to simulate its eventual load of Pu-239.

To refuel the core, the hold-down device was first lifted off the top of the core. The fuel-transfer rotor was turned by a motor until it clicked into place with an empty slot under the handling mechanism, which was a long, motor-driven arm hanging down next to the reactor. The arm then turned to the core, gently snatching a designated rod assembly, picking it up and out of the reactor, swinging over to the transfer rotor, and lowering it down into the open slot. The rotor then rotated clockwise and clicked to the next open slot. After the rotor became completely full, the refueling car would bring up the rod assemblies one at a time and put them into shielded storage. To put new fuel or breeding material in the reactor, the process was reversed, with the empty rotor first loaded with fresh fuel from the refueling car. The rotor would index around, presenting the new rod assemblies one at a time to the arm, which would pick them up, swing them through the core, and insert them. All this happened under liquid sodium. It was a totally blind operation, depending only on mechanical precision to find rod assemblies where they were supposed to be and transporting them with great accuracy. When a refueling operation was completed, the hold-down device would descend and cover the top of the core. To be able to automatically refuel this reactor from the comfort of a swivel chair by pushing a few buttons was a marvel to behold.[147]

Given the recent tribulations at the SRE, great attention was given to the design of the coolant pumps. Made of pure 304 stainless steel, one primary pump was the size of a Buick. It was a simple centrifugal impeller, driven by a 1,000-horsepower wound-rotor electric motor, with a vertically mounted drive shaft 18 feet long. It turned at 900 RPM to move 11,800 gallons of thick, heavy, 1,000°F liquid sodium per minute and lift it 360 feet. The motor speed was controlled using a simple 19th century invention, a liquid rheostat, as once used to dim the lights in theaters.

147 When I was in school, fast breeder reactors were all the rage, and a wild story about the refueling machinery in the Fermi 1 was going around. It was said that a spin-off of the breeder reactor project was the AMF pin-spotting machine, now used in many bowling alleys. The method used for collecting and placing fuel assemblies in an automatic, utterly blind situation under liquid sodium was found useful for doing a similar job with bowling pins. I am sad to say that I can find no connection between bowling alley machinery and fast reactor refueling devices, much as I would like to. The patent for the AMF "candlepin" pin spotting machine was filed on April 14, 1964, which is close to when the Fermi 1 was designed, but it was based on the original AMF 82-10 pin spotter, first demonstrated in 1946.

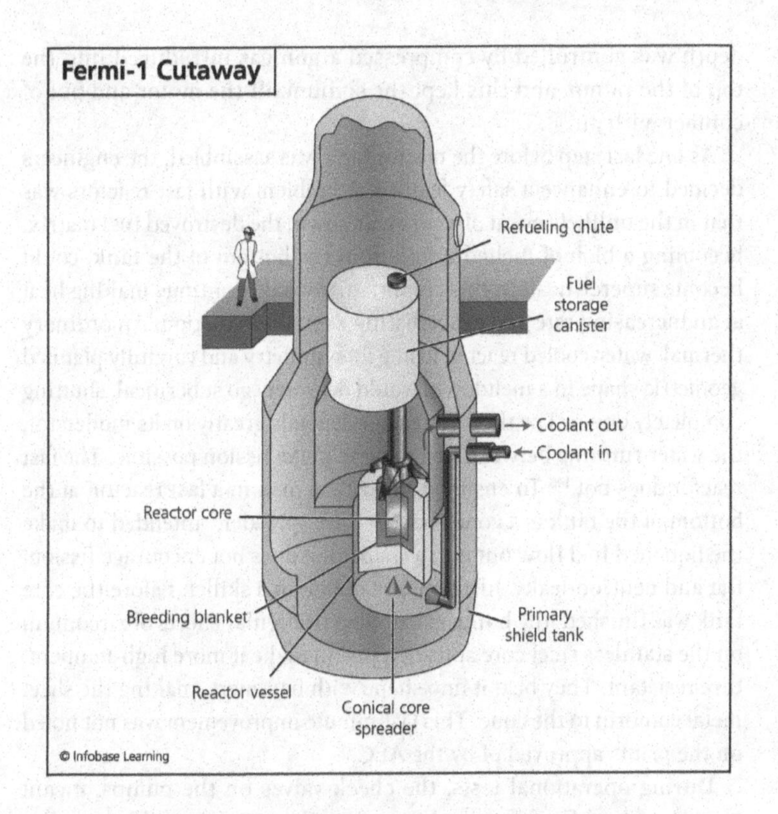

The Fermi-1 fast breeder reactor was a bold move in the development of commercial power reactors, ensuring a future with no problems of obtaining fuel to generate power in a growing economy. Unfortunately, the fuel-supply problem failed to materialize, and the complexity of its sodium-based cooling system proved to be its downfall.

There were no liquid-cooled ball bearings to worry about, and no seal to keep sodium out of the pump motor. There was no tetralin seal coolant, of course. The main bearing at the impeller end was a simple metal sleeve, with the stainless shaft running loose inside it. It was a "hydrostatic bearing," like the bearing on a grocery-cart wheel, but instead of having oily grease to keep it from squeaking, the lubricant was liquid sodium, supplied out of the fluid that the impeller was supposed to pump. The sodium was allowed to puddle up in the long shaft gallery, rising several feet above the impeller housing. Its depth was controlled by compressed argon gas introduced into the top of the pump, and this kept the sodium off the motor and out of contact with air.

As one last step before the reactor tank was assembled, the engineers decided to enhance a safety feature. A problem with fast reactors was that in the unlikely event of a core meltdown, the destroyed fuel matrix, becoming a blob of melted uranium in the bottom of the tank, could become supercritical. In this condition, it would continue making heat at an increasing rate and exacerbating a terrible situation. An ordinary thermal, water-cooled reactor, losing its symmetry and carefully planned geometric shape in a meltdown, would definitely go subcritical, shutting completely down. The thermal reactor depends greatly on its moderator, the water running between fuel rods, to make fission possible. The fast reactor does not.[148] To ensure a subcritical melt in a fast reactor, at the bottom of the tank is a cone-shaped "core spreader," intended to make the liquefied fuel flow out into a shape that does not encourage fission, flat and neutron-leaky, like pancake dough in a skillet. Before the core tank was finished, the builders installed triangular sheets of zirconium on the stainless steel core spreader, just to make it more high-temperature resistant. They beat it into shape with hammers, making the sheet metal conform to the cone. This last-minute improvement was not noted on the prints approved of by the AEC.

During operational tests, the check valves on the pumps, meant to prevent backflow if something stopped working, would slam shut instead of closing gently but firmly, as they were supposed to. This flaw was re-engineered, and further extensive testing would ensure no problems from the pumps.

Still, working with a large volume of pure sodium would prove challenging. Tests were being performed in an abandoned gravel pit about 20 miles north of the building site on August 24, 1959, when a load of sodium exploded in air. Houses in the nearest neighborhoods, Trenton and Riverview, were damaged by the blast and six people were hospitalized. By December 12, 1962, there was enough of the

148 An enduring question of fast reactor design is: "What is the sodium void coefficient?" If the sodium coolant leaks out of a fast reactor, does it go subcritical or supercritical? The correct answer if this question should come up on your Ph.D. qualifying exam is: "That depends." There are many factors that come to bear, including the geometrical shape of the core.

reactor assembled to test the main cooling loop. An operator was at the control console watching the instruments as the sodium circulated at full speed through the system. The temperature started reading high on a thermocouple gauge. There was no fuel in the reactor, so where was the heat coming from all of a sudden? Were sodium and water mixing due to a flaw in the steam generator? Thinking this through, the operator reached over and hit the red water-dump button. Water gushed out of the secondary cooling loop into a holding tank, taking it quickly out of the steam generator and away from possible contact with the sodium loop. Unfortunately, this caused a sudden vacuum in the system, which was designed to hold high-pressure steam and not an airless void. A safety disk blew open, and sodium started oozing out a relief vent, hit the air in the reactor building, and made a ghastly mess.

No one was hurt. When such a complicated system is built using so many new ideas and mechanisms, there will be unexpected turns, and this was one of them. The reactor was in a double-hulled stainless steel container, and it and the entire sodium loop were encased in a domed metal building, designed to remain sealed if a 500-pound box of TNT were exploded on the main floor. It was honestly felt that Detroit was not in danger, no matter what happened.

Cost of the Fermi 1 project reached $100 million, and it was too far along to turn back. The fuel was loaded on July 13, 1963, and the fuel car was not acting well.[149] The first startup was a few weeks later, on August 23 at 12:35 P.M. A system shakedown at low power would continue until June of 1964. A few problems surfaced. The number 4 control rod delatched from the drive mechanism, leaving it stuck in the core. The large, rotating plug in the top of the reactor vessel, used to move the refueling arm between the fuel rotor and the core, jammed and wouldn't move. A sodium pump had to be repaired, and the cap on the reactor vessel had to be rebuilt so that it would fit correctly. Some electrical connections and cable runs were defective, causing instrument problems. These glitches were all knocked down.

By January 1966, the Fermi 1 plant was wrung out and ready to go to full 200 megawatts of heat in several cautious steps. August 6,

---

149 It is difficult to find a cross-section view of the Fermi 1 reactor that does not have a big X drawn through the refueling car. It was not a popular accessory.

1966, was a day of triumph. The thermal power was brought up to 100 megawatts, enough to make 33 megawatts of electricity, or about half what the backup diesel generators could produce. The project cost had also reached a high point, at a cool $120 million, and critics pointed out that the reactor had so far been able to generate measurable electricity for a total of only 52 hours.

At this half-power level, unusually high temperatures were indicated in fuel assemblies M-091 and M-140, the steam generator started leaking steam, and control rod no. 3 seemed to stick in the guides. The next day, August 7, the positions of the hot fuel assemblies were swapped with trouble-free fuel assemblies, to see if the problem moved. There seemed no correlation between the specific fuel assemblies and overheating. The problem seemed to be the position in the core, and not the fuel.[150] Could it be that the thermocouples in those locations were just reading wrong?

The operating crew was ready to try another cautious power-up on October 4, 1966. The Fermi 1 reached criticality at 11:08 P.M., and it idled at low power while every little thing was checked. At 8:00 A.M. the next day, the operators were ready to bring the reactor to half-power, but a steam-generator valve seemed stuck. It took until 2:00 P.M. to resolve that problem, and then the feed-water pump in the secondary loop was not working. They powered down and worked on it. By 3:05 P.M. they had resolved the problem and the power was increased to 34 megawatts and rising.

Something was not right. The neutron activity in the reactor core was erratic and bouncy. There was no reason for the neutron level to be anything but smooth and steady. The power ascension was halted while Mike Wilbur, the assistant nuclear engineer in the control room, contemplated the meaning of these instrument readings. Based on previous problems, Wilbur had a hunch. He stepped behind the main control panel to take a look at the thermocouple readouts on the fuel assembly outlet nozzles. These instruments were not considered essential for running the power plant, so they were not mounted on the main panels. They were included in the hundreds of instrument

150 Geek-joke from 1975: How does a Digital Equipment Corporation computer-repair technician fix a flat tire? He swaps the wheels to see if the problem moves. You had to be there. . . .

readouts, lights, and switches in the control room for diagnostics, and here was a problem that required a deeper look.

Fuel assemblies M-140 and M-098 were both running hot. At this power level, the temperature of coolant flowing out the top of fuel assembly M-140, which had given trouble in the past, should have been 580°F. It was reading 700°F. As Wilbur was taking this in, at 3:08 P.M. the building radiation alarms started sounding. It was a rude, air-horn sound: two mind-numbing blasts every three seconds. There were several possible explanations for the radiation alarms, but the assistant nuclear engineer knew deep in his heart that one was likely. Fuel had melted, spreading fission products into the coolant. The only thing that was not clear at all was: Why?

The crew executed emergency procedures as specified in the operations manual. All doors were closed, and all fresh-air intakes were closed in the building. Detecting radiation in the building was an emergency condition. It was supremely important to not let it leak out into the world outside Lagoona Beach, which would make it a big, public emergency. They executed a manual scram at 3:20 P.M., shutting Fermi 1 down with the floor-trembling shudder of all controls dropped in at once. One rod would not go in all the way. Not good. Was a fuel assembly warped? They tried another scram. This time, it went in. With the neutron-poisoning control rods all in, there was no fear of the core being jostled or melted into a critical condition. There could be no supercritical runaway accident, and if the core were completely collapsed and flowing onto the spreader cone, the uranium would be mixed with melted control material, which would definitely discourage fission. News of the accident, specifying that the engineers did not know what had happened, spread across the land.

The reactor had never run at full power, and only for a short time trying to get to half power, so there was no worry that it could melt down any further in the shutdown condition. There were not enough delayed fissions and fission product decays to cause havoc with high temperatures in the fuel. Over the next few weeks, the operators and engineers tried to find the extent of the damage without being able to see inside the reactor core. One at a time, they pulled control rods and noted the increase in neutron activity in the core. A few control rod locations did not return the activity they expected. Near these, the fuel may have sagged out of shape.

Next, they attached a contact microphone to a control-rod extension and listened as the liquid sodium was pumped around the primary loop. They heard a clapping sound. They slowed the pumps. The clapping sound slowed. Was there something loose in the core? It was impossible to see.

They had to feel around using the fuel-loading devices to determine the state of the core. Proceeding slowly and cautiously, it would take four months, into January 1967, to confirm that fuel had melted. First they raised the hold-down column on top of the core, and they found that it was not welded to the reactor. This was good. Next, they swung the refueling arm over the core and tried to lift the fuel assemblies, one at a time. The strain gauge on the arm would weigh each assembly. Two seemed light. Fuel had dropped out the bottom, apparently. Two were stuck together and could not be moved without breaking something. It took five months, until May 1967, to remove the fuel using the automatic equipment. Finally, seeing the fuel assemblies in the light of day, it was clear that two had melted and one had warped. There was still no explanation as to why.

The sodium was drained out of the reactor tank, although there was no provision in the design for doing so. A periscope, 40 feet long with a quartz light attached, was specially built to be lowered into the darkness from the top of the reactor vessel. Finally, the engineers could see the bottom of the reactor tank. It looked clean and neat. There was no melted uranium dripped onto the spreader cone. No loose fuel slugs were scattered around. All of the damaged fuel had collected on the support plate for the bottom of the reactor core. There was, however, something out of place. It looked like . . . a stepped-on beer can, lying on the floor of the reactor tank? That could explain the core melt and the clapping noise. Caught in the maelstrom of coolant forcing its way through the bottom of the reactor core, this piece of metal had slapped up against an inlet nozzle and blocked sodium from flowing past a couple of red-hot fuel assemblies. But what was it, and how did it get in the reactor?

To answer that question, the metal thing would have to be removed from the reactor, and that was not easy. Cisler stood firm under a hail of abuse from anti-nuclear factions as the Fermi 1 engineers accomplished the impossible. They built a special remote-manipulating tool and lowered it down into the floor of the reactor tank through the sodium inlet pipe. It had to make two 90-degree turns to get there.

With the new tool in place, they were able to move the metal thing closer to the periscope, flip it over, and take pictures.

Still, the experts could not tell what it was. They all agreed on one thing: it was not a component for the reactor as shown on the design prints. It would have to be removed. With enormous effort and a great deal of money, another special remote operator was built and inserted through the sodium inlet. One worker swung the quartz light on the end of the periscope to bat the object into the grabber at the end of the new tool. Another man, working 30 feet away, manipulated the grabber tool blindly from instructions over an intercom. Finally, a year and a half after the accident, on a Friday night at 6:10 P.M., the mystery object fell into the grip of the tool. Slowly, taking 90 minutes, they snaked it up through the pipe and into the hands of the awaiting engineers.

They looked at it, turned it over, looked again, and finally truth dawned. It was a piece of the zirconium cover that they had attached to the stainless steel spreader cone, nine years ago. They had not bothered to have it approved and put on the final prints. It had cost an additional $12 million to figure this out.

By May 1970, all repairs had been made and Fermi 1 was ready for a restart. AEC inspectors were on hand, making the operators nervous at the close monitoring of their every action. Things were tense as 200 pounds of sodium suddenly broke loose in the primary transfer tank room, tearing out a water-pipe run and causing a loud thud as the mixture of water, air, and sodium exploded. This embarrassing incident cost another two months of down-time to repair the damage, but in October the plant finally reached its designed power level, making 200 megawatts of heat. For the next year of operation, the plant was able to remain online for only 3.4% of the time.[151] Denied an extension to its operating license in August 1972, its operation ended on September 22, 1972. The plant was officially decommissioned on

151 This number, 3.4%, is the *capacity factor* for Fermi 1. In 2011 the average capacity factor for a nuclear power plant in the United States was 89%. A wind-turbine plant, such as the Burton Wold Wind Farm, consisting of ten Enercon E70-E4 wind turbines, had a capacity factor of 25% in 2008. The Hoover hydroelectric dam has an average capacity factor of 23%. The capacity factor of a power plant is the amount of energy produced over a set period of time divided by the energy it could have produced if working at full capacity during the same period.

December 31, 1975, the fuel and the sodium were removed, and it still sits quietly at Lagoona Beach, next to Fermi 2, a General Electric boiling-water reactor that is currently making power for DTE Energy. All things considered, Fermi 1 failed at its mission, to spearhead the age of commercial plutonium breeding in the United States. Admiral Rickover had summed it up clearly back in '57 in one sentence, saying that sodium-cooled reactors were "expensive to build, complex to operate, susceptible to prolonged shutdown as a result of even minor malfunctions, and difficult and time-consuming to repair."

Did we almost lose Detroit? No. There was no water in the reactor vessel to destructively explode into steam with a suddenly overheated reactor core. No steam meant that there was no source of force to break open the containment dome and spread fission products from the core.[152] The water and steam were out in another building, and the worst that could happen was to mix them with the secondary sodium loop and not with the primary loop containing radioactive sodium. The results of a massive breakdown in the secondary loop could dissolve every aluminum drink can within five miles with vaporized sodium hydroxide, but it would spread no radioactivity. The reactor was too feeble to build up enough fission product to justify the thousands of casualties predicted if the core were to somehow explode. In a maximum accident, the entire core would overheat and melt into the bottom of the reactor vessel, but it would melt the controls and the non-fissile core structure along with it, making it unable to maintain a critical mass. The dire predictions and warnings had been dramatic, but hardly realistic.

The school of business at Northern Michigan University was renamed the Walker L. Cisler College of Business. He died in 1994 at the age of 97.

Not even slightly discouraged, the AEC proceeded to secure funding for yet another stab at a commercial sodium-cooled contraption in 1970, the Clinch River Breeder Reactor Project, to be built inside the city limits of Oak Ridge, Tennessee. This would be a full-sized power

152 No steam not only meant that there could be no steam explosion, it also meant that there could be no explosive hydrogen buildup. Hydrogen gas can be made from steam in a reactor accident by heat-induced corrosion of metals or radiolysis. More on this topic in the Fukushima chapter.

The Power of Radium at Your Disposal
Twenty-three years ago radium was unknown. Today, thanks to constant laboratory work, the power of this most unusual of elements is at your disposal. Through the medium of Undark, radium serves you safely and surely.
Does Undark really contain radium? Most assuredly. It is radium, combined in exactly the proper manner with zinc sulphide, which gives Undark its ability to shine *continuously* in the dark.
Manufacturers have been quick to recognize the value of Undark. They apply it to the dials of watches and clocks, to electric push buttons, to the buckles of bed room slippers, to house numbers, flashlights, compasses, gasoline gauges, autometers and many other articles which you frequently wish to see in the dark.
The next time you fumble for a lighting switch, bark your shins on furniture, wonder vainly what time it is *because of the dark*—remember Undark. *It shines in the dark.* Dealers can supply you with Undarked articles.
For interesting little folder telling of the production of radium and the uses of Undark address
RADIUM LUMINOUS MATERIAL CORPORATION
NEW YORK CITY
UNDARK
Radium Luminous Material
Shines in the Dark
321
To Manufacturers
The number of manufactured articles to which Undark will add increased usefulness is manifold. From a sales standpoint, it has many obvious advantages. We gladly answer inquiries from manufacturers and, when it seems advisable, will carry on experimental work for them. Undark may be applied either at your plant, or at our own.
The application of Undark is simple. It is furnished as a powder, which is mixed with an adhesive. The paste thus formed is painted on with a brush. It adheres firmly to any surface.

TOP: Chapter 1. Tho-Radia was one of many radioactive cosmetic lines promising to cure boils, pimples, redness, pigmentation, and the increasing diameter of pores. This particular product, "RACHEL No. 1" face powder, contained both thorium chloride and radium bromide, both radioactive and nothing that one should rub into the skin. Dr. Alfred Curie, named on the can, was not related to Pierre or Marie Curie, co-discoverers of radium, but the name recognition helped sell the product. Tho-Radia face cream, powder, and soap were introduced in France in 1933, and these products were sold in Europe until the early 1960s.

BOTTOM: Chapter 1. A magazine ad for Undark radium-charged luminous material. It suggests that radium would be good on your bedroom slippers as well as the push-button light switch on the wall in your bedroom. The use of radium in consumer products is now banned in most countries.

Chapter 2. This is a mock-up of the Mk-2 bomb-core setup that Harry Daghlian was testing in 1945. The shiny, rectangular blocks of material surrounding the plutonium sphere are tungsten carbide (WC) bricks, and Daghlian was stacking them around the center-mounted sphere to see how much WC the assembly could stand before the neutron-reflection effect caused it to go critical. He accidentally dropped a brick right on top, and the plutonium went prompt supercritical. This was not an atomic bomb configuration, where the plutonium sphere would be crushed down to the size of a large marble, but it was a functioning nuclear reactor, out of control.

Chapter 2. Not long after Daghlian's accident, Louis Slotin was demonstrating the criticality effect to his replacement. The screwdriver, shown in the mock-up picture, shimming up the hemispherical reflector on top, slipped, and the assembly came together suddenly. Slotin died from the radiation pulse caused by the prompt startup of the nuclear reactor that he had accidentally assembled. The mock-up was as accurate as possible, down to Slotin's empty Coke bottle on the setup table.

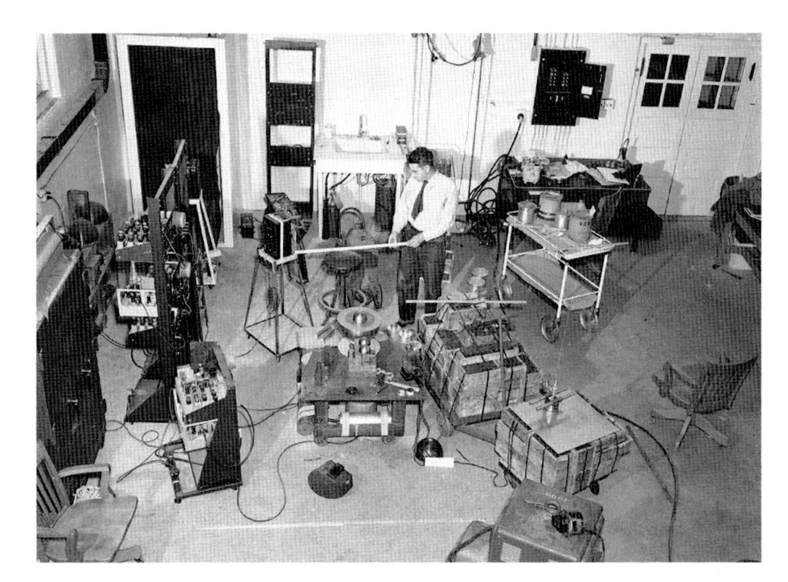

Chapter 2. This is a rarely seen shot of the entire space around Slotin's setup table, showing many interesting details. Note the old bank vault at the left, used to store the plutonium sphere, the radiation-counting equipment racked up on the left, the welder's helmet on the floor, and the 400 Hz motor-generator in the foreground. The motor-generator was used to simulate the power environment on a B-29 strategic bomber, for testing equipment that would be attached to the bomb and using aircraft power. The gliders on the floor are piled with lead bricks for radiation shield applications, and there is an active neutron source atop the brick pile closest to the motor-generator set. I'm not sure what the welder's helmet was for.

Chapter 3. The Castle Bravo test in the Pacific in 1954 used this ground-level thermonuclear device, named "Shrimp" for its modest size. It was a new bomb design, using a stock RACER IV plutonium atomic bomb adjacent to a cylindrical assembly containing lithium deuteride powder. It was predicted to yield 5 megatons of explosive energy, but gave 22 instead. It was a surprise. Note the NO SMOKING sign at the lower left.

Chapter 3. The NRX heavy-water reactor in Chalk River, Canada, in 1955, after a complete rebuild due to the unfortunate incident in 1952. The world's first reactor core meltdown occurred accidentally in this reactor, soon after which the first radiation-induced hydrogen explosion happened. Ensign James Earl Carter from Plains, Georgia, participated in the cleanup of the site.

Chapter 3. Samples of various materials were placed in the radiation environment of the NRX reactor core to be tested for stamina under high-flux conditions. The sampling ports on test reactors are usually driven by compressed air and controlled remotely, but the NRX system, called the "self-serve unit," seemed to be manually operated. A health physicist is holding a "cutie pie" ion chamber to closely monitor the radiation during this operation.

Chapter 3. The NRU heavy-water reactor under construction at Chalk River, Canada, in 1956. This reactor was used to test concepts that are in use today in CANDU reactors all over the world. It is in use today, and it may be the oldest reactor in the world that is still running. If you have had a medical test performed using technetium-99m, then that nuclide was produced in this reactor.

Chapter 4. The instant of Sam Untermyer's BORAX-I explosion, a controlled test of the very worst that could happen to a boiling water reactor. The experiment did not disappoint, as it sent the contents of the reactor vessel flying. This is a still from the 16mm movie that was made of the test. The movie camera stopped functioning soon after this frame was exposed, as its power cable was blown away in the explosion.

Chapter 4. Early in the analysis of the SL-1 explosion incident, a water sample was needed from the coolant spill on the reactor-room floor. Under normal circumstances, this was a simple task, but in this case the inside of the building was contaminated with highly radioactive fission products, and extraordinary measures were necessary. The crane setup shown in the picture is going to sample remotely, through the refueling door.

LEST WE FORGET

SL-1 1-3-61

TOP: Chapter 4. The poster made to go on nuclear engineers' walls. It refers to the explosion of the SL-1 power reactor on January 3, 1961, reminding all engineers that designing a reactor with a single control rod that can increase reactivity to the point of criticality is not an acceptable concept. The view is looking straight down into the SL-1 core with the top removed. The insides are so scrambled, it's hard to tell what you are looking at. The rods sticking out are connected to three of four peripheral flux-shaping controls. The controls themselves are normally cross-shaped, and here the crosses are flattened. What looks like strips of metal were once vertically mounted fuel assemblies.

BOTTOM: I took this picture looking up at the reactor refueling face of the X-10 graphite reactor at Oak Ridge, Tennessee. What appear to be dots on the white wall are plugs. To refuel, a worker pulls out a plug and pushes in a fuel canister using a steel rod. This reactor was built during World War II, and to the postwar world this was what a nuclear reactor looked like.

Chapter 5. A close-up of the back of the back of the two Windscale plutonium production reactors, showing the concrete exhaust stacks. The rectangular building in the foreground is one of two air-blower buildings per reactor, disconnected from the reactor buildings. What look like windows are intake louvers for the air that is blown through the reactor, up the stack, and over the surrounding dairy farms.

Chapter 5. A wider shot of the Windscale reactors, showing the lack of other tall structures in 1957. The landscape filled in with other buildings in the following decades. I was unable to gain permission to include photos of the internal structure of the Windscale reactors, but look on pages 11 through 13 in a lecture slide show from the University of Manchester, available on the Internet here: http://web.up.ac.za/sitefiles/file/44/2063/Nuclear_Graphite_Course/A%20-%20Graphite%20Core%20Design%20Air%20&%20Magnox.pdf. Or search on "graphite core design air & magnox" for the PDF file. The photo on page 13 of the slide show was taken looking straight into an open port on the core face. You are looking through the 5-foot-thick concrete shield, across the air void, and at an aluminum "charge pan." There are five holes. Four are for fuel, and the center hole is for isotope-production cartridges. The fire started, as is clearly described in a plant worker's deposition, in an isotope cartridge and not in the graphite or the fuel, as has been long assumed.

Chapter 5. This aerial shot of Windscale Unit 1 shows the two air-blower buildings and the air-filter assembly at the top of the exhaust stack. A label identifying the fire hoses entering the building has been blanked out.

Chapter 6. The Sodium Reactor Experiment building and auxiliary buildings at the Santa Susana Field Laboratory in Simi Hills, California, about 50 miles north of Los Angeles. The reactor is located in the middle of the floor of the tall building on the right. The smaller building with a peaked roof in front of the reactor building is the helium control station, and behind it with a flat roof is the air-blast heat exchanger. The steam generator is in the maze of pipes across the road, on the left.

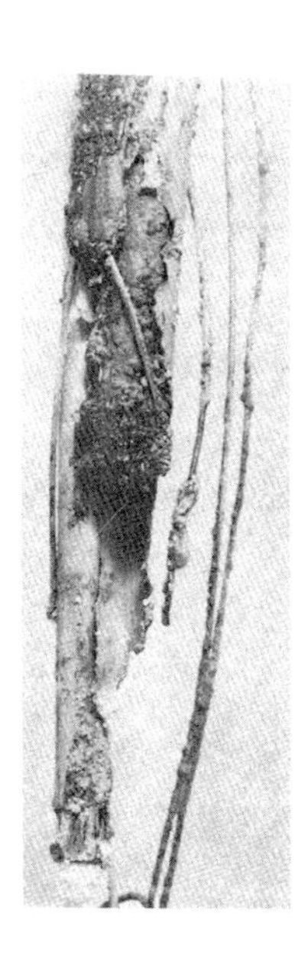

LEFT: Chapter 6. The bottom of a heat-damaged fuel rod in channel 55 in the Sodium Reactor Experiment at Santa Susana. The stainless steel tube containing a column of uranium fuel slugs has melted away, allowing fuel to drop into the bottom of the reactor core. The stainless steel wire that spirals around the tube is to prevent it from touching other tubes in a fuel element cluster. The location guide and orifice plate at the bottom of the rod are completely gone. RIGHT: Chapter 7. Americium extraction hood WT-2 in the 242-Z Building, Americium Recovery Process, at the Hanford site in southeastern Washington. Behind the long vertical window at the top left was the resin column that exploded, blowing out the glass in it and the diamond-shaped window below it. Harold McCluskey, the "Atomic Man," was standing on the step-stool at the far left.

Chapter 7. Room 180 in Building 771 at the Rocky Flats atomic bomb plant in Colorado. This is where the fire started on September 11, 1957, in the glove box, middle left in the picture. Gloves have been turned inside out and are hanging down.

Chapter 7. This is the second floor in Building 771 at Rocky Flats, with HEPA filters in racks from floor to ceiling. They were supposed to keep radioactive dust from escaping the workspaces and being blown into the environment by the ventilation fans. In the fire of 1957, two men opened the door to see if the filters were on fire, and the rush of fresh air caused the plutonium dust that had built up in the room for years to ignite quite suddenly. The explosion destroyed the filter banks, and radioactive dust started going up the exhaust stack.

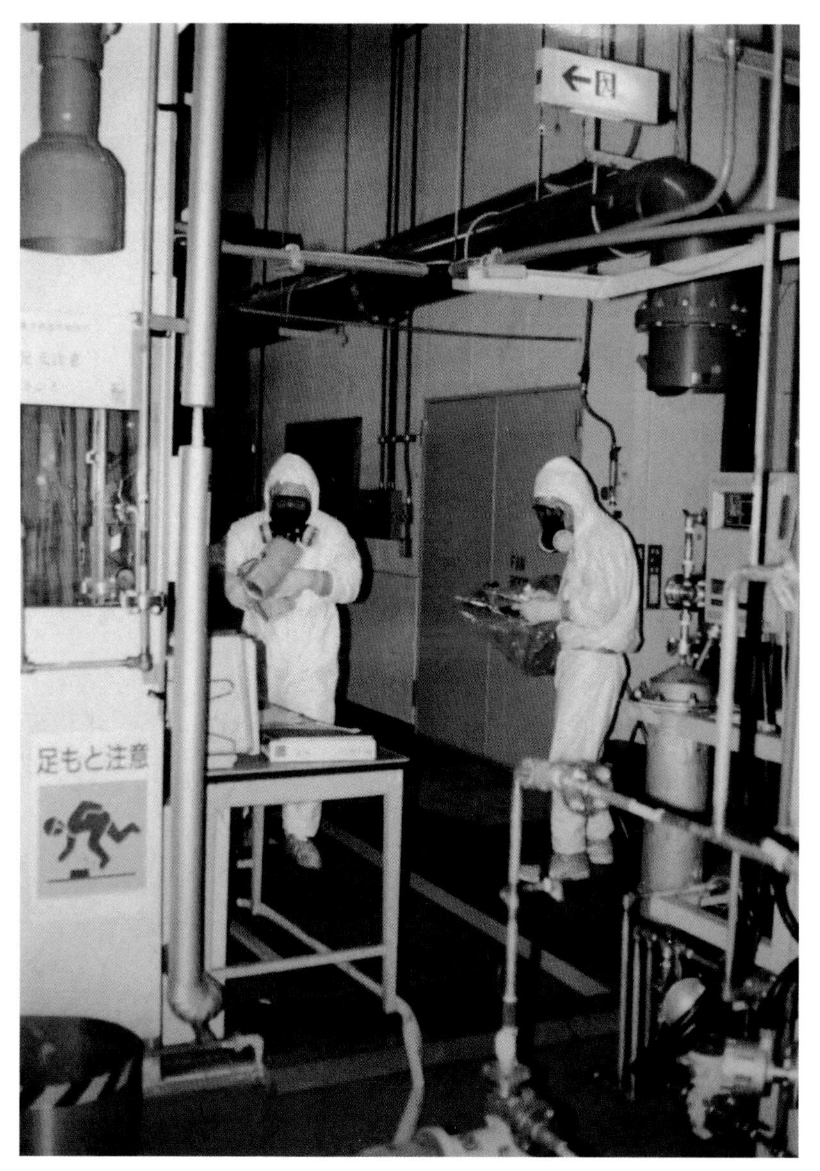

Chapter 7. Inside the Fuel Conversion Test Building at the JCO plant in Tokaimura, Japan, in 1999. Workers are suited up and evaluating the radiation lingering just after the criticality in Precipitation Tank B had been brought under control. They are looking at the desk at which Yutaka Yokodawa was sitting, doing paperwork, when Tank B became a nuclear reactor out of control. Tank B is located just out of the frame on the right.

Chapter 8. An MK-28FI thermonuclear weapon being unloaded from a B-52H strategic bomber by a crew of three at Ellsworth Air Force Base, South Dakota, in 1984.

Chapter 9. The control room at TMI-2 near Harrisburg, Pennsylvania, on April 1, 1979. President Jimmy Carter with his wife, Rosalynn, are being briefed by James R. Floyd, supervisor of TMI-2 operations, who is the only one not wearing anti-contamination booties. Harold R. Denton, director of the Office of Nuclear Reactor Regulation in the Nuclear Regulatory Commission, is standing in the foreground. Hidden behind Denton is Richard L. Thornburgh, governor of Pennsylvania. Carter demonstrated his knowledge by asking the right questions concerning the buildup of hydrogen in the containment building.

Chapter 9. An aerial shot of the Chernobyl-4 power reactor in Ukraine, USSR, after it had exploded in 1986 and the smoke had cleared. It is completely destroyed and is unrecognizable as a power plant.

Chapter 9. Chernobyl-4 after it has exploded in 1986 but as the fire still burns in what remains of the graphite moderator. This photo was taken by a helicopter flying near and risking a dangerous radiation dose.

This is the author, operating the Emergency Response Data System end of the Safety Parameter Display System (SPDS) in the control room of the E. I. Hatch Nuclear Power Station. It was built to withstand a force-9 earthquake.

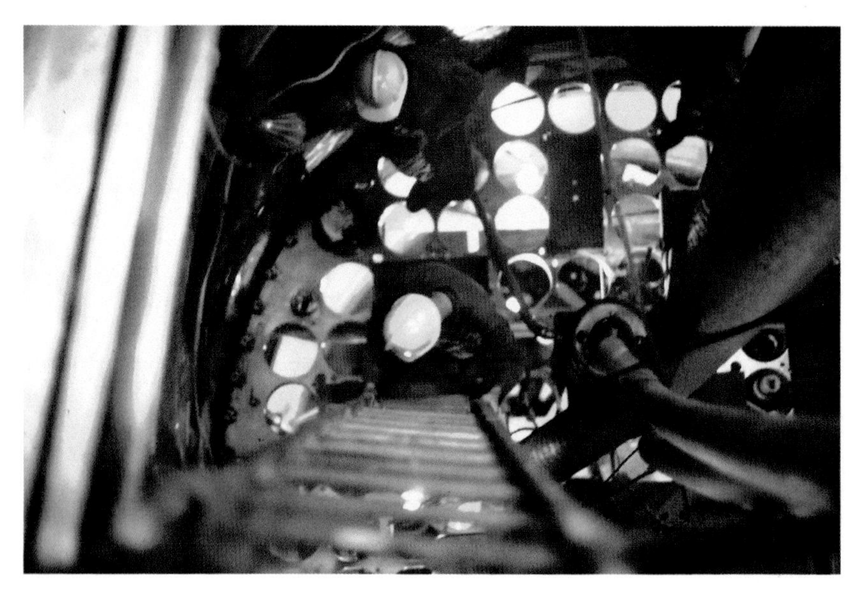

Chapter 10. I made this picture inside the reactor vessel of a General Electric BWR/4 with a Mark I containment, exactly as used at the Fukushima Daiichi power plant in Japan. I was standing on the steam separator, shooting down at the upper fuel-support plate. The fuel assemblies plug into the round openings in the plate, and cooling water flows upward through the holes. The large hoses are ventilation so we can breathe, and there is a large vacuum cleaner on the support plate to remove the dust we track in.

Chapter 10. I made this picture standing in the wet well of a General Electric Mark I containment structure, exactly as used in reactors at Fukushima Daiichi. You can see the wall curving upward into a large sphere. In the middle is the outlet diffuser for one of eight vent lines, intended to conduct a blast of suddenly escaped steam downward into the torus, where it will be quenched in the pool of cool water. The ladder in the middle gives some perspective.

Chapter 10. I made this picture standing in the torus (wet well) of a General Electric Mark I containment structure. As you can see, it's as big as a subway tunnel. We were standing on a catwalk, below which is the pool of water. The large tubular structure on the right is the vent header, built to distribute a massive steam pulse coming down the vent lines into 96 smaller pipes that release the steam under the water.

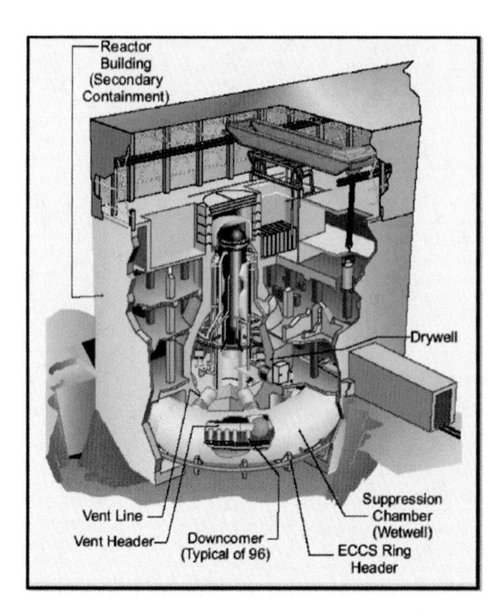

Chapter 10. This cutaway diagram shows the relationships among the dry well "inverted lightbulb," the wet well or "torus," and the reactor vessel, which stands upright inside the dry well. The refueling machine runs on rails in the building's top floor. It was this top floor, having thin walls, that was destroyed in the hydrogen explosions, and not the heavy concrete building that extends one floor underground.

Chapter 10. The Fukushima Daiichi power plant on the east coast of Japan after the earthquake in 2011. This picture was shot by an unmanned drone aircraft flying over the site on March 24, 2011. The reactor buildings of Units 3 and 4 are shown, dismantled by hydrogen gas explosions. The damage, which appears devastating, is not quite as bad as it looks. The top floors of the reactor buildings, which were built only to keep rain off the refueling equipment, have been blown away, but the solid concrete structures that hold the reactors and the fuel pools are all intact.

reactor, making a billion watts of heat, turning out 350 megawatts of electricity, and producing more plutonium than it burned. Lessons learned from Fermi 1 were applied to the design, including multiple coolant intake passages per fuel assembly to make inlet blockages "impossible."

Estimated cost of the plant was $400 million, with $256 million to be paid by private industry. Being built right in the middle of the Great Atomic Downturn in the mid-seventies, the project spun out of budgetary control and was plagued with contracting abuse charges, including bribery and fraud. By the time the Senate drove a stake through its heart in 1983, $8 billion had washed away, and plans for a commercial breeder economy in the United States went with it.[153]

The U.S. was not alone in an early quest for a sodium-cooled plutonium breeder. In 1964, the Soviet Union under Minister of Atomic Energy Yefim P. Slavsky began construction of what would become the world's first and only nuclear-heated desalination unit making more fuel than it used, the BN-350 power station. The site was two miles in from the shore of the Caspian Sea on the Mangyshlak Peninsula. The reactor was designed to run at 750 megawatts, driving five sodium loops at the same time with one spare. It was first started up in 1972, and although it was not able to make its designed power level, it ran for 26 years. For 22 of those years, it actually had an operating license, and the fact that it kept going for so long speaks well of the operating staff. They were an exceptionally tough bunch, eventually developing an immunity to frequent sodium fires and explosions.

Startup tests immediately found weld problems in the steam generators. Both the endcaps and the tubesheets would tend to break and mix water with sodium. In the first three years of operation, there were eight sodium explosions. For some odd reason, loop No. 4 never came apart. In 1974, after two major leaks and three small ones, it was decided to rebuild all the steam generators except No. 4. By February 1975, three loops had been fixed, and they decided to start up. Seven

153 Meanwhile, in 1975 the Shippingport Atomic Power Station in Pennsylvania, a pressurized water reactor based on a Westinghouse aircraft carrier engine, was converted into a thermal breeder using a thorium-uranium-233 fuel cycle. The Shippingport reactor had been running since May 26, 1958, and was the first commercial reactor making significant electrical power in the U.S. The reactor ran smooth and quiet for 25 years, and it never gave any trouble.

days later, loop No. 5 disintegrated. The steam generator in loop No. 5 was replaced with one made in Czechoslovakia.

As of December 16, 1991, the BN-350 was no longer in the Soviet Union. It was in the newly formed Republic of Kazakhstan, and it was re-named the Aktau Nuclear Power Plant. It continued to supply 120,000 cubic meters of fresh water for the city of Aktau way past its projected lifetime of 1993. BN-350 fissioned its last nucleus in 1999, due to a lack of funds to buy more fuel.[154]

Before the BN-350 was started up, the Soviet government was building an even larger sodium-cooled fast breeder reactor, the BN-600 in Zarechny, Sverdlovsk Oblast, Russia. It successfully started up in 1980, but a larger Russian sodium-cooled reactor means larger sodium explosions. As of 1997, there had been 27 sodium leaks, 14 of which caused serious fires, and the largest accident released over a ton of sodium. Fortunately, each steam generator is in its own blast-wall-protected cubicle, and any one can be reconstructed by the on-site workers while the reactor is running. That is one way to solve the problem of bad welds. A BN-800 and a BN-1600, still larger breeder reactors of the same type, are currently under construction in Russia.

We move on to France, a country firmly committed to a nuclear economy.

France saw the same writing on the wall that warned others about the limited supply of uranium, and in 1962 started construction of a modest-sized, experimental sodium-cooled fast breeder reactor named *Rapsodie*, in Cardache. It began operations on January 28, 1967, making 20 megawatts of heat. It ran okay, and in 1970 the core was redesigned and the power was increased to 40 megawatts. Under the stress of the higher output, the reactor vessel developed cracks, and it was reduced to 24 megawatts. By April 1983, the sodium leakage in *Rapsodie* was too costly to fix, and it was put to sleep. On March 31, 1994, a highly experienced, specialized,

154 Even a breeder reactor has to buy fuel. If the supply services had been operating as planned, the plant would exchange its spent fuel and breeding material for a fresh load of plutonium and depleted uranium. Plutonium would be extracted chemically from the spent load, and there would be enough extra in the breeding stock to more than pay for the chemical separation and fuel-rod fabrication. The BN-350 started out with an expensive load of 20% enriched uranium with some mixed uranium-plutonium surplus from the Soviet nuclear weapons being demilitarized. By 1999, the now-obsolete fuel-rods would have to be custom-built.

59-year-old CEA engineer was killed in an explosion while cleaning the sodium out of a tank at *Rapsodie*. Four people were injured.

In February 1968, after *Rapsodie* had run for a year, ground was broken in Marcoule for a bigger, 563-megawatt sodium-cooled breeder named Phénix. Ownership and construction costs were shared by the French Atomic Energy Commission (CEA) and *Electricité de France* (EDF), the government-owned electrical utility. It was big and ambitious. The starting core was 22,351 rods of pure plutonium, with 17,360 depleted uranium rods in the breeding blanket and an awesome 285,789 around the periphery for neutron reflection back into the blanket and radiation shielding. Its 250 megawatts of electrical power were connected to the grid on December 13, 1973, two months after the Organization of Petroleum Exporting Countries (OPEC) halted oil deliveries to countries that supported Israel and increased the market price of crude oil by a factor of four. The first sodium leak occurred in September 1974. In March and July there were two more, causing slow spontaneous combustion in pipe insulation.

The engineering innovation of *Phénix* was a free-standing or "free-flowering" core restraint, allowing it to expand or contract as it wished due to thermal, mechanical, or irradiation effects without bending or breaking anything. It ran perfectly for 16 years, as the spot price for uranium went from $6 per pound in 1973 to $40 per pound in 1976. Running on plutonium seemed a splendid idea. On July 11, 1976, a sodium fire broke out at the intermediate heat exchanger. Another one on October 5, and three more by the end of 1988. In May 1979, the fuel cladding failed, releasing radioactive xenon-135.

On August 6, 1989, something very odd happened to *Phénix*. A sudden very negative reactivity excursion triggered a scram. The reactor simply quit making neutrons, and the power level fell like a lead brick. The engineers had no idea why. The reactor restarted without a problem, but 18 days later it happened again. This time, the instruments were blamed for both negative excursions, but nothing wrong could be found. The ability to make electrical power dropped to near zero as the incidents continued.[155]

155 Subsequent investigations found that similar incidents had occurred in April 1976 and June 1978. They were explained away as "control rod slippage," which turned out to be wrong.

On September 14, 1989, the power again went to zero. The cause was believed to be a gas bubble in the core periphery. The problem was solved with mechanical maintenance, and *Phénix* was restarted in December. It happened again on September 9, 1990, and the gas-bubble hypothesis went out the window.

For the next 12 years, panels of experts pondered the problem, and the plant was tested, taken apart, put back together, repaired, modified, and refurbished, not generating any power to speak of. No one specific scenario or cause of the strange incidents was identified, but the only thing that made sense, given the speed and extent of the events, was that the core was in motion, moving in different directions as it generated power and disturbing the critical configuration of the reactor as designed. Lesson learned: avoid free-flowering core restraints in future reactor designs. In June 2003, *Phénix* was restarted and ran at reduced power, 130 megawatts, until final shutdown in 2009.

In 1974, a European fast-neutron reactor consortium, NERSA, was established by France, Germany, and Italy to build the biggest plutonium-fueled breeder reactor in the world. This extraordinary allegiance of countries lasted about a year. Germany spun off and decided to make their own breeder, the SNR-300 in Kalkar, and in the middle of 1976 President Valéry Giscard d'Estaing of France proclaimed that his country would build the *Superphénix*, a 3-gigawatt sodium-cooled breeder, at Creys-Malville, 45 kilometers east of Lyon. Italy stood still and did not make a sound.

Minister of Industry André Giraud announced to a spellbound crowd at the American Nuclear Society meeting in Washington, D.C., that there would be 540 *Superphénix*-sized breeder reactors in the world by the year 2000, and 20 of them would be in France. Meanwhile, 20,000 people occupied the building site to protest the thought of another reactor project. On July 31, 1977, the protest got serious, with 50,000 participants, and riot police went in armed with grenades. A local teacher, Vital Michalon, was killed, another protester lost a foot, and a third lost a hand in the battle. Ongoing protest and sabotage made construction work difficult, and on the night of January 18, 1982, militant Swiss eco-pacifists fired five rocket-propelled grenades at the containment building, causing cosmetic damage.[156]

Construction continued under bombardment, and the completed *Superphénix* went critical on September 7, 1985. It was connected to the grid on January 14, 1986. A series of administrative hurdles and incidents prevented any significant power production until March 8, 1987, when a massive liquid-sodium leak was discovered issuing forth from the refueling rotor tank (storage carousel). The reactor was down until April 1989 as an alternate refueling scheme was designed. The rotor tank was too far down in the guts of the reactor, and there was no way to repair the leak, a crack 24 inches long, without dismantling the entire plant.

Operation at very low power proceeded until July 1990, when a defective compressor was found to be blowing air into the liquid sodium line and making solid sodium oxide. In December the roof of the turbine hall collapsed in a heavy snowstorm. *Superphénix* seemed cursed.

The reactor was shut down for good by government decree on December 30, 1998. It had remained offline for two years because of technical difficulties and four and a half years for administrative debating. It never ran at its designed power level. *Superphénix* is scheduled to be completely dismantled by 2025. No new breeder reactor is planned in Europe.

The Germans' 327-megawatt SNR-300 sodium-cooled fast breeder reactor, costing a scant $4 billion to build, was completed, filled with sodium, and ready to be started up in 1985. Political hand-wringing kept it in standby condition for six years, costing about $6 million per month to keep the sodium liquefied using electric heaters. On March 21, 1991, the project was officially cancelled. The SNR-300 and the ground it was sitting on were sold at auction for half the monthly upkeep cost to a Dutch developer, who turned it into an amusement park named *Kernwasser Wunderland*.

---

156 The unnamed group of terrorists obtained a Soviet RPG-7 shoulder-fired rocket launcher and eight rockets ("bonbons") from the German "Red Army Faction" via the Belgian counterpart *Cellules Communistes Combattantes*. They lost three of the missiles in the dark. Chaïm Nissim, elected to the Swiss Geneva cantonal government for the Green Party in 1985, admitted 22 years later to leading the attack. He should stay out of France.

Of all the countries building sodium-cooled fast breeder reactors, India would be voted least likely to pull it off without blowing something to kingdom come. A list of sodium fires, explosions, and inexplicable power excursions in their Fast Breeder Test Reactor (FBTR) would be monotonous, but one accident stands out as unique. The reactor was first started up in October 1985. FBTR is a copy of the French *Rapsodie*, built at Kalpakkam using elephants for heavy lifting. Like every other fast breeder made since 1957, the blind fuel-handling machinery was based on the ingenious and complex Fermi 1 design.

In May 1987, a fuel assembly was being transferred from the reactor core to the radiation shield at the periphery. Each assembly held 217 fuel rods in a square metal matrix. For reasons unknown, one fuel assembly was sticking up one foot above the core as the handling arm swept across to find the one it was supposed to pick up. The electrical interlock that normally prevents the arm from moving if anything is in the way had been bypassed. Crunch. The arm rammed the protruding assembly, bending it out of shape, and then knocked the heads off 28 assemblies in the reflector as it tried to back away. Trying to make everything back like it was, the arm mechanism accidentally ejected an assembly in the reflector and put a 12.6-inch bend in a substantial guide tube. It took two years to sort out the damage and repair the core. Reasons for the accident were never really understood.

India is presently building two larger sodium-cooled fast breeder reactors. Be afraid.

The Japan Atomic Energy Commission published its first Long Term Plan in 1956, and at the top of the list of technologies to be developed were a sodium-cooled fast breeder and an associated fuel cycle using plutonium extraction. Of all countries in the game, Japan had the strongest incentive to build breeder reactors and become energy-independent. Domestic power options were few, and they had recently fought a war over energy resources and lost.

Two breeder reactor projects were started in parallel. The smaller unit, *Jōyō* (Eternal Sun), achieved criticality on April 24, 1977, developing 50 megawatts of heat energy. Located in Ōnari, Ibaraki, this reactor is now on its third core-loading, and the power has been increased to 140 megawatts. It has been used as a test-bed for fuel mixtures and materials for use in future breeder reactors.

The second breeder, *Monju*, was built in Tsuruga, Fukui Prefecture, and was first brought to criticality in April 1994.[157] It is less of a test reactor than *Jōyō*, producing 280 megawatts of electricity from 714 megawatts of heat. It has three double-loop core coolers, A, B, and C. The inside loop is full of radioactive sodium, and the outside loop, of course, has non-radioactive sodium, connected to the steam generator.

All was well with *Monju* until December 8, 1995. It was operating at 43% power when a smoke alarm went off near the hot-leg pipe for outside loop C, about where it exited the reactor vessel. It was 7:47 P.M. High-temperature alarms went off. The operating crew started a slow controlled shutdown 13 minutes later. It was beginning to look like a sodium leak, so at 9:20 P.M. they went ahead and hit the scram button. By 12:15 A.M. they had successfully drained the outside C-loop of sodium.

Turbulent flow in the sodium pipe had caused intense vibration, which broke off a sealed thermocouple well inside the pipe and bent the thermocouple 45°. Hot, liquefied sodium oozed through the now-opened thermocouple penetration in the pipe to the electrical connection and started dripping on the floor. Eventually, three tons of sodium collected in a clump on the concrete, and it caught fire, hot enough to warp and melt steel structures in the room. No one was hurt, and no radioactivity was released, but there was a lot of eternal shame.

The government-established Power Reactor and Nuclear Fuel Development Corporation (PNC) tried to cover up the accident, but when accounts broke free there was a public and political uproar and questions as to what else they had not divulged. A restart after repairs was delayed until May 8, 2010. Fuel was replaced, but on August 26, 2010, a 3.6-ton in-vessel transfer machine being hoisted into place slipped its bindings and did a free dive into the reactor vessel. It was

157 *Monju* is the Japanese word for Manjusri, a bodhisattva or enlightenment-being in *Mahāyāna* Buddhism. The Japanese tradition holds that *Monju* invented male homosexual love. What this has to do with a sodium-cooled fast breeder reactor, I'm not sure.

somewhat mangled and could not be retrieved from the opening through which it had fallen. A lot of engineering work by the Japan Atomic Energy Agency achieved removal on June 23, 2011. To this date, *Monja* has managed to generate power and put it on the grid for one hour. Another, larger fast breeder in Japan is planned, but plans have gone awry in recent years.

Given this interesting set of mini-disasters in which there were no injuries, one would have to consider the truth in Admiral Rickover's terse assessment of liquid-metal-cooled reactors. Take a step back for the longer gaze, and it starts to look like the most awful way to build a machine that has ever been designed. The people who championed and worked hard for this new and dangerous technology were visionaries. There was actually no particular need for an advanced power system that would last forever and free entire nations from a dependence on others. In the 1950s, there was enough burnable material, oil, coal, and gas, to go around, and enormous reserves of uranium were discovered throughout the decade. These individuals were forcing us into the future with all the speed we could handle, developing novel and outrageous concepts, materials, and machinery, to a place where mankind would eventually have no choice but to go. They were just too early. Was this wrong?

Next, we will explore an area in which the man-machine interface is pushed to the limit, and people die.[158]

158 I had to leave out the other participant in the sodium-cooled circus, Great Britain, not because they never had a coolant leak, but because the British incidents were too routine to stir the reader. There were two notable British fast reactors built at Dounreay, near the chilly, northernmost frontier of Scotland. There was the Dounreay Fast Reactor (DFR), which was nack-cooled, and the Prototype Fast Reactor (PFR), cooled with sodium. DFR was started up on November 14, 1959, and PFR was the last to shut down permanently in 1994. The U.K. push for a plutonium economy has since been declared dead.

Chapter 7

THE ATOMIC MAN AND LESSONS IN FUEL PROCESSING

"Expect to have a fire."

—the concluding sentence in AEC Accident and Fire Prevention issue no. 21, 28 October 1955, "Plutonium Fires"

AT 64 YEARS OF AGE, Harold McCluskey was hovering near retirement vintage, but he wanted to pass along his skills as a chemical operator to the dwindling supply of youngsters eager to enter the exciting life of a plutonium production engineer. He worked for the Atlantic Richfield Hanford Company of Richland, located in the middle of absolute nowhere in southeastern Washington State, and he was not in particularly good shape to be punching a clock at Hanford. McCluskey had recovered from a near-fatal heart attack two years before, and three years before that he had an aortic aneurism, treated with a prosthetic graft.

It was August 30, 1976, 2:45 A.M. on the graveyard shift when McCluskey entered the Americium Recovery Room, Building 242-Z of the Plutonium Finishing Plant. The operator on duty turned the recovery task over to him and left, and a few minutes later the junior chemical operator showed up to assist, observe, and learn.

Americium-241 is a transuranic byproduct of plutonium production. Rather than throw it away into the atomic landfill, Atlantic Richfield found that it was a valuable substance that could be sold and offset the cost of making fissile bomb material. This particular nuclide decays with a powerful alpha-particle emission, much like polonium-210, but unlike polonium with its short half-life, this species would outlast a Galapagos tortoise. Only half of it is gone after 433 years, and this makes it an attractive material for use in air-ionizing smoke detectors. Eventually every house in the country will have at least one smoke detector stuck to the ceiling, and that will require a lot of americium. It decays into neptunium-237 with a spray of gamma rays accompanying the alphas. The neptunium stays around for about two million years.

It is removed from fission waste products in very small batches, about 10 grams each, by dissolving it in 7-molar nitric acid and dribbling it down through a long cylindrical column filled with Dowex 50W-X8 ion-exchange resin beads. This process is observed, controlled, and adjusted using a large glove box, which allows the worker to insert his hands into two long-sleeved rubber gloves, bolted to the front of a metal workstation. The worker can manipulate objects in the glove box as if they were on a tabletop, yet he or she is completely isolated from the materials in the tightly sealed box, never directly touching anything. No radioactive contamination can get through the gloves or the box, and the worker is perfectly safe. He or she can see the work through the lead-glass window.

A general-purpose glove box is a simple affair, having a wide, tilted window on the front and two gloves at elbow level, but the americium recovery glove box was tall and rather strange. Each station in the line of americium boxes had not two but six glove positions, so that you could get hold of something at the top, the middle, or the bottom of the 6-foot ion-exchange column. There were seven windows. One long, thin window, right in front of the column, came down from the top, halfway to the floor. Five diamond-shaped windows were spaced all around, and a triangular window was at the top on the right side, all giving special viewing to certain parts of the chemical extraction process. Each window was made of laminated safety glass, to prevent shattering, covered with quarter-inch lead glass for gamma-ray protection. The inside of the box was dominated by the metal column, festooned with a confusion of tubes, valves, and conduits. There was a two-step metal ladder to help you see the top of the column and put your hands into the upper set of gloves. To work in the Americium Recovery Room, you had to love complexity and the fact that not just anybody could do this.

It was unfortunate that the americium recovery had been shut down for five months due to a labor dispute, and starting it up again was not going to be trivial, which is why an old-timer like McCluskey was on board. As it turned out, when the workers struck, the column was left in mid ion-exchange with acid on top of the resin and the drain valve at the bottom closed. In five months of sitting there, the resin beads had compacted into an unnatural configuration, saturated with americium.

McCluskey stood on the top ladder step and opened the valves on the column to get the process started, with the americium dissolved in acid dripping into the resin. Satisfied that it was started, he retraced his steps through a narrow corridor and back to the control panel desk, where the junior engineer was sitting.

He had just sat down, verbally downloading wisdom to his young associate, when the junior engineer interrupted the conversation. He heard something. Hissing. Like, a steam leak? McCluskey got up and went back to the recovery station, listening and following the sound. The entire glove box was filled with dense brown smoke. Uh oh. He shouted to the junior operator, telling him to call the control room up on the fourth floor and ask for help. Junior had just arrived at the glove box. He took one quick look and ran to the intercom back at the desk.

McCluskey climbed the stepladder and put his hands into the top gloves. They felt strangely warm. Had he forgotten to open the drain valve? He tried it. It was already open. He could not see the pressure gauge because the fumes were so dense, but now he could hear a new hiss, out the bottom of the column. He turned his face to the left and called out to junior, "It's gonna blow!"

WHAM tinkle tinkle. The operator in the control room heard it over the intercom as the resin column disintegrated in a heavy blast, and the junior operator turned around to see the cloud of debris make it through the corridor maze and to the desk. As soon as his ears cleared, he heard McCluskey's voice. "I can't see! I can't see!"

Junior ran to him and found him knocked to the floor, covered with blood, and the room socked in with americium fog. The windows in the glove box were blown all over the room, two gloves had turned inside out, and three gloves were simply gone. He tried to hold his breath as he rolled McCluskey over and helped him crawl back past the desk and to the outer door. Just then, the control-room operator, having heard and felt the explosion, was scrambling down the stairs to see what was going on, and he saw Junior and McCluskey near the door. He called back to the man following behind him, "A tank has blown up. Call the ambulance and shut down the plant as quick as you can." It was 2:55 A.M., only 10 minutes after McCluskey had clocked in.

At that instant, a health physicist, trained to monitor radiation and assist in any emergency and called "HP," ambled through the door and immediately perceived that an explosive radiation release had occurred. The potential for further contamination was obvious to him. He held up both hands and said, "Stay right there. I'll come and get you." He turned to the control room operator and told him to back off. No sense getting more people contaminated. He opened the emergency cabinet, pulled a respirator mask over his head and tossed one to Junior and one to the control room operator, telling them to put them on.

As Junior tried to adjust his mask so he could breathe, he heard HP's muffled voice saying, "We've got to get him under some water." McCluskey looked like he was about to faint. It had to sting like hell, particularly with the nitric acid in his eyes. HP and Junior took him to the emergency shower in the next room, but they hesitated to put him in it. The water would be ice-cold. They were afraid that with his well-known heart condition, the temperature shock could kill him. They stripped off his clothes, took him to the sink on the opposite wall, and sat him down on a stool. HP wiped McCluskey's face with wet rags while Junior scrounged more rags and some soap.

They tried to keep McCluskey conscious and talking. The glove box had blown up, he said. The last thing he saw was a blue-white flame. HP knew that no criticality alarm had gone off, so it was not a rogue chain reaction with plutonium, as was constantly the concern. It had to have been a chemical reaction, but unfortunately it was a big one. McCluskey's face and neck, particularly the right side, were perforated with bits of glass, heavily contaminated with radioactive americium-241. His eyes were swollen shut. The right one looked particularly bad, with "black stuff" around it and his right ear, and there was a cut above on his forehead.

At 3:00 A.M., the ambulance arrived and a nurse, fully decked out in radiation protection clothing, took over. Workers were already putting down plastic sheets on the floor leading to the outer door. The physician-on-call arrived, intending to treat McCluskey for radiation poisoning on the spot, but as he examined the patient it became obvious that a great deal more attention would be required. He and two nurses cleaned him of any obvious contamination, loaded him into the ambulance, and pulled out with full siren and flashing lights at 4:37 A.M. It was 25 miles to the hospital in Richland. The Geiger counters on board were tapping out gamma-ray hits as fast as possible, sounding like radio static. There was no way to estimate the extent of his americium contamination, because the radiation instruments ran off scale when held to his face or his neck. McCluskey was hot as a pistol. It had to be somewhere between 1 and 5 curies. There had been 17 previous incidents of human americium contamination at Hanford, starting in 1956, but those were all microcuries or nanocuries, a billion times less activity. This was new, unexplored territory.

It was similar to the contamination that the three workers had received at SL-1 back in 1961 in Idaho. They had been standing on the reactor vessel when it steam-exploded, and fuel and fission products, reduced to a fine aerosol, had been driven deeply and irretrievably into their bodies by proximity to the blast. McCluskey's condition was similar, inoculated by the americium-241. The main difference was that he had survived the explosion. The men at SL-1 had died instantly of the mechanical trauma, before the radiation could have any effect. McCluskey would have to be decontaminated, as if he were a truck sandblasted by an above-ground weapon test, only much more gently.

He was lucky. In 1967 the AEC had paid for an extensive Emergency Decontamination Facility (EDF) to be built at the Richland Hospital, all in response to the SL-1 incident. Everything was made to minimize radiation exposure to the health professionals from a heavily radioactive patient, while preventing a spread of the contamination by radioactive fluids, dust, or gases. An electrically operated hoist and monorail system was in place to move the patient from the ambulance to the operating room, which was equipped with an operating table shielded on all sides with lead. Heavy concrete walls and labyrinthine entrance halls stopped gamma rays from dosing anyone not attending the patient, with movable sheets of lead hanging from the monorail to shield those who were. Lead holding tanks were used for all waste collection, and a two-stage HEPA-filtered exhaust blower took care of the air.[159] Closed-circuit television cameras were used for remote viewing, and there was even a room dedicated to medical-equipment decontamination. There were long-term sleeping quarters for contaminated patients. It was not exactly homey, but to have it in place was good planning. HP had called ahead at 3:08 A.M. to activate the facility.

The ambulance arrived at the EDF at 5:14 A.M., and treatment for McCluskey's americium contamination began promptly. Bits of glass, metal, resin, acid, and plastic, mostly too small to see and all covered with americium, were embedded in his right face and shoulder.

159 HEPA stands for High Efficiency Particulate Air (or "Arresting"). It was developed for the Manhattan Project during World War II for preventing the spread of airborne radioactive contaminants, and it has become a set of industry standards, a trademark, and a generic term for the best air filters available. It is now used in aerospace, pharmaceutical plants, hospitals, computer chip manufacturing, and all nuclear industries. By specification it must remove 99.97% of all particles larger than 0.3 micrometers from the air that passes through it.

The danger was from it leaking into his bloodstream, which would take it to his bones and liver. That would kill him if left untreated. Fortunately there was a medicine available that would chemically capture any americium in the blood stream and excrete it through the kidneys. This treatment is called chelation therapy, and it had been extensively tested using beagle dogs. The drug, calcium diethylene-triaminepentaacetate or Ca-DTPA, was quite efficient at cleansing the bloodstream of americium, but in large doses it was toxic. The calcium would displace zinc in the delicate human metabolism. There was Zn-DTPA made to eliminate this problem, but there were very limited supplies of it and it had yet to achieve FDA approval. McCluskey would require megadoses of unprecedented size. For the first five days he would receive large doses of Ca-DTPA with zinc sulfide tablets to counteract the zinc depletion. He was immediately given a gram of Ca-DTPA through a 21-gauge needle, and the gate to the road of recovery swung open.

It would be a long journey, but McCluskey was in the most capable hands. By the end of the first day of treatment, the thorough cleanup and chelation start had reduced his level of contamination to 6 millicuries, or reduction by a factor of 1,000. By day five, chemists at the Pacific Northwest Lab in Richland had produced some Zn-DTPA, and permission to use it was granted by the FDA over the phone. X-rays revealed a galaxy of tiny pieces of debris embedded in his face and neck, but there was no way to go after them with surgery. You cannot remove what is too small to see and pick out with tweezers. On day nine McCluskey was up and walking, and on day 12 he was able to walk around outside. He was taking a big dose of Zn-DTPA twice a day, and his only complaint was eye irritation from the acid burns. On day 22 a splinter was removed from his right cornea, and that helped.

By day 45 he was becoming depressed with the living conditions in the long-term sleeping quarters, and plans were devised to put him in a house-trailer adjacent to the building. The physicians did not feel that they could send him home yet, because highly radioactive debris particles were gradually working their way to the surface in his face and were being left on his pillow at night. Dealing with that level of contamination spread would be a problem if he were at home. McCluskey, his wife, and his dog moved into their new mobile home on day 79. On day 103 he was allowed a trip by automobile to his home in Prosser, Washington, 30 miles away, where he was pleased with a six-hour stay. His contamination level was falling steadily with chelation, and there were no measurable health effects from his continuous and extremely close bombardment with alpha and gamma radiation. On day 125 he attended church, and here the story of Harold McCluskey took an interesting turn.

All aspects of McCluskey's injuries had been treated with appropriate care, including for the first time the psychological effects of being in an industrial accident and being a mobile, talking piece of radioactive contamination. Everywhere he went, he carried millicuries of unshielded radioisotope with him. To carry that much radiation in a lead bucket, you have to have a federal license. He harbored a deep concern of contaminating his home, his loved ones, and everything he touched. The psychologists worked on this, assuring him that he would do no harm, but his friends and fellow church members were not so certain. They were overjoyed that he had survived and that he was back in the world with them, but they did not want to get too close to him. They would rather wave to him at a distance and move on. He heard it over and over: "Harold, I like you, but I can never come to your house." Even though they lived in proximity to a sprawling plant that manufactured plutonium-239 by the ton and had been steeped in the atomic culture all their lives, they were terrified of the fact that he was radioactive. McCluskey had become the Atomic Man. He felt shunned.[160]

His growing despair was somewhat lifted when his church pastor gave an impassioned sermon on his behalf, convincing people that it was both Christian and physically safe to be around Harold

160 To be fair, I must point out that wherever he went, McCluskey carried a radiation counter with a speaker to broadcast the amplified sound of gamma rays crashing through the detector tube. If he held it up to his face, it made quite a noise. He thought that doing this would show that although he was still heavily contaminated, he was a normal, healthy human being with whom you could shake hands without dropping dead from radiation poisoning. This message did not get through as he may have hoped. The fear of anything radioactive, even a family friend, still runs deep in the civilized, educated world, and the buzzing Geiger counter could spook a horse.

McCluskey. The senior physician in charge of his case at Hanford, Dr. Bryce D. Breitenstein, gave several lectures concerning his now-famous and unprecedented case of human contamination, with McCluskey along as the actual specimen.

On day 150 of treatment, he was able to return home permanently, visiting the EDF at least twice a week for continued chelation treatment. On day 885 he was taken off Zn-DTPA, and on day 1,115 a slowly decreasing platelet count was noticed, probably due to the continuous radiation exposure. He was put back onto Zn-DTPA on day 1,254, and on day 1,596 his blood platelets were still trending low, but by that time he had bigger health problems. McCluskey died of coronary artery disease, not attributable to radiation, on August 17, 1987. He was 75 years old, and to the last he was in favor of all things nuclear, including nuclear power. He had insisted to all who would get near him that his injuries were a result of an industrial accident and nothing more.

The Atomic Man incident is a representation of the nuclear industry in the late 20th century. It was, even into the 1990s, obsessed with building nuclear weapons. This mission, protected from deep public scrutiny by the often-cited need for national security, seemed to be given priority above any peaceful application of nuclear energy release, and the impression given to the general public continued to erode and distort individual beliefs about the dangerous and the not-dangerous aspects of the industry. It shows that the plutonium production plant in Hanford was better equipped than one might have thought to deal with extreme accidents involving radiation. It also shows that many improvements had been made to the national labs involved in defense nuclear work after the SL-1 explosion, which was a definite wakeup call. McCluskey's medical treatment was the most advanced state-of-the-art and was expertly applied.

Although it seems illogical, McCluskey's level of lingering contamination was higher than that of any one person on Earth, yet he was not as affected by it as he was by nitric acid burns on his corneas and clogged arteries in his heart. The nuclear production plants were run with reasonable industrial safety measures in place, but nothing was foolproof, and there were plenty of figurative land mines to step on. Was it any more safe to work there than in a peanut butter factory?

The Hanford Plant made raw plutonium-239, a fissile nuclear fuel, delivered in roughly cast "buttons," about the size of hockey pucks. It was up to other plants in other states to make it into something useful, and it was always shipped to them in very small batches. Care was taken at every step not to let too much of it bunch up and become an impromptu nuclear reactor, enthusiastically making a great deal of heat and flesh-withering radiation. The next stop in making it into weapons was the Rocky Flats Plutonium Component Fabrication Plant, where it would be formed into shiny, barely subcritical spheres.[161]

Rocky Flats was a flat mesa covered with rocks, devoid of trees, about 15 miles northwest of Denver, Colorado, bought by Henry Church for $1.25 an acre back in 1869. It was good for grazing cows if you spread them out. Then came World War II, which the United States brought to an end with its new and unique weapons, catapulting technology abruptly forward. Shortly after came the Korean Civil Conflict in 1950, and the United States, weary of war, found itself trying to prevent North Korea from invading South Korea.

President Harry S Truman sought to gather his options. "If we wanted to drop atomic bombs on somebody, how many do we have in stockpile?" he asked.

"Well," he was told, "at Los Alamos if we use all the parts that are lying around, we can probably put together two of them."

President Truman found this answer disturbing. The other nations are looking to us as the benevolent, all-powerful force, able to crush any aggression with a single, white-hot fireball, and we don't have atomic bombs piled up in a warehouse somewhere? And so began the extended bomb crisis, soon becoming the H-bomb development scramble. The AEC commenced Project Apple to build a special factory to produce the core or "pit" for plutonium-fueled

161 Plutonium is a dull-looking metal that quickly corrodes in atmosphere, so the two hemispheres used to make a bomb core were coated with something to keep the air out. The most-used coating was nickel plating, which gave the finely machined parts an attractive metallic shine. You did not want to scratch the plating, as doing so would result in heavy white smoke as the plutonium caught fire.

implosion weapons and hydrogen bomb triggers, operating under enforced secrecy. This would take the strain off the Los Alamos Lab so that it could devote most of its effort to improving bomb designs. Dow Chemical of Michigan was awarded the contract, and Senator Edwin "Big Ed" Johnson of Colorado pushed really hard for his state to be the site of this new venture.[162]

The U.S. Army Corps of Engineers, real-estate division, acquired the 2,560 acres of Rocky Flats from Marcus Church by the 5th Amendment of the Constitution and took possession on July 10, 1951. They offered $15 an acre, but the purchase price bounced around in the courts for a couple of decades. In the meantime, bulldozers gouged out the foundation of Building D, and construction proceeded at a rapid pace. Building D was for final bomb-core assembly using parts made of plutonium, uranium, and stainless steel built in other buildings. It was eventually renamed Building 991. Building C, or 771, was where the plutonium parts were made.

Everybody who worked at or on the plant, more than 1,000 people, had to have a Q clearance from the AEC, requiring extensive background checks of each person, his or her relatives, and everyone he or she knew. A guard shack was built at the entrance to the property. Three layers of barbed-wire-topped fencing and concertina wire were installed, and security holes were closed. By 1953 the plant was tuned up for full bomb production with 15 shielded, windowless buildings. By 1957, there would be 67 buildings on the site, with only its general mission known to those

162 Big Ed (six foot two) was one of nine members of the Joint Committee on Atomic Energy (JCAE) as well as senior member of the Senate Military Affairs Committee. He and the other senator from Colorado, Eugene Millikan, were able to divert some big, important projects to their state, including the North American Air Defense Command (NORAD, an A-bomb-proof headquarters in Cheyenne Mountain), the United States Air Force Academy in Colorado Springs, and a lot of uranium mining. The Rocky Flats plant was their crowning achievement. Big Ed eventually made a disastrous slip of the tongue in a live television show, "Court of Current Issues," in New York on November 1, 1949. Johnson casually mentioned that the United States was developing the hydrogen bomb, which would be 1,000 times more powerful than what we had dropped on Japan. Television being what it was in 1949, it took a while for this incredible announcement to sink in, but the *Washington Post* took hold of it on November 18. The project was now subjected to the blinding glare of public opinion, and there would be no more quiet examination of the issues. President Truman was not pleased, and he wrote a one-sentence "Dear Ed" letter to Senator Johnson on December 17.

who did not work there. The Rocky Flats facility was thus a prime example of a Cold War battlefield, where front-line fighting was done in locked rooms, paranoia was actively encouraged, and not a shot was ever fired.

Plutonium is an inherently dangerous material to work with, but there are worse, more radioactive substances. The main problem with plutonium or any of the transuranic elements such as uranium or neptunium is its pyrophoric tendency, or the enthusiasm with which it oxidizes. It is similar to wood, in that a fresh-cut log will burn, but ignition is not necessarily easy. You can waste a lot of matches trying to set fire to a log, even though you know it will burn. It is too massive to heat to combustion temperature easily, and the surface area, where burning takes place, is tiny compared to the volume of the heavy piece of wood. If you really want to set fire to it, then carve it into slivers with a knife. Each thin slice of wood is all surface area, without much mass that has to be heated up. Strike a match to a big pile of shavings, fluffed up and full of oxygen-bearing air, and it will burn like gasoline. The same principle applies to metallic plutonium. A billet of it weighing over a pound (under critical mass!) will sit there in air and smolder. Work it down in a lathe, peeling off a pile of curly shavings, and you have made a fire. No match is necessary. Machining and close work with plutonium components must be done in an inert atmosphere, such as argon.

The problem with a plutonium fire is putting it out. Exposure to the usual extinguisher substances, such as water, carbon dioxide, foam, soda-acid, carbon tetrachloride, or dry chemical, can cause an explosion as the extreme chemical reduction scavenges oxygen or chlorine wherever the plutonium can find it. The predominant nuclide, Pu-239, is an alpha-gamma emitter, but it radiates at a slow, 24,000-year half-life, and it is not difficult to shield workers from the radiation. The smoke from burning plutonium is, however, another matter. If you breathe it, the tiny, alpha-active vapor particles lodge in your lungs, and subsequent cancer by genetic scrambling is practically unavoidable. However, the smoke is extremely heavy, and it drops to the ground quickly. It is not something that will rise into the air, drift with a breeze, and contaminate a large city 15 miles away, or even an adjacent building. It travels beyond the burn site on the bottoms of your shoes.

One cannot work in an argon atmosphere for very long without passing out for lack of oxygen, so the plutonium workpieces and the workers must be insulated from each other. At Rocky Flats in Building 771, long lines of stainless steel glove boxes, raised three feet off the ground, were welded together. A worker standing in front of a glove box would insert his hands in the gloves and perform whatever fabrication task was assigned to that position in the line. The small, always subcritical plutonium units were carried on a continuous conveyor belt, made of small platforms linked together. A plutonium thing would move down the line, from glove box to glove box filled with inert argon gas, and the workers could perform close, precision work using the touch-sensitive gloves, looking through Plexiglas windows, with practically zero exposure to radiation. The conveyor could then turn 180 degrees and continue down the next line in the room, with the plutonium object finally being removed from the line once the nickel plating had been applied.

The room was so crowded with continuous lines of glove boxes, the only way to get from one side of the room to the other was to go under the boxes. For this purpose, each line had a sloping valley dug out under it in the middle, where a person could stoop over slightly and get under the boxes without having to crawl.[163] The air in each assembly room was kept at below atmospheric pressure using blowers and racks of expensive HEPA filters. Air could therefore not leak out of the building and carry any plutonium oxide dust with it. Air could only leak into the building from outside, through imperfections in the airtight structure or when a door was opened. The world outside the building and certainly beyond the fence line was thus protected from plutonium contamination. Of even greater importance, no enemy could tell what was going on inside the buildings or tell how much was going on by reading the radiation signature at a distance. There was no radiation signature.

163 The workers named the underpass ravine "sheep dip." There was no drain at the lowest point in the dip, and anything dropped on the floor would eventually end up there. Visiting physicists renamed this feature the "Lamb dip," having to do with the spectral hole in the HeNe laser cavity at 1.1 microns. The humor went over most heads.

Plutonium was shipped to Rocky Flats from Hanford as liquid plutonium nitrate in small stainless steel flasks. Each flask, which held seven ounces of plutonium, was purposely isolated from all other flasks by putting each in the center of a cylinder the size of a truck tire. In Building 771, the flasks were emptied using a tube connected to a vacuum pump, and the liquid was transferred to a tall glass cylinder in a special glove box that looked very similar to the one that blew up at Hanford. Hydrogen peroxide was added to the plutonium nitrate, and a solid, plutonium peroxide or "green cake," precipitated to the bottom. Moving through the glove-box line, the solid material was washed with alcohol, desiccated using a hair dryer, and pressed into 1.1-pound biscuits.

Continuing down the line, the biscuits were sent down the "chem line" to the G furnace, were they were baked into plutonium dioxide and mixed with hydrogen fluoride to make "pink cake," or plutonium tetrafluoride. On they went on the conveyor to another furnace, where the pink biscuits were reduced to 10.5-ounce buttons of pure plutonium metal. In a day, Building 771 could produce 26.4 pounds of plutonium.[164] This raw material was then transferred to the fabrication line, where it was cast and machined into bomb parts, touched only with rubber gloves.

In 1955, a radical design change for atomic bomb cores came up from Los Alamos. Instead of a solid sphere of plutonium with a small cutout at the center for a modulated neutron source, the new cores would be thin, hollow spheres of alternating uranium-235 and plutonium-239. This design change would result in a lighter bomb that could be lobbed by a compact missile aboard a submarine, a ground-launched anti-aircraft missile, or an air-to-air missile on a fighter plane. It could also be boosted by injecting into the empty space in the core a mixture of deuterium and tritium gases, giving the atomic bomb a kick from hydrogen fusion. The management at Rocky Flats was delighted by the news and said that it should not

164 This was about enough to make two bombs. The old Mark I "Fat Man" used 13.6 pounds of plutonium in the core, but the newer designs used less plutonium, and some had U-235 components as well, reducing the amount of needed Pu-239, increasing the bomb yield, and reducing the weight.

take more than two years of construction and $21 million to make the necessary improvements to the physical plant.

The improvement schedule was fine, but in the meantime Rocky Flats would have to accommodate the new core design with whatever they could put together quickly. The timing was critical, because the Soviets seemed to be improving their arsenal at an alarming rate, and it was critically important to stay ahead of whatever they were doing. The new hollow core would require a lot more complicated machining and casting. There would be many times more machine cuttings and shavings to be recycled back, and larger lathes would be necessary. The fabrication room in Building 771 became extremely crowded as new box-lines were installed. It was hard to walk in the modified building with all the new glove boxes in the way, and the new lathe boxes were made of Plexiglas on all sides, instead of being a metal box with a Plexiglas window. The machinists had to be able to see the work from any angle, and to make a box out of leaded glass was too expensive. The Plexiglas could catch fire easily, and it was against AEC policy to use it in a plutonium line. The policy was waived under these emergency conditions, as it had been when the windows were installed in the original glove boxes.

With the time-critical, crowded work now at the factory, accidents were inevitable. It was June 1957, and the facility was still dealing with several times more plutonium than it was designed for.

The first problem occurred in the glass column used to mix plutonium nitrate with hydrogen peroxide. When you do too much of it too fast, oxygen builds up and pressurizes the column. One day without warning it exploded, blowing the side off the tall, metal glove box and wetting down two workers with plutonium nitrate. It was similar to the Atomic Man incident that happened two decades later at Hanford, but it was minor compared to the next accident. On September 11 the fabrication space, Room 180, in Building 771 caught fire. It was the result of production stress, exactly a month before the plutonium-conversion reactor at Windscale, England, caught fire for basically the same reason. Resolution of the Rocky Flats fire would be eerily similar to that of the Windscale disaster.

A box in the middle of the room was filled with leftover pieces of plutonium and machine shavings, contained in six steel cans and amounting to 22 pounds of fissile, flammable material. In one can was the result of a casting operation. The plutonium metal had been melted as usual in a hemispherical cast-iron crucible, shaped like a punch ladle. To make a plutonium casting, you melt it over a flame in the ladle, holding it by the long handle with the glove. When the metal goes liquid, you carefully pour it into the mold and let it cool and become solid. There is a thin film of plutonium left in the ladle. You peel it off and put it in a can for recycling. It is very thin and perfect for starting a fire. It is called a "skull."

Just on its own, the skull caught fire at about 10:00 P.M. Plutonium burns with a brilliant white heat. Soon the all-Plexiglas glove box was on fire, and Room 180 was filled with smoke. Two security guards discovered it, seeing flames out of the glove boxes reaching for the ceiling. One went for a CO_2 fire extinguisher, and the other called the Rocky Flats fire department.

By 10:15 P.M., the firemen under Verle "Lefty" Eminger were unreeling hose and going in, but Bob Vandegriff, the production supervisor, and Bruce Owen, the night radiation monitor, advised that no water be used. There were 137.5 pounds of plutonium in the room. Water covering it would amount to a neutron moderator, and the fissile plutonium could go critical and become a problem even larger than a raging conflagration. The two men quickly dressed down in Chemox breathing apparatus and went in with CO_2 extinguishers, which they quickly emptied into the mass of flames. Nothing happened. They retreated, ran down the hall, and found the large fire extinguisher cart, dragged it into the room, which was now fully engaged, and opened the release valve. The firemen watched as the massive blast of carbon dioxide filled the room and engulfed the fire. It had absolutely no effect on the flames. Where was all the fresh air coming from to feed the blaze?

The 771 supervisor, Bud Venable, had been called at home at 10:23 P.M. and told that his building was on fire. Venable worried that the men going into Room 180 would suffocate or pass out in the heat. He ordered that the fans be turned up to the highest speed.

There were four huge exhaust fans up on the second floor, designed to keep the first floor at negative pressure. They blew the air into a concrete tunnel and up a smokestack, 142 feet high. To keep any plutonium dust particles from making their way into the atmosphere, the air from the first floor passed through a massive bank of HEPA filters on the second floor before exiting the building. The filter bank ran the entire length of the floor, over 200 feet long, made of 620 paper filters. Each filter was a foot thick and presented four square feet to the airflow. The fans, pulling 200,000 cubic feet of air per minute at high speed, sucked fresh air into the blazing fire and sent the flames up to the second floor, where by 10:28 P.M. they set fire to the paper filter elements. As was exactly the case with Windscale, thoughtful consideration for the men who were fighting the fire was keeping it going and spreading it.

The blowers should have automatically shut down by now because of the heat buildup in the filter bank, but the fire-detection equipment had been disabled earlier because it was always going off and disrupting production. Fire was obviously being sucked into the air vents in the corner of Room 180. Owen was still apprehensive about causing a criticality in the room, but by now Vandegriff was willing to allow the fire department to hose it down with water. He suggested fog nozzles on the inch-and-a-half hose that the firemen had already laid, and told them not to aim at the glove boxes. The flames went down almost immediately.

Eminger and Vandegriff rushed to the second floor to see if the filters were on fire. The filters had not been changed in four years, and there was a large buildup of finely divided plutonium detritus from all the machining. Just as they opened the door, the dust exploded, knocking Vandegriff to the floor and Eminger back through the double doors. The filtering function was destroyed, and plutonium dust was forced to where it would not go if left on its own, up the stack and into the air over Colorado, by the four powerful blowers. The blast was intense enough to dislodge the lead cap on top of the smokestack. By 10:40 P.M., the remains of the filter bank were engulfed in flame, and at 11:10 P.M. the electrical power failed, finally turning the blowers off. By 11:28 the fire was officially extinguished.

Building 771 was back in business by December 1, 1957, but Room 180 would not be decontaminated until April 1960. The cleanup and replacement of the filter bank cost $818,000. Of the plutonium known to be in the building, 18.3 pounds of it could not be found.[165]

That was the small fire. The big one started in connected Buildings 776-777 on Sunday, May 11, 1969. Mother's Day.

Small, controllable plutonium fires had become a way of life at Rocky Flats. The fire trucks had been called to hundreds of fires since Building 771 was smoked back in '57. The worst small fire broke out in 1966, when workers were trying to unclog a drain. In that barely mentionable incident 400 people were contaminated, with most of them inhaling plutonium smoke. Countless fires were smothered in the day-to-day work without calling the fire department or mentioning it to superiors.

Plutonium is a very strange element, and some of its characteristics are not understood. It has seven allotopes, each with a different crystal structure, density, and internal energy, and it can switch from one state to another very quickly, depending on temperature, pressure, or surrounding chemistry. This makes a billet of plutonium difficult to machine, as the simple act of peeling off shavings in a lathe can cause an allotropic change as it sits clamped in the chuck. Its machining characteristic can shift from that of cast iron to that of polyethylene, and at the same time its size can change. You can safely hold a billet in the palm of your hand, but only if its mass and even more importantly its shape does not encourage it to start fissioning at an exponentially increasing rate. The inert blob of metal can become deadly just because you picked it up, using the hydrogen in the structure of your hand as a moderator and reflecting thermalized neutrons back into it and making it go supercritical. The ignition temperature of plutonium has never been established. In some form, it can burst into white-hot flame sitting in a freezer.

Unfortunately, the frequent plutonium fires did not make everyone wary of this bad-behaving material. The effect was just

165 Dow Chemical's report of this accident was classified SECRET until 1993. The investigation found that somewhere between 1.8 ounces and 1.1 pounds (50 to 500 grams) of plutonium made it up the smokestack and landed somewhere in Colorado. No trace of it has ever shown up in radiation surveys of the surrounding land. Note that the safety exposure limit for a worker at Rocky Flats was 0.0000005 grams of plutonium.

the opposite. Everyone in the plant, starting with the top management, became convinced, at least subconsciously, that a plutonium fire was easy to control and was no big problem. Starting in 1965, there were more than 7,000 pounds of plutonium in Building 776-777 at a given time, and it was not given a second thought. There was a list of dangerous procedures, equipment configurations, and building materials at Rocky Flats that converged on Mother's Day 1969.

First, there was the Benelex. It was a type of synthetic wallboard, no longer made, that resembled Masonite. It had a high density, it could be put together with glue, and unfortunately it was flammable. It was used to make boxes to hold plutonium, shielding for entire glove-box lines, and even the walls of fabrication rooms. Compared to other versatile materials, it was inexpensive. In 1968, 1.17 million pounds of flammable Benelex and Plexiglas were added to Building 776-777.[166]

The walls and the glove boxes were flammable, and to top that the hangers on the overhead conveyor belt, used to hook heavy parts and take them to another building, were made of magnesium. Magnesium cut down on the weight, but if it caught fire, it would burn white-hot, like a flare.

Another problem was the cleanup of the machine tools, which made a lot of plutonium chips, scraps, and even dust. These remnants were always oily from the coolant that was sprayed on the plutonium parts as they were shaved down into shape on the lathes

166 Why Benelex? It turns out, Benelex is an excellent neutron shield, and in the 1960s it was used extensively in nuclear research for shielding neutron collimators and interferometers. A component of Benelex is wood fibers, and harvested wood in the U.S. grew up soaked with borax, a wood preservative and insect-damage preventative. Plutonium-239 can spontaneously fission on occasion and send neutrons careering through the room, and the contaminant plutonium-240 is particularly apt to do this. When hit by neutrons of any speed, these particles are slowed to thermal speed by the hydrogen in the cellulose wood fibers and are summarily absorbed by the boron-10 remnants from the borax treatment. The classified purpose of using Benelex was to kill any neutron activity in the building, preventing the plutonium pieces in the building from cross-connecting by neutron flight and causing the building to become one enormous nuclear reactor, running uncontrolled with people inside it. On Mother's Day 1969 there were 7,641 pounds of plutonium in the building. The first power reactors in the world, the plutonium production reactors at the Hanford Site, used Masonite, a similar material, for neutron shielding beginning in 1944.

and milling machine. When this stuff started building up under the machine, it was gathered, put in a can with a lid, and sent by conveyer to Room 134. Here it was supposed to be degreased using carbon tetrachloride and then pressed into a rough briquette, three inches in diameter and an inch thick, weighing about 3.3 pounds. The degreasing was unofficially dropped from the procedure, as the carbon tetrachloride treatment would too often result in a fire or an annoying explosion. When the press squeezed down on the non-degreased plutonium debris, oil flowed onto the floor, taking with it little pieces of plutonium. Workers sopped it up off the floor using rags.

On that Mother's Day in '69, late in the morning, a heap of oily, plutonium-enriched rags beneath the briquette press spontaneously caught fire. There was nobody at work on the plutonium line that day. The Benelex glove box above the burning rags had a ventilation fan, feeding air into the big filters on the second floor. It pulled the hot air from the fire into the box, where there was a can holding a briquette. Someone had neglected to put the lid on the can. The plutonium briquette caught fire, and it burned white hot. The Benelex glove box started smoldering. It lighted some more briquettes. At this point, the fire alarms should have been blaring and automatically calling the fire department, but the detection equipment had been removed to make room for all the new Benelex shielding. More plutonium ignited. The Plexiglass windows and the rubber gloves caught fire, flaming up, and this left the arm-holes open. Air rushed into the glove boxes and fanned the burning plutonium. The fire moved down the north-south conveyer line, away from the connected building 777, taking everything that would burn.

At 2:27 P.M., the heat sensors in the building triggered, alarms sounded, and the fire department rushed to the scene. The firemen found the north plutonium foundry in Building 776 fully engaged. Captain Wayne Jesser ordered a man to discharge a hand-held carbon dioxide extinguisher at the fire to try to scare it while he rolled a fifty-pound extinguisher to the east end and emptied it into the flames. The fire was not impressed. Aware of all the dangers of using water on a plutonium fire, the risk of a hydrogen explosion from oxygen being pulled out of the water and a possible criticality from the moderation effect, he could see no choice. At 2:34 P.M., he ordered his men to deploy the fire hoses and wet it down.

It was not your average industrial fire. In the black smoke the plutonium and the magnesium were burning furiously, and the room was lighted up like a new shopping center opening, even as the fluorescent lights melted and came crashing down on the firemen. Molten lead from the gamma-ray shielding was hitting the floor in globs, and even the glue used to hold the Benelex together was on fire. The Styrofoam in the roof started melting, and the tar on top was getting soft. The powerful blowers were pulling air through the wall on the second floor, just as they were supposed to do, and the filters were already burning. All seemed lost when, in a miraculous stroke of good fortune, a fireman accidentally backed a truck into a power pole and killed the electricity to the building. The fans squeaked to a stop.

Firefighters in full radiation gear were now able to enter the room without being incinerated by the air-fed flames.[167] They came into the crowded space spraying water, and when they moved from line to line, they had to go under, using the sheep dips. The ravines were now filled with water, so it was like wading through a creek in a space suit, up to your chest. The firemen noticed that the oxidized plutonium powder stuck together when wet, like gray Play-Doh. They tried to corral it into a corner using the high-pressure hoses. Fortunately it was too sticky, and it clung to the floor. If they had been able to push it with the hoses as they desired, it would have assumed a critical shape and killed them all with an unshielded supercriticality.

By 8:00 P.M., the fire was declared not burning, but the wet plutonium was still flaring up as late as Monday morning. The AEC investigated the cause of the accident, and its report, criticizing both itself and Dow Chemical for allowing obvious safety lapses,

167 The danger from radiation in most situations aside from criticality is that broken skin, lungs, or the gastro-intestinal system can be contaminated with radioactive dust. You do not want to be inoculated with a long-term radioactive nuclide in small quantities. Wading into a mixed radiation field for a short time is not really what causes health problems, but having dust decay inside you for a long time is. For this reason, a radiation protection suit seals you up against any interaction with the environment. An airtight coverall with tape-sealed gloves and booties is standard, along with self-contained breathing apparatus and a full head-covering. It is not pleasant. You cannot scratch your nose.

was classified SECRET. With $70.7 million in damages, it broke the record for industrial loss in a fire in the United States. The report from the fire in 1957 had apparently not been read, possibly because it too was SECRET, and no lessons had been learned and applied to preventing further plutonium fires. There were no injuries, and no wind-borne contamination of the surrounding area was detected. Connected Buildings 776-777 were turned into a large parking lot.

By 1992, the mission of Rocky Flats, to fabricate nuclear weapon components, had completely dried up. The Cold War was over, and the last remaining fabrication job, making cores for the W88 "Trident II" submarine missiles, reached the end-of-contract. Of the 8,500 workers, 4,500 were laid off permanently and 4,000 were kept on for cleanup. In 2001, the Rocky Flats National Wildlife Refuge Act of Congress turned the bomb factory site into a nature park. It was amazing, but in three decades of difficult work at Rocky Flats using a great deal of a very dangerous material, plutonium, nobody died in an accident. The work was so secret, negative publicity from it could not do much damage to the public perceptions of radiation and nuclear energy release.

The parts built at Rocky Flats, Colorado, were shipped to the Pantex bomb assembly plant, about 17 miles northeast of Amarillo, Texas, to be made into nuclear weapons. Before 1942, Pantex was a large, 16,000-acre, utterly flat wheat field; but with the pressing need for bombs to drop in World War II, an ordnance factory was built on the site. For the next three years, thousands of workers loaded gravity bombs and artillery shells with TNT. It was dangerous work, and an unusual amount of care was needed to keep from accidentally blowing everything up. The pay was good, but there was no smoking. All nicotine-delivery systems had to be held in the mouth, closed, and chewed. At the end of the war, on VJ Day, the plant was deactivated, and the workers had to find something else to do.

In 1951, the AEC had a sudden need for a centralized plant in the middle of nowhere to manufacture atomic bombs, and the old Pantex site was ideal. The original buildings were expanded and refurbished for $25 million. Proctor & Gamble, expert at making shampoo and laundry detergent, was put in charge.[168] The metal parts for the bombs, including the fissile material, were built elsewhere, but at Pantex the chemical explosives used to set off the nuclear detonations were cast and machined to fit, as if they were made of plastic.

The nuclear weapons used by the United States were almost all "implosion" types, in which the fissile material is forced into a hypercritical configuration using a hollow sphere of high explosives. The shell of explosive material is assembled from curved segments, like a soccer ball. In a normal explosion of a sphere, such as a hand grenade, a detonator at the center of the explosive sets it off. The explosion starts at the point of detonation and moves rapidly outward in a spherical wave-front until the entire mass of explosive is burned up. The wave-front then proceeds outward, as a sphere of compressed gas moving outward very fast. The spherical wave grows larger and larger, and the energy imparted to it by the brief explosion becomes stretched thinner and thinner. Far enough from the explosion, it is no longer strong enough to knock you down, and the destructive explosion is reduced to a loud noise as the energy pulse is spread out over the entire surface area of the sphere.

The implosion works in reverse. Instead of being detonated from a point in the middle, the sphere of explosive must be detonated from its entire outer surface. There is still a spherical wave-front that starts at the surface heading away from the bomb, but this is a waste of energy. There are two wave-fronts. One heads out, and one heads in. Inside the sphere, the wave-front grows smaller and smaller as it heads for the center. The energy from the brief explosion, instead of being dispersed and losing impact, is concentrated down, losing nothing. At the center of the sphere is a small ball of plutonium. The converging wave-front hits it so hard, it bends the molecular forces that are maintaining its density, and it shrinks to a concentrated, smaller size. This new configuration makes it hypercritical. Flying neutrons do not have as far to travel to hit a

168 The Maintenance and Operations (M&O) contract for Pantex was reassigned to Mason & Hanger—Silas Mason Co., Inc., Mason Technologies, Inc., in 1956. Mason & Hanger was the oldest engineering and construction company in the United States, dating back to 1827 in Virginia, and they had a lot of experience in managing ammunition plants.

nucleus, and with the nuclei crowded closer together, it becomes hard to miss. The ball explodes with a sudden release of nuclear power.

The problem is detonating the entire outer surface of the sphere all at once. It is not really possible to do so. The best you can do is to set off about 40 point-detonators placed around the outside of the explosive shell. These can be made to all go off at once, but the explosion does not make a perfectly spherical wave-front. To form this unacceptable, knobby wave-front into a perfect sphere does not require expertise in making hand grenades. It requires optics.

Optics is the art of taking a wave-front of light and warping it into a desired shape using the fact that light travels at different speeds in air and glass. When a light wave traveling in air hits clear glass, it must slow down, and if it hits at an angle, then it is bent, or refracted. Controlling the angle at which the light encounters the glass controls the refraction. This is accomplished by grinding the glass into a curved surface, specifically designed to bend the incoming wave-front into the desired shape. This is how telescopes, microscopes, and vision-correcting glasses are made.

The chemical explosives in an atomic bomb work exactly the same way. There are actually two, nested explosive shells. The outer shell, having the point-detonators, is a fast explosive, producing a high-speed shock wave. The inner shell is a slower explosive, making a lower-speed shock wave. The interface between the two shells is shaped in very specific ways to refract the segmented knobs of the outer shell explosion into a perfectly spherical shock wave in the inner shell, based on the difference in wave-speed in the two media. Explosion in the outer shell is caused by little firecracker-like detonators, and the inner shell is set off by the shock-wave from the outer shell hitting it. Shaping of the solid explosive segments to make this happen must be precise, and one must be careful not to impart a shock to the explosive while machining it.

Post-war improvements on the World War II atomic bombs were numerous and rapidly applied. The old Fat Man nuclear device that wiped out Nagasaki was five feet in diameter and contained 5,300 pounds of high explosive. That seemed clumsy, but by the late 1950s the bomb engineers had it down to 44 pounds of explosive in a bomb that was many times more powerful. At one point, they got it down to 15 pounds. This improvement meant lighter, smaller atomic bombs that could be put in cruise missiles, air-to-air missiles, or a rocket fired from a recoilless rifle bolted to a jeep. The antique formulas having baratol or RDX explosives were supplanted with such exotics as cyclotetramethylenetetranitromine (HMX) and triaminotrinitrobenzine (TATB).[169]

Making bombs less bulky was all well and good, but as the energy from the explosives became concentrated into smaller spaces, they got touchy, or very sensitive to being slapped. There were three accidental explosion events in the 1960s, when improper handling procedures led to detonations, but there were no deaths. All operations at the plant were carefully sequestered, with strong blast walls separating an operation from all other operations and not letting an accident become a catastrophe, setting off adjacent explosives or even setting off a nuclear event. All steps of explosive manufacture were done in the smallest possible batches.

On March 30, 1977, the luck ran out at Pantex in Building 11-14A, Bay 8. A machinist had chucked a billet of high explosive in the lathe chuck and turned it by hand to see how it would spin. It was slightly out of alignment, running a bit wobbly on the lathe spindle, and at cutting speed it would vibrate. This is a common occurrence when using a gear-chuck on a lathe. As careful as you are, the work-piece will not necessarily sit right in the chuck when you tighten it up. To remedy this, an experienced machinist will pick up his much-used wooden mallet and tap the piece into alignment, hitting it on the edge that causes the most "run-out." The last thing the machinist saw was the mallet coming down on the edge of the explosive work-piece. He and two others died instantly in the blast.

The Energy Research and Development Administration report on the accident was issued on March 1, 1979. The sensitive PBX-9404 fast explosive was replaced by less sensitive PBX-9502, and a movement to change out all the aging, increasingly sensitive explosives

169 In the 1960s at the apex of nuclear weapon development, the most favored chemical explosive formulas were PBX-9404 (93% HMX, 6.5% nitrocellulose, 0.5% wax) and LX-17 (92.5% TATB, 7.5% wax). Wax was used as a binder that would melt in heat and re-solidify when cooled. The "exploding wire" detonators used pentaerythritoltetranitrate (PETN).

in weapons on the shelf gained attention. A *Department of Energy Explosive Safety Manual*, DOE M 440.1-1A, was in place by the mid-1990s. Pantex is still in business, refurbishing and repairing our aging inventory of weapons, which is probably about 2,200 units.

These misfortunes in the production of nuclear devices are interesting, but none were true atomic accidents. They were industrial accidents of types that could occur anywhere in the technosphere. Authentic nuclear accidents in fuel processing, usually but not always for bomb manufacture, did occur, unlike anything in the history of technology. Some were predictable, and some were not. There have been 22 documented cases of process accidents in which an unexpected criticality occurred in the United States, Russia, Great Britain, and Japan. In these incidents, there were nine fatalities due to close exposure to radiation from self-sustained fission. Accidents occurred with the fissile material in a solution or slurry in 21 cases, and one occurred in a pile of metal ingots. No criticalities were the result of powered fissile material. No accident has occurred in the transportation of fissile material or while it was being stored. Of the many survivors of criticality accidents, three had limbs amputated due to vascular system collapse. Only one incident exposed the public to radiation. There was a clump of 17 accidents between 1957 and 1971, and only two have occurred since.

The first atomic bomb was conceived, designed, and built at the Los Alamos Scientific Laboratory in New Mexico, and after the war it was expanded to one of the largest and most versatile facilities in the galaxy of national labs. In 1958 they were still doing chemical separation of plutonium at Los Alamos, even though most of this was being carried out elsewhere. Somewhere in the above-ground portion of Los Alamos was a dreary, windowless concrete room packed neatly with 264-gallon stainless steel tanks, about three feet in diameter, each held off the floor with four stubby legs and seemingly connected together in all kinds of ways by a maze of pipes, tubes, and cables. They looked like short water heaters. There was a tall sight-glass bolted to the side, so that an observer could see the liquid in the tank and tell how full it was. On top was a push-button switch. Press the switch, and an electric motor would spin a stirring impeller at the very bottom of the tank, mixing the contents into a homogenous fluid.

The tanks were part of the chemical separation system, meant to recover plutonium from machine-shop waste, leftovers in melting crucibles, or slag from casting. The tanks typically held aqueous solutions that were about 0.1 gram of plutonium per liter, which was way below anything that could be made critical, but the tanks, which had been in daily use for the past seven years, were obsolete, and they were scheduled to be replaced soon. They were still in fine condition, but they had been made in a perfect shape for accidental criticality. They had the surface-area-minimizing shape of a soup can, and the ends were rounded. By now it was realized that this was a dangerous shape, even though the procedures were designed to absolutely prohibit there ever being enough plutonium in one tank to go critical. The replacement tanks would be 10 feet high and six inches in diameter, which would discourage anything less than solid plutonium from becoming a runaway reactor.

It was 4:35 P.M. on December 30, 1958, a little before quitting time on the last shift before the New Year's holiday. A load of 129 gallons of a murky fluid consisting of plutonium, nitric acid, water, and an organic solvent had been drained out of two other vessels and transferred to this particular tank.[170] Allowed to sit for a while, the liquid had separated into 87 gallons of water in the bottom of the tank with 42 gallons of oily solvent sitting on top of the water. This was to be expected, which is why there was an aggressive stirring mechanism built into the tank. Unknown to anyone, plutonium solids, built up from years of processing, had dissolved off the insides of the tanks upstream and landed in this tank. The water in the bottom had only 2 ounces of plutonium dissolved in it, but the thin, disc-like layer of solvent on top contained a barely subcritical 6.8 pounds of plutonium, helped along in its quest to go critical by being homogeneously mixed with a hydrocarbon liquid, an excellent neutron moderator.

170 The official reports of this incident always refer to an "organic solvent" without specifying what exactly was in the mix. It was surely a 30-percent solution of tributyl phosphate in kerosene, the active ingredient in the ion-exchange process known as PUREX. Invented during the Manhattan Project at the University of Chicago, PUREX (Plutonium URanium EXtraction) was the fuel-processing method of choice through the 1970s. It was classified SECRET at the time of this incident, and report writers were careful not to divulge any information that was not necessary in explaining an accident.

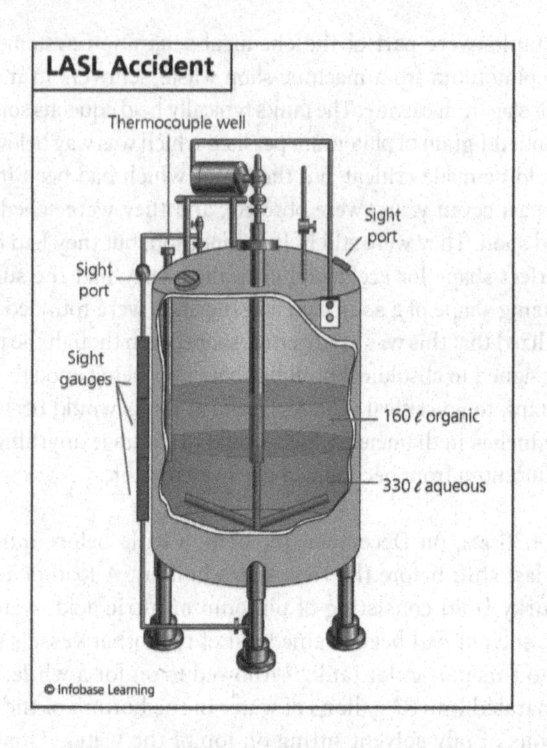

It was hard to predict this accident, except for the fact that the steel tank, built to use as little metal as possible, was an ideal shape for nuclear criticality using uranium dissolved in a liquid. When the electric motor was started to stir the two solutions together, it formed a whirlpool in the center. Instead of mixing immediately, the organic solution containing uranium was suddenly reduced in diameter and surrounded by water, making it a supercritical nuclear reactor.

Cecil Kelley had spent the last 11.5 years as a plutonium-process operator at Los Alamos, and he had almost seen it all. He stepped up on a footladder to look at the contents of the tank through a glass porthole on top, cupping his left hand to shut out ambient light. The ceiling fluorescents were illumining the surface of the liquid through another porthole on the other side. It was time to mix the water and the light solvent together. Leaning on the tank, he reached for the stir button with his right hand and pressed it. It took one second for the impeller to reach speed at 60 RPM.

A blast of heat washed over the front of Kelley's body, going clean through him and coming out the back. It was like being in a microwave oven, as fast neutrons saturated his insides, exchanging momentum with his comparatively still hydrogen nuclei. He felt the strange tingling from gamma rays ionizing the sensitive nerve endings. A rushing noise was coming from the tank, over the whir of the stirring motor. Boiling? There was a slight tremor, moving the tank sideways very slightly, one centimeter, as it walked across the floor on its four legs. He fell backward onto the concrete floor. Dazed and confused, he got to his feet, turned off the stirrer with another push of the button, then turned it back on and ran out of the building.

Two other process workers in the same room saw a flash of blue light on the ceiling, as if a photoflash had gone off, and then they heard a dull thud. No criticality alarms went off, but they both knew that something bad had happened. They rushed to help Kelley and found him outside. He had lost control of his limbs. "I'm burning up!" he cried. "I'm burning up!" They hustled him to the emergency shower, turning off the stirrer as they passed it.

In a few minutes the medical emergency and radiation monitoring staff arrived. Kelley was in deep shock, phasing in and out of consciousness. He looked sunburned all over. By 4:53 they had him in the ambulance and headed for the lab hospital. The radiation monitors ran their Geiger counters over the tank. It was hot—tens of rads per hour. It was the remnant of a criticality in the tank, but how?

When Kelley started the stirring impeller at the bottom of the tank, it was supposed to mix the oily layer on top with the water on the bottom, and it would eventually do this, but first it started the water spinning in a circle, independent of the disc-shaped, plutonium-heavy solvent stratum. The water assumed the shape of a whirlpool, a cone-shaped depression in the middle of the tank. The solvent fell into the cone, losing its large surface area and becoming a shape favorable to fission with the neutron-reflective water surrounding it in a circle. Instantly the cone of solvent became prompt supercritical, releasing a blast of fast neutrons and gamma rays.[171] The criticality only lasted for 0.2 seconds,

171 I remind the reader that "prompt" supercriticality means that the mass of plutonium plus moderator is sufficiently supercritical to begin increasing the rate of fission exponentially without waiting for the delayed fission neutrons to contribute.

but in that brief spike there were 1.5×10^{17} fissions. When the two fluids mixed together under the continued influence of the impeller, the plutonium-laden solvent was diluted by the water. The plutonium nuclei became too separated from one another for adequate neutron exchange, and the criticality died off as quickly as it had started.

Kelley's condition was dire. He was semiconscious, retching, vomiting, and hyperventilating. His lips were blue, his skin was dusky red-violet, and his pulse and blood pressure were unobtainable. He was shaking, and his muscles were convulsing uncontrollably. His body was radioactive from neutron activation.

After an hour and forty minutes, he settled down and was perfectly coherent. He was moved to a private room. The staff drew blood and tried to get an estimate of his radiation dose from counting the activated sodium-24 in the sample. He had absorbed about 900 rad from fast neutrons and somewhere between 3,000 and 4,000 rad from gamma rays. A dose of 1,000 rad was thought fatal.[172] His bone marrow had changed to inert, fatty tissue. He started having severe, uncontrollable pain in his abdomen, and he turned an ashen gray. At 35 hours after he had touched the stirrer switch, Cecil Kelley died.

The plutonium process was shut down for six weeks and the tanks were ripped out and replaced with the six-foot columns, as had been planned but put off.

Back in the 1960s, all the fuel reprocessing was not for weapons work, and all was not government-owned. There were also privately owned plants. The early startups were not large operations, but the ultimate goal was to take the spent fuel from power company reactors, extract the unused uranium, sell plutonium waste to the government, compact the fission products for efficient burial, and deprive the Canadians of a monopoly on the manufacture of medical isotopes, such as technetium-99M. Spent reactor fuel was seen as a cash cow, and not as a burden on the power industry. Fuel reprocessing was

172 The radiation dose specification of "rad" (Radiation Acquired Dose) used in the official reports is now considered obsolete. It is often expressed in "rem" (Roentgen Equivalent Man," because radiation counters are calibrated in rem, but that specification has been replaced with the sievert by the *Système international d'unités,* and the rad has been replaced by the "gray." The estimate of Kelley's total dose has been revised a few times, and it may have been as high as 18,000 rems, or 180 sieverts. The only important point of these numbers is that that much radiation could have killed him 18 times.

also considered a necessity for commercial breeder reactor operations, and breeders were expected to start coming online later in the decade.

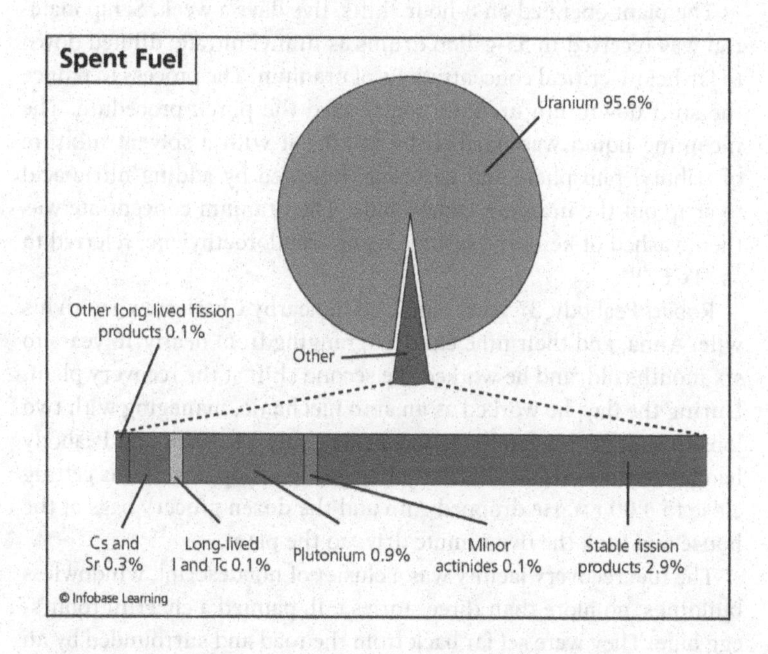

Spent reactor fuel is mostly unused uranium, which is a natural part of the Earth's crust and not particularly dangerous. If it could be processed down to the 1.4% that is highly radioactive, nuclear waste disposal would be much easier and reusable fuel would not be wasted.

The United Nuclear Corporation alone owned five plants. There was one in downtown New Haven, Connecticut, one in White Plains, New York, one in Hematite, Missouri, a research lab complete with a nuclear reactor in Pawling, New York, and a brand-new scrap uranium recovery facility in Wood River Junction, Rhode Island.

Before March 16, 1964, Wood River Junction was known for two things. It was the site of a railroad trestle washout leading to a passenger train disaster on April 19, 1873, and it was regarded the coldest spot in Rhode Island. On that day in March, the sparkling new United Nuclear Fuels Recovery Plant began operations. Its first contract was to recover highly enriched uranium from manufacturing scraps left on the floor at a government-owned fuel-element factory.[173]

The plant operated on 8-hour shifts, five days a week. Scrap material was received in 55-gallon drums as uranyl nitrate, diluted down to far-below-critical concentrations of uranium. The process to reduce the stuff down into uranium metal used the purex procedure. The incoming liquid was purified by mixing it with a solvent mixture of tributyl phosphate and kerosene, followed by adding nitric acid to strip out the uranium compounds. The uranium concentrate was then washed of kerosene residue using trichloroethylene, referred to as "TCE."[174]

Robert Peabody, 37 years old, lived in nearby Charlestown with his wife, Anna, and their nine children, ranging from nearly 16 years to six months old, and he worked the second shift at the recovery plant. During the day, he worked as an auto mechanic, managing with two jobs to support his family. It was Friday, July 24, 1964, and Peabody had taken time off from his day job to go food shopping. It was getting close to 4:00 P.M. He dropped Ann and the dozen grocery bags at the house and took the five-minute drive to the plant.

The fuel recovery facility was a cluster of nondescript, windowless buildings, no more than three stories tall, painted a cheerful robin's-egg blue. They were set far back from the road and surrounded by an

---

173 This material was probably MTR fuel scraps from Oak Ridge. The Materials Test Reactor (MTR) was built at the Nuclear Reactor Test Station in Idaho and started up in March 1951. Its fuel was a unique design, made of bomb-grade uranium metal, enriched to 93% U-235. The uranium was mixed with pure aluminum to make an alloy, formed into a long rod, and clad in a layer of pure aluminum, or "aluminum 1000." The rod was then flattened between two steel rollers and bowed slightly along the major axis. This simple fuel-element design became an international standard, and "MTR fuel" was used in dozens of research reactors all over the world. As these aged reactors are decommissioned, particular care is taken to see that the highly enriched uranium fuel does not fall into the wrong hands.

174 The trichloroethylene (TCE) used in the wash-out step is incorrectly referred to as trichlorethane (chlorothene) in *A Review of Criticality Accidents: 2000 Revision* (LA-13638) by the Los Alamos National Laboratory. Trichlorethylene was originally formulated in Great Britain as a general anesthetic to replace chloroform, but by 1956 better, less-toxic substitutes were found. Since the 1970s it has been widely banned from use in the food and medical industries, and is considered a carcinogen. Exposure to it seems to lead to Parkinson's disease.

imposing chain-link fence on a flat, 1,200-acre plot. Peabody clocked in, as usual, and changed into his coveralls.

It had been a nerve-racking week at the plant, mainly due to false criticality alarms. When processing nuclear materials, working with highly enriched uranium in aqueous solutions was about as dangerous as it could get, and an accidental criticality was something to be avoided at all costs. Wednesday he had been working on the second floor, washing down some equipment, when the criticality alarm sounded. Having it blare off nearby was like having a tooth drilled. He and the other four workers left the building in a hard sprint. It took a while to figure out that there was no danger, and that water had splashed into some electrical contacts. The radiation-detection equipment was set to be nervous and sensitive to the slightest provocation.

Later on the same shift, they discovered a substance technically designated "black goo" collecting at the end of the line where uranium was supposed to be coming out. The next day, the line had to be shut down, and everything had to be taken apart and cleaned, and the labeling conventions got a little scrambled as parts and wet rags and bottles were scattered around.

Most of chemical engineering practice had been fully automated by 1964, but this part of nuclear engineering, which was actually similar to other parts of nuclear engineering, seemed based in the nineteenth century. Processing nuclear fuel at this level was an astonishingly manual operation, requiring human beings to carry bottles of liquid material by hand and empty them into vats, tanks, or funnels at the tops of vertical pipes. The system was mostly gravity-driven, with liquid flowing naturally from the top floor to the ground floor in the plant. Somehow it seemed safer to have men carry around batches of uranium in small quantities, knowing exactly where and when it was to be transferred, than letting uncaring machinery do it.

The basic transfer units were specially built bottles, made of polyethylene, four feet tall and five inches in diameter, with plastic screw-on tops. Being tall and skinny, they discouraged a critical mass. No matter how concentrated was the uranium solution, it was impossible to get enough into one bottle to cause a chain reaction. However, it would be possible to stack them together in a corner and make a working nuclear reactor out of filled bottles, as stray neutrons flying back and forth from bottle to bottle would cross-connect them. That possibility was eliminated by having the anti-criticality bottles rolled around from place to place in "safe carts." Each cart was built to hold the bottle vertically at the center of an open-framed, three-by-three-foot cube, made of angle irons and fitted with four casters at the bottom so it would roll. There was no way that all the bottles in the plant filled with highly enriched uranium solution could go critical as long as they were sitting in safe carts. The bottles were always separated by at least three feet of open space.

The problem with the polyethylene bottles was labeling them to identify the contents. Paper labels stuck on with Scotch tape would come off easily, because the bottles were frequently covered with slippery kerosene residue. The only way to get a label to stick was to hold it on with two rubber bands. On Thursday, the day shift was cleaning out the black goo in the system, and they found a plug of uranium nitrate crystals clogging a pipe. They cleaned it out with steam and drained the highly concentrated, bright yellow solution into polyethylene anti-criticality bottles. Paper labels were attached with rubber bands identifying them as containing a great deal of highly enriched uranium.

Five people at any one time ran the entire plant. On the night shift it was three young technicians, Peabody, George Spencer, and Robert Mastriani, the supervisor, a 30-year-old chemist named Clifford Smith, and the security guard. The plant superintendent, Richard Holthaus, was usually there during the day.

The back-breaking task of the evening was to clean the TCE, which had been used to wash the kerosene out of the uranium concentrate. It was expensive stuff, and it had to be recycled back into the process, but it always picked up a little uranium oxide when washing out the oil. The uranium was separated out of the TCE by adding some sodium carbonate and precipitating it to the bottom of the vessel. This process, like others in the plant, was carried out in small batches, and a shift-load of bottles loaded with dirty TCE was bunched up in safe carts. Peabody was expected to pick up each 35-pound bottle of solution, pour in some carbonate, and shake it for 20 minutes to ensure mixing. There had to be a better way.

Another way of agitating the TCE had been worked out in a previous shift. Weary of manipulating the heavy bottles, a technician had noticed that on the third floor was a perfectly good mixing bowl with a motorized stirrer, and why couldn't we use that to slosh the TCE? It is not recorded, but I am sure he got the standard nuclear-work answer from the supervisor: "No! Give me a few minutes, and I will think of why you can't do that." Technicians could not be allowed to improve operating procedures on a whim. Eventually, the technician was able to wear down the supervisor, and word of an undocumented labor-saving procedure traveled through the plant with the speed of sound. The vessel in question, the carbonate make-up mixer, was about 18 inches in diameter and 26 inches tall, or the size and shape of a very efficient submarine reactor core. It was okay to use it, as long as the uranium content in what it was mixing was less than 800 parts per million, or very, very dilute.

It was nearly 6:00 P.M. Peabody rolled the safe cart with the first bottle in the cluster of what he assumed were bottles of TCE to be cleaned to the base of the stairs. The cart would not make it up the stairs, so he hefted the bottle to his shoulder. The label slipped out of the rubber bands and fluttered to the floor. The contents of this bottle looked about like the stuff in all the bottles. It was yellow, due to the extreme fluorescence of uranium salt, but this was not a bottle of contaminated TCE: it was uranium nitrate dissolved in water, from the black goo cleanout.

It was about as much work to get it up the stairs as it would be to shake the bottle, but Peabody arrived on the third floor, dragged the bottle over to the mixer, and unscrewed the top. The mixer was against the north wall of the room, held a couple of feet off the floor by metal legs, making the rim five feet high. The stirrer motor was hanging over the open top. Workers were protected from falling off the third floor and to the ground floor by a railing on either side of the narrow platform. Leaned against the railing on the right side, very near the mixer, was a folded two-section ladder, lying on its side.

The mixer already had 41 liters of sodium carbonate in it, and the motor was running. Peabody, who was only six inches taller than the lip of the vessel, stepped up on the sideways ladder and tilted the bottle into the mixer. Glug, glug, glug. As the last dregs emptied into the mixer, there was a bright blue flash and the sound of an enormous water balloon being slammed against the wall. As the geometry improved from long and thin to short and round, 6.2 pounds of nearly pure U-235 homogeneously mixed with water went prompt critical. Instantly, the contents of the mixer boiled violently, sending a vertical geyser hitting the ceiling, the walls, and thoroughly soaking Peabody with the products of 1×10^{17} fissions. He fell backwards off the ladder, jumped to his feet, and lunged for the stair well, screaming "Oh, my God!" The criticality alarm went off, and this time it meant it.

Peabody ran full tilt down the stairs, out the door, and was quickly making for the emergency shack, 450 feet away. His fellow workers were right behind him, fleeing the criticality alarm and watching Peabody tear his clothes off. He almost made it, but he fell to the ground naked, vomiting, and bleeding from the mouth and ears. Smith, the supervisor, ran to call Holthaus while Spencer and Mastriani grabbed a blanket from the shack and tried to wrap the injured man on the ground. He got up twice and tried to walk around, but he sank back to the ground with severe stomach cramps.

Soon the company officials and the police were backed up at the gate, and Peabody was loaded into the ambulance. His wife and eldest son, Charles "Chickie" Peabody, were found by a police officer. "There's been an accident," he began. "We'll take you to the hospital."

At 7:15 P.M., Richard Holthaus arrived at the plant, waving a radiation counter. Peabody had been the only person anywhere near the criticality, and he was the only one affected by the radiation burst. There was no radiation evidence on the ground floor that anything had happened, but nobody had turned off the criticality alarm, and the klaxon was still screaming. At 7:45 Smith joined him, and they cautiously climbed the staircase to the third floor, radiation probe held in front. There was no hint of a continuing criticality. Clearly, enough material had immediately boiled out to stop the chain reaction, but the walls, floor, and mixer showed fission-product contamination and were painted a brightly fluorescent yellow. Peabody's bottle was still upended in the mixer. Holthaus went over to the mixer, removed the bottle, flipped the switch to turn off the stirrer, and quickly turned to go out the door. Smith took one last look and was right behind him.

They had to quickly go downstairs and drain the contents of the mixer into anti-criticality bottles.

The stirring motor coasted to a stop, and the deep, funnel-shaped maelstrom in the mixer vessel relaxed to a momentarily flat surface, before the mixture started another furious boil. The radiation caught Smith in the back as he was hurrying through the doorway. Fortunately for Holthaus, Smith's body shielded him from the neutrons and he only got a 60-rad dose. Smith at least was not standing directly over the mixer, but he got a serious 100-rad blast of mixed radiation, head to toe.

In its spinning configuration, the uranium-water mixture was a good configuration only when there was a great deal of excess reactivity (uranium) in the mixer. It boiled away the excess until the contents went barely subcritical and the reaction stopped. Holthaus then removed the empty bottle, which was a non-productive void in the would-be reactor, and he stopped the spinning. The surface area of the geometric shape in the mixer went down as the stuff stopped spinning, and the lack of a bottle-shaped void made it complete. The mixture once again went supercritical.[175] The two men were unaware of it, as they were both looking down into the stairwell when it happened, and the alarm was still blaring from the first criticality. Feeling a little strange, they returned to the ground floor, turned off the alarm, and took half an hour draining the mixer.

---

175 At a glance, this incident bears a close resemblance to the fatal accident at Los Alamos detailed in the previous sketch. The two are different in subtle ways, but both are examples of how an unexpected concentration of fissile material can be dangerous if it is shaped just the wrong way. At Los Alamos, the tank containing a benign mass of plutonium was made critical by powerful mixing action that first stirred the water beneath the oily solution. For just a second, before the oil and water were able to mix, the funnel in the water caused by the stirring forced the oily plutonium solution to assume a shape with less surface-to-volume ratio, which reduced the non-productive escape of stray neutrons, and the mass went supercritical. In the case of Wood River Junction, a solution of sodium carbonate (washing soda) was already turning slowly at the bottom of the tank before Peabody poured in the uranium dissolved in water. Going from the high-surface-area bottle to the low-surface-area mixing tank is what made the uranium-235 go supercritical. The change in shape made such a huge difference, it did not matter that the uranium solution was diluted when it hit what was already in the tank. In both cases, the plutonium criticality and the uranium criticality, there are always a few stray neutrons bouncing around from spontaneous fissions in the fissile material. Unless there is a critical mass for the given shape, spontaneous fission leads to nothing.

At the hospital, Anna and Chickie were cautioned to stand at the foot of the bed. Peabody was radioactive, conscious, lucid, and restless. He was given a sedative. "Somebody put a bottle of uranium where it wasn't supposed to be," he told them. By Sunday morning he was starting to slip away. His left hand, the one that had held the front of the bottle, was swelling up, and his wedding band had to be sawed off. He drifted off into a coma, and that evening, 49 hours after he saw the blue flash, Robert Peabody died. His exposure had been 10,000 rads, or enough to kill him ten times.

Smith and Holthaus survived with no lasting effects, but they had to give up the silver coins in their pockets, which had been partly activated into radioactive silver isotopes by the neutron bombardment and were quickly decaying into stable cadmium. They were saved only by the distance between them and the supercriticality event and not by any cautious prescience. The walls on the third floor were decontaminated, and production resumed by February 1965. Contracts gradually dwindled away, and the plant closed for good in 1980. Robert Peabody was the first civilian to die from acute radiation exposure in the United States. So far, he was also the only one.

Impressed by the flash-bang end to World War II, the Soviet Union was quick to replicate the nuclear materials production facilities used by the United States. The U.S., in an unprecedented show of openness and generosity, published the final report for the atomic bomb development project, *Atomic Energy for Military Purposes*, or *The Smyth Report*, in hardback three days after the Empire of Japan surrendered. It included a map of the Hanford Works, a detailed photo, and an explanation as to how we manufactured the synthetic fissile nuclide plutonium-239. It was available to anyone in the world with $1.25 to invest, and many copies were bought for use by Soviet scientists and engineers, eager to get started.[176]

176 The first edition did not have the photographs, but there were at least eight editions, and the helpful photos were added. There were over 170,000 copies of the U.S. edition alone, and it was on the *New York Times* bestseller list until late January 1946. A British edition was published in 1945, and eventually it was published in 40 languages all around the world. At the top of page iv is written: "Reproduction in whole or part is authorized and permitted."

The robust Soviet building program produced the Tomsk-7 Reprocessing Plant, the Novosibirsk Chemical Concentration Camp, the Siberian Chemical Combine, and, most impressive of all, the Mayak Production Association, covering 35 square miles of flat wilderness.

The production reactors and plutonium extraction plants were built and running by 1948, and the site was treated as the deepest military secret in the Union of Soviet Socialist Republics. Not trusting anyone with anything, the Soviet government was careful not to divulge what was going on at Mayak, particularly to the thousands of people who worked there. This policy resulted in a lack of essential knowledge among the workers, and studies have blamed this for the 19 severe radiation accidents at the site occurring from 1948 through 1958. Among the 59 people who suffered from the effects of radiation exposure, six men and one woman died in criticality accidents. Since the cluster of accidents in those early years of nuclear weapons production, there have been 26 more accidents at Mayak that we know of.

Mayak was an irritating black hole in the intelligence community. It was literally a blank spot on the map of the Soviet Union, and it seemed important to know what was going on there. On May 1, 1960, an outstanding jet pilot named Francis Gary Powers flew a Lockheed U-2 spy plane 70,000 feet above Mayak. It was a covert CIA mission, the existence of which would be vehemently denied by the President of the United States, Dwight D. Eisenhower. The specially built airplane carried a terrain-recording high-resolution camera in its belly, clicking off frame after frame as Powers guided it over the plutonium plant.[177]

Unfortunately for Powers, Eisenhower, and the CIA reconnaissance-photo analysts, the Soviets sent up everything they had against the U-2 flying over their most secret installation. An entire battery of eight S-75 *Dvina* surface-to-air missiles to blow it up, a MiG-19 fighter jet to shoot it down, and a *Sukhoi* Su-9 interceptor just to ram into it were

177 This was actually the second flight over Mayak. The first had been made in April, with the camera running as the plane flew over a 300-kilometer line from Kyshtym to Pionerski in the East Ural Mountains. The reason for these expeditions will become clear at the end of this chapter.

deployed in anger. The first S-75 blew up somewhere behind the U-2. It did not hit the plane, but the U-2 was fragile, built only to take pictures and not to withstand roughhousing. The shock wave from the missile destructing in air folded up the U-2 like a wadded piece of junk mail. A second missile shot down the MiG-19, another one caused the Su-9 to auger in, and the remaining five missiles were simply wasted.

Powers bailed out and was immediately captured. His plane was spread out over square miles, but it was gathered up and glued back together as evidence of espionage on the part of the Eisenhower administration. The cover story that it was a weather plane that had strayed off course did not work, and peace talks between Premier Khrushchev and President Eisenhower were cancelled. Powers was eventually repatriated in a prisoner exchange in Berlin, Germany, with the Soviets getting back their ace spy, Rudolf Abel.[178] Mayak remained a mystery until 1992, when the Soviet Union fell apart and true *glasnost*, or openness, spilled it all.

Of the many ghastly accidents at Mayak, one stands out as unusual and worth a detailed look. Mayak was run under war footing, as were the atomic bomb labs in the United States during World War II, and most workers were undertrained. Carelessness and minimal safety considerations led to many problems, but in this case the participants were nuclear experts near the top of the food chain, and they knew exactly what they were doing.

On December 10, 1968, the night-shift supervisor and a couple of highly placed plant operators conspired to set up an experiment in the basement of the plutonium extraction building. It was an unauthorized research project, breaking the rules and protocols, but they wanted to investigate the purification properties of some organic compounds. They

---

178 Powers returned home to a cold reception. The CIA was upset because he had not hit the self-destruct button in the U-2, nor had he injected himself with the suicide needle, hidden in a coin. He got a job as a test pilot for Lockheed, and later as a news helicopter pilot. He died in 1977 when covering brush fires in Santa Barbara County when his helicopter ran out of fuel. Although he was awarded the Order of the Red Banner back in the USSR, Rudolf Abel (Willie Fisher) failed to recruit or even identify a single Soviet agent in his eight-year deployment in New York City. His cover began to unravel in 1953 when his assistant, a Finnish alcoholic, accidentally spent a nickel containing a microphotograph of a coded message. A newsboy dropped the nickel, and the hollowed-out coin split in half, revealing the strange film negative inside. It was downhill from there, and Abel was captured in 1957.

were sure that they would get points for coming up with something better than kerosene and tributyl phosphate as the extraction solvent.

It was 7:00 P.M. In the basement was a long, narrow room, having two 1,000-liter tanks bolted to the concrete floor. Four and a half feet above the concrete was built a raised floor with the tops of the two tanks protruding through it. The tanks were used to temporarily hold very dilute mixtures of plutonium salt and water originating upstairs. The pipes had been changed, and there was some resulting confusion as to what the tanks were connected to. Along the walls were two shelves. You entered the room by climbing seven wooden steps to the open doorway, and on the shelf to your left sat an unauthorized stainless steel bucket with two handles. It looked like a cookpot stolen from the cafeteria kitchen, probably used to make soup. It had no business being in the same room with plutonium extract. It could hold 60 liters of fluid.

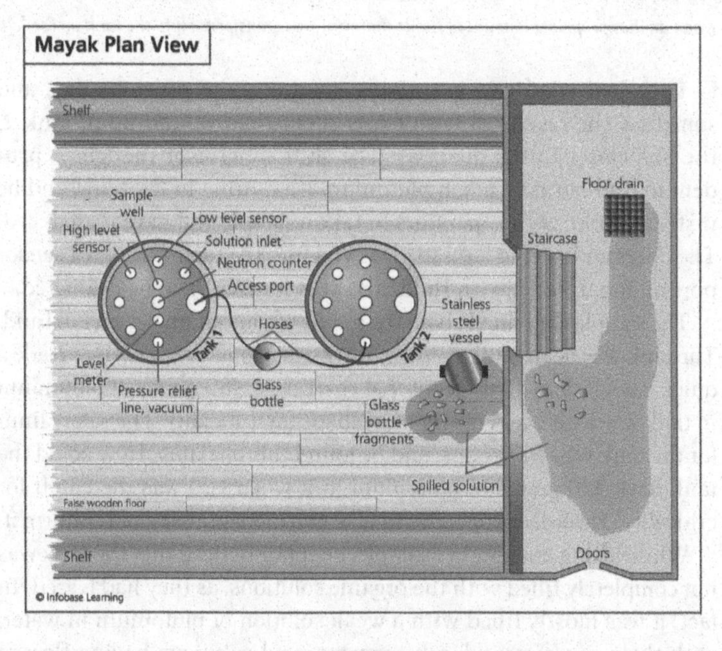

The unauthorized Mayak fuel reprocessing experiment, shown from the top after the second criticality excursion. The stainless steel cook-pot, large enough to contain a liquid nuclear reactor core, should not have been anywhere near this room.

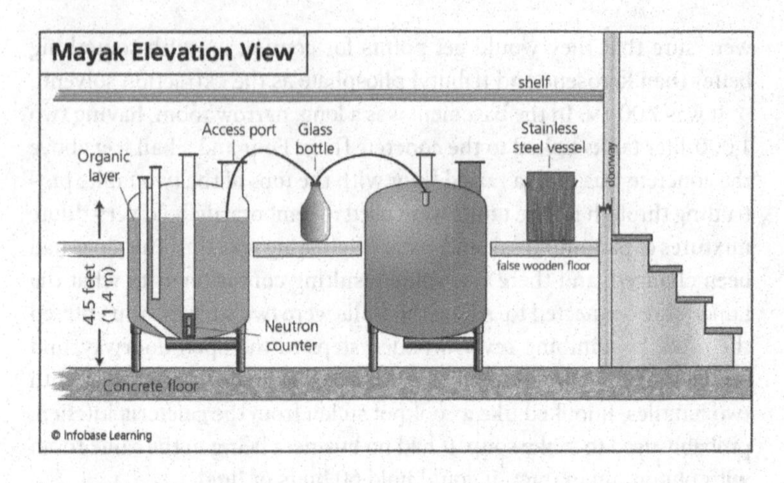

Scene of the Mayak criticality accident, shown from the side. The first criticality occurred when an inappropriate amount of fissile material was transferred into the tank on the left.

They had tried two extracting solutions the previous day, and somehow the results of both experiments had wound up in tank 2, the one closer to the doorway. The shift supervisor thought it prudent to find out how much plutonium was sitting in the tank, and he instructed one of the operators to take a sample and have it analyzed. The operator lowered a small glass vial into the tank through a sensor port on top, filled it with fluid, and sent it upstairs to be evaluated.

The sample contained about 0.6 grams of plutonium per liter of fluid. The tank was nearly full, holding 800 liters of liquid. Supervisor made a quick calculation. That meant that there were 480 grams of plutonium in tank 2, which was a nearly ideal shape for criticality. The safety limit for the tank was 400 grams, and he found this puzzling. How could the tank have 480 grams of plutonium in it when 400 was the cutoff for criticality? He ordered operator to take two more samples and confirm it.

While lifting out another sample, operator noticed that the tank was not completely filled with the organic solutions, as they had feared. In fact, it was mostly filled with a weak solution of plutonium in water, with their experimental, oily, concentrated solutions having floated to the top and sitting in a thin disk-shaped layer in the tank. This explained why the tank had not gone supercritical when filled, and supervisor felt slightly better. There may have been enough plutonium in the organic layer to go critical, but only if it were shaped like the entire tank and not spread into a thin layer on top of the water. The process downstream of tank 2 was not set up to work on organic solvents, so they had better decant it off the top of the water before something happened. They found a 20-liter glass bottle, normally used to hold chemical reagents, a two-hole stopper, and two rubber hoses. Operator put the bottle on the shelf, connected the hoses to the stopper, and connected one hose to an active vacuum port on top of tank 1, which was completely empty. He connected the second hose to the stopper and lowered the open end of it into the wide access port on tank 2, being careful to let it suck fluid only from the thin layer of organic solvent on top. The oily stuff was dark brown, indicating that it was thoroughly loaded with plutonium.

Satisfied that all was well, the supervisor left his two operators with the task and went to see how the rest of the shift was performing. When the bottle looked about full, the operators pulled the stopper loose, hefted the 17 liters, and poured it into the big pot, still up on the shelf. One operator had to go to his other duties, and the other found the supervisor to ask him, what next? The supervisor told him to make sure it was all gone out of the tank. By being mindful of the depth of the end of the rubber hose, operator was able to suck out another 20 liters into the bottle, filling it to the top.

Operator pulled the stopper and started emptying the bottle into the big pot up on the shelf. The liquid, a mixture of two plutonium-extraction experiments, was thick, with globs in it. It poured slowly, but the bottle was gradually getting lighter and easier to hold. BANG! A flash of blue light filled the room, and the operator felt the sudden blast of heat hit him right in the face. He instinctively dropped the bottle. It crashed to the floor, sending shards of broken glass and dissolved plutonium scattering across the floor, out the door, dripping down the steps, and to the floor drain. The criticality alarm in the building went off immediately, along with the criticality alarm in the next building, 50 yards away. Everybody in both buildings dropped what they were doing and hastened for the escape tunnels.

The building radiation-control supervisor was the only one not leaving. He switched into emergency mode and made sure that everyone surrendered his or her personal dosimeter before getting away. He then ran into the operator, looking frightened and in shock, and he directed the injured man to decontamination and medical care. Everyone was ordered to not go back into the building until the reason for the criticality alarm could be determined and made safe.

The oily mixture in the stainless steel pot had heated instantly to the boiling point upon going critical, and the resulting thermal expansion turned off the chain reaction. The supercritical condition in these very small, unplanned reactors teeters on a knife-edge, and the slightest modification of the density, the total mass, or the shape of the reactor can shut it down as quickly as it came into being. There it sat, undisturbed, until 11:30 P.M., when, having cooled, it suddenly lapsed again into criticality and faded out when the liquid re-heated. This time, the reaction was too weak to set off the criticality alarm, and nobody was there to be harmed by the mixed radiation pulse.

The shift supervisor was feeling dread about how completely he had botched his experimental program. He had a strong urge to erase what he had done, resetting the situation to normal conditions, but to set everything right he had to get back into the room. The radiation supervisor would hear none of it, but the shift supervisor was adamant. Finally, radiation agreed to follow shift to the area and scope out the extent of the contamination. As they neared the doorway to the tank room, the rate meter on the gamma-sensitive "cutie pie" instrument slid off scale. The room was hot with fission products, and it would be crazy to go in there. Wait for the clean-up team to dress out in radiation suits and come up with a plan to disable the reactor.

Shift supervisor was too impatient for that. Somehow, he talked the radiation supervisor into leaving him standing in front of the room while he went to check something.[179] Shift supervisor, seeing radia-

---

179 The existing record of what happened at Mayak was given to the team at Los Alamos compiling *A Review of Criticality Accidents* in 1993 by Gennadiy S. Stardubtsev and A. P. Suslov. This particular report is unusual, in that no written papers or articles were referenced, and the account is apparently taken from memory and notes. This documentation does not specify how the shift supervisor was able to talk the radiation supervisor into leaving him in front of the room. In my imagining of the situation, the shift supervisor probably told the radiation supervisor that he needed to make sure that the heavily dosed operator had not spread radiation all over the floors and walls on his way to decontamination. He would wait quietly here while the radiation supervisor checked it out with his instrument. Radiation supervisor left, reminding shift supervisor not to move.

tion supervisor turn a corner, raced up the steps into the tank room. He saw the pot on the shelf and quickly scoped the problem. The pot was cooling down, and he had to do something very soon, before it had a chance to go critical again. He took the handles in both hands and lifted the pot, planning to dump its contents down the steps and into the floor drain. The plutonium mixture would be so spread out into a thin puddle, it could not possibly regain criticality. The thing was a lot heavier than it looked, and he managed a controlled fall to the floor. It hit with a wet thud, right in the middle of the puddle of plutonium solution.

This time, the supercriticality pulsed like it meant it. As the barely subcritical pot hit the floor, its tendency to fission was extended by the flat field of plutonium-239-bearing solvent now under it. Not only were neutrons reflected back into the pot, they were multiplied by causing fissions outside the critical mass, giving back as many as two neutrons for every one lost by leakage from the surface of the pot. Alarms in both buildings went off again, and supervisor was drenched in fluid as the reactor boiled explosively. Supervisor staggered down the steps and made it to decontamination. He had absorbed 2,450 rem of mixed radiation, and he was a dead man walking.[180]

The operator who had made the first criticality suffered from acute, severe radiation sickness. He had absorbed 700 rems of mixed radiation. His vascular system collapsed, and eventually one hand and both legs had to be amputated, but he was still alive 34 years after the accident. A little over a month after the accident, the shift supervisor died.

Mayak is still in business, and safety conditions improved over the decades from "medieval" to levels in keeping with 21st-century handling of radioactive and potentially critical materials. There has not been a criticality incident since the fatal accident in 1968, and the

---

180 We may never know his name, but the night shift supervisor at the Mayak plutonium extraction building was awarded the not-coveted 1994 Darwin Award. The Darwin is given to that individual who has taken his or her (usually his) self out of the gene pool by doing something really stupid, therefore proving that evolution works by not allowing people who should not reproduce to do so. http://darwinawards.com/darwin/darwin1994-25.html

last death at the plant was in 1990, caused by a chemical explosion in a reagent tank.

The last fatal accident due to an unplanned criticality occurred in Japan in 1999, in a publicly owned nuclear-fuel-processing plant. This accident was unusual in that the criticality was not over in a flash, but would continue to react for an impressive 20 hours, and the two men who died broke the records for length of survival after receiving lethal radiation dosage. It was similar to the previous accident in Rhode Island, in that a break with the standard procedure to make the work easier led to the criticality, and even in 1999 the fuel processing incorporated a surprising amount of manual labor. It was also the first and only criticality accident in which members of the public not involved with uranium processing were exposed to measurable radiation.

The Japan Nuclear Fuel Conversion Co. Ltd. was established in 1979 as a subsidiary of the Sumitomo Metal Mining Co. Ltd. The Fuel Fabrication plant was built in Tokaimura, Ibarakin Prefecture, Japan, on a 37-acre, inner-city plot of ground. Unlike the United States or Russia, where a nuclear plant of any purpose was built in a lonely, isolated place, in Japan it was put in a highly congested, tightly packed city of over 35,000 people. In two large buildings, incoming source material, uranium hexafluoride gas, was converted to either uranium oxide powder or uranyl nitrate dissolved in water. The plant handled uranium used in light-water commercial power reactors. It was a large-scale plant, handling 540 tons of uranium per year at the peak in 1993, but it was only licensed to process low-enriched fuel, about five percent U-235.[181] Competition with foreign companies doing the same thing was stiff and production efficiency always needed tight-

181 The company wished to process Mixed OXide fuel (MOX), which is a combination of uranium and plutonium, derived from reprocessing spent power-reactor fuel. The United States opposed this plan under a new nuclear nonproliferation policy, fearing that Japan would either secretly build up a nuclear weapons stockpile or sell plutonium under the table to some other Asian tiger, neither of which seemed likely. Heated negotiations went on for three years, beginning in 1977. The U.S. finally gave in, agreeing to a proliferation-resistant process for mixing plutonium and uranium devised by the Japanese Power Reactor and Nuclear Fuel Development Corporation (PNC). Several power reactors, including Fukushima Daiichi 6, have been operated using MOX fuel, saving uranium and burning off otherwise unusable plutonium.

ening, but in 1993 the company sold ¥3,276,000,000, or $32,760,000, worth of product.

In 1983, a small facility, the Fuel Conversion Test Building, was erected to be used for special products. The plant's license was modified to allow the processing of uranium enriched to up to 20 percent U-235 so that startup fuel for the *Jōyō* fast breeder reactor could be produced. *Jōyō* needed fuel enriched to 18.8 percent U-235. Care was supposedly taken in the building's design to ensure that no enriched uranium would ever be in a critical-sized or -shaped container, so no criticality alarms were called for in the license. An accidental criticality of any kind in this facility, run by highly disciplined Japanese laborers, was not a credible scenario.[182] Gamma-ray detectors were bolted to the walls in all the buildings, in case some mildly radioactive fuel was somehow misplaced.

A step in the licensed procedure for making highly enriched uranyl nitrate was to mix uranium oxide powder and nitric acid together in a dissolver tank. As the nitrate product dripped through the dissolver, it was conveyed by a stainless steel pipe to a long, thin stainless steel holding vessel, specifically designed not to allow a critical mass of liquefied uranium solution to exist in it. The uranyl nitrate solution was then drained out the bottom of the vessel into small polyethylene bottles, each holding a non-critical four liters of solution. A little petcock on the bottom of the vessel controlled the flow into a bottle held under it. Just follow the procedure, being careful not to stack the bottles close together, of course, and nothing can happen.

In 1998 the company's name was shortened to JCO, requiring less ink to print. By then the fuel-conversion business had fallen to 53 percent of the peak back in 1993, but in September 1999 JCO won a contract to convert 16.8 kilograms of uranium into uranyl nitrate for *Jōyō*. On September 29, three operators, Masato Shinohara, Yutaka Yokokawa, and Hisashi Ouchi, were assigned the task of dissolving the uranium oxide in nitric acid in seven batches of 2.4 kilograms of uranium each. With each run of uranium being only 2.4 kilograms, there was no chance of criticality.

---

182 This "credible accident" criterion would bedevil the Japanese nuclear industry on March 11, 2011, when the Fukushima Daiichi power plant was knocked out of service by an earthquake and tsunami wave. This was not a credible scenario, so no preparations were made to prevent damage during such an event.

There was an immediate problem. The drain petcock on the bottom of the long, thin holding vessel was only four inches off the floor. There was no way to fit a bottle under it. The resourceful workers decided to mix the uranium oxide with acid in a 10-liter stainless steel bucket instead. They could then tip the bucket, pour the solution into a five-liter glass Erlenmeyer flask, and then dump it directly into Precipitation Tank B, which had an electrically driven stirrer. This would save time by not having the solution sit around in little four-liter bottles, and the stirrer in Tank B would do the job a lot faster than just letting it drip through the dissolver. This plan indicated a weak understanding of the factors that lead to criticality. True, 45 liters of 18.8-percent enriched uranium solution is not critical, but only if it is in a geometry that does not encourage criticality, such as the long, thin tank. The 100-liter Precipitation Tank B was round and short, meant to incorporate as little expensive stainless steel as possible in its design, and it was therefore an ideal reactor vessel.

Ouchi stood on the metal platform surrounding the top of the tank, holding a glass funnel with his body draped over it. Shinohara climbed the metal steps to the platform, carefully cradling the flask full of solution, and poured it slowly into the funnel. Yokokawa sat at a desk nearby and completed the paperwork. By quitting time, they had successfully processed four batches, now sitting in Precipitation Tank B.

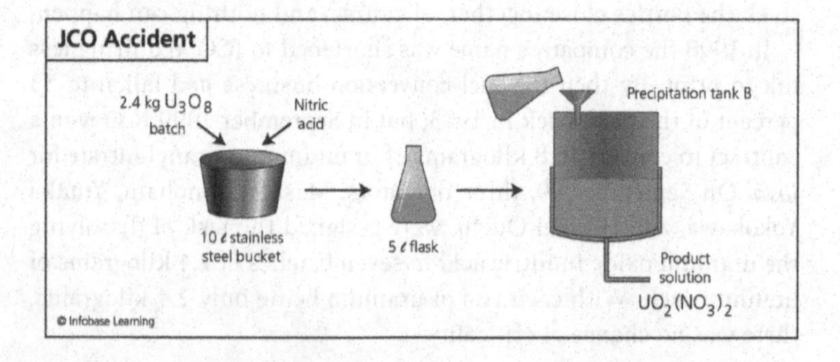

It is always dangerous to have a liquid containing uranium and a vessel of the right size and shape to make a reactor in the same building. By simplifying a transfer process, workers at JCO in Japan managed to make a supercritical reactor.

Next morning, it was more of the same. By 10:35 A.M. they had done two more batches, and they were almost through pouring the last of batch number seven into the tank. There were 0.183 liters left in the flask. Drip. Drip. There was a blue flash out the open port, lighting up the ceiling. Shinohara and Ouchi staggered down the steps, starting to feel strange. Then came extreme abdominal pain, waves of nausea, and difficulty taking a breath. Yokogawa looked up from his paperwork and turned in his seat, quizzical. The three workers had no idea what had happened, but the gamma-ray alarms were sounding. Ouchi had lost control of his muscles and was sinking into incoherence. His two fellow workers helped him out of the building. Someone had released gamma radiation somewhere in the plant, and they had to get out of the building. The unshielded reactor they had assembled in Precipitation Tank B was still running at power, boiling the uranium solution and broadcasting a deadly mix of gamma rays and neutrons in all directions.

Workers in all three buildings were streaming out and going to the emergency mustering point as the gamma-ray alarms rang everywhere. A worker from the building next door noticed that three guys from the Fuel Conversion Test Building looked injured and confused. He summoned an ambulance, and they were quickly removed to the nearest hospital.

People high up in the organization began to realize that there had to be a criticality on site, and it looked as if it was in the Fuel Conversion Test Building from the high gamma readings near it. No concentration of the gamma rays from uranium could make radiation with this intensity. Nuclear fuel is radioactive, but not *that* radioactive. Somewhere, a reactor was running. Studying the pen-chart recordings, they could see that it had started up with a large surge of supercriticality, then settled down into quasi-stable critical condition, and the power level dropped gradually by about half in the next 17 hours. They had to find a way to shut it down.

After 4.5 hours, radiation was detected beyond the plant's fence, with gamma rays and neutrons streaming into the streets of Tokaimura. The mayor suggested that people living within 0.2 miles of the plant should probably go somewhere else. After 12 hours, government authorities stepped in and suggested that people within 7.5 miles of the plant should stay indoors and not take deep breaths. The solution in the reactor was apparently boiling, with steam coming out the open port on top of the tank. Highly radioactive fission products, the scourge of nuclear power, were falling all around in a light, invisible mist.

A plan was worked out to kill the chain reaction, and workers volunteered to execute it shortly after midnight on October 1. Precipitation Tank B was water-cooled by a jacket encircling it. All they had to do was drain the water out of the cooling jacket and the reaction would stop. Neutrons reflected back into the uranium solution from the jacket were all that was maintaining the fission. It seemed simple, but it turned out that it was a lot easier to put water into the jacket than it was to remove it. The piping would have to be disassembled, and it could be done from outside the walls of the plant, but the radiation was still too high for men to work on the plumbing. They had to approach the pipes in relays, with each man allowed to absorb no more than 10 rem (0.1 sievert) of radiation.

The last drop of water was drained from the cooling jacket after the reactor had been running 17 hours. The power level dropped by a factor of four, but it leveled off. The thing was still critical. There was still water trapped in the system. It took three more hours, but the plant workers were finally able to shut it down by blowing out the water using pressurized argon gas. Just to make sure, they pumped a borax solution into the tank through a rubber hose.[183]

Hisashi Ouchi, 35 years old, had received about 1,700 rem of mixed radiation. He was burned over most of his body, and his white-blood-cell count had dropped to near zero. He died 82 days later of multiple organ failure. Masato Shinohara, 40 years old, expired on April 27, 2000, 210 days after the accident. He had absorbed 1,000 rem, teetering on the border of a fatal dose. Yutaka Yokokawa was hit with

---

183 Nothing shuts down a chain reaction faster than boric acid in the coolant. The boron-10 nuclide absorbs thermal neutrons voraciously. There is a legend in nuclear engineering about some janitorial workers who were tasked with cleaning up the inside of an aluminum research reactor vessel. They did a marvelous job, making it sparkle and shine, but they used 20 Mule Team Borax as the detergent. The reactor never again achieved criticality, with the boron scrubbed into the inside surface of the vessel. It doesn't take much.

300 rem. He left the hospital on December 20, 1999, and he is still alive and well.[184]

At least 439 plant workers, firemen, and emergency responders were exposed to high levels of radiation, as were 207 residents near the plant. Although their exposures were probably 1,000 times the normal background radiation, there have been no unusual sickness or radiation effects reported from these people. The mindset at all levels of the JCO organization and the government regulators had been that no such accident was possible, and therefore there were no accident plans, no review of work procedures, and little thought was put into the equipment layout. The workers were minimally trained, and the primary goal was for everyone to work more efficiently. The Japanese work ethic, for all its strengths, would have to be modified for this peculiar line of endeavor. The JCO uranium-conversion activities ceased in 2003, due to regulatory pressures and dwindling profits, and Japan's high hopes for nuclear power suffered along with the rest of the economy in a decades-long recession.

A summary of production disasters would be incomplete without mention of the Kyshtym catastrophe near the Mayak plant in Russia. It may go down in history as the worst release of radioactive fission products to have ever occurred, or it may not. Of all the significant nuclear accidents, this one was a black hole in the firmament of knowledge for many decades, locked up tightly by both the Soviet KGB intelligence service and the Central Intelligence Agency of the United States.[185] With so little information to go on, speculation

184 I am expressing radiation doses received in the archaic "rem" notation to try to keep them in context with the earlier accidents. To convert to sieverts, or Sv, divide the rem by 100. Radiation levels, dosage, or exposure is expressed in many ways, on technical grounds, and this can make it difficult to simplify explanations of the effects of radiation on human beings. Please bear with me.

185 Why did the CIA keep this locked up? It may have just been a product of the secret mindset of intelligence organizations. Having big secrets was important to the CIA, but if everybody knew about it, it would not be a secret anymore. More likely it was fear that if information concerning an accidental, massive contamination of a large patch of populated territory was released to the public, there would be mass hysteria and a popular call to bring an end to nuclear power. This would ruin the careful campaign of the AEC to promote nuclear power, get it off the ground, and transfer it to the public sector. If so, this was a case of one government agency looking out for the welfare of another.

ran rampant and wild theories rushed into the vacuum. All we had was a trickle of partial, confusing reports taken third-hand from some excitable defectors or exiles. We could not even tell when this contaminating incident had occurred, with dates ranging from 1954 to 1961. It looked as if several lesser incidents may have been woven together into a combined story.

What type of accident was it? It was variously described as an earthquake, a landslide, an accidental A-bomb detonation, a test-drop from a Soviet bomber, a reactor explosion, a graphite fire in a reactor, a meteor hit, and a steam explosion in a holding tank. No explanation made sense, and overflights by U-2 spy planes trying to find visual confirmation of a nuclear catastrophe were curtailed by the loss of Gary Powers's plane over Mayak. Articles, papers, and even books were written about it, but the mystery of what happened at Kyshtym would not be solved until the beginning of the end of the Soviet Union in 1989.

The first published inkling of a radiological problem in the East Ural Mountains claims to be the June 1958 edition of *Cosmic Voice*, the monthly journal of The Aetherius Society.[186] On April 18, 1958, George King, founder of the society in 1955, was contacted telepathically by two beings riding around in a UFO.[187] The first message was from an individual identified only by origin, Mars Sector 6:

> Owing to an atomic accident just recently in the USSR, a great amount of radioactivity in the shape of radioactive iodine, strontium 90, radioactive nitrogen and radioactive sodium have been released into the atmosphere of Terra.

186 I would prefer to think that there was an earlier leak of the story, possibly in the Danish newspaper *Berlingske Tidende* in April 1958. The event was not widely noticed by the Western press.

187 UFO is an Air Force term, meaning Unidentified Flying Object, or an apparently controlled machine moving through the atmosphere that cannot be classified by type, country of origin, manufacturer, or serial number. The Aetherius Society is technically a "UFO religion," in that it depends upon a belief in extraterrestrial entities operating UFOs. There are many such religions, the largest of which is Scientology.

This message, relayed through King's larynx and recorded on a reel-to-reel tape, was followed by a second pronouncement from The Master Aetherius from Venus:

> All forms of reception from Interplanetary sources will become a little more difficult during the next few weeks because of the foolish actions of Russia. They have not yet declared to the world as a whole, exactly what happened in one of their atomic research establishments. Neither have they declared how many people were killed there. Neither have they declared that they were really frightened by the tremendous release of radioactive materials from this particular establishment during the accident. . . .

The report goes on to claim that the Interplanetary Parliament will have to use an enormous amount of energy to clean up this mess. They were, however, able by Divine Intervention to save 17,000,000 people from having been "forced to vacate their physical bodies," which may be a euphemism meaning to die of acute radiation poisoning. They were given permission by the Lords of Karma to intervene on behalf of Terra, presumably in cooperation with Divine Intervention.

This brief announcement was followed by mention of the contamination accident in a book, *You Are Responsible!*, published by The Aetherius Society in 1961, and there it sat for another 15 years, with no further mention outside certain secretive compartments in the CIA.

On November 4, 1976, an article, "Two Decades of Dissidence," was published in the *New Scientist* magazine for its 20th anniversary issue. The author, Zhores A. Medvedev, a Soviet biologist, had gone into exile in London with his family in 1972 and gotten a job as senior research scientist at the National Institute for Medical Research. In his article, he mentioned that in 1957 or 1958 an explosive accident had contaminated a thousand square kilometers of territory in the Ural Mountains with radioactive debris. Hundreds of people were killed, thousands had to be evacuated, and the area would be a danger zone from now on. The *New Scientist*, aware of the stale claim by The Aetherious Society, proclaimed

"Scooped by a UFO!" on a back page. Medvedev's story was roundly denounced as "science fiction," and was thrown into the same box with the Aetherions.

To be fair to the naysayers, the story was unbelievable. If a ground-level nuclear device had accidentally detonated, its signature of high-thrown radioactive dust would have been picked up by Western governments within hours of the blast, and the same was true if a bomber-dropped weapon had been used to test a simulated city on the ground. It was hard to hide such tests or accidents.[188] It was true that the Soviets had six graphite reactors at Mayak, which was then known as Chelyabinsk-40, but it would be hard for a graphite reactor fire or even a graphite steam explosion to kill hundreds of people unless they were standing on top of it, and how were thousands injured, needing immediate hospitalization? Not even Soviet engineering would place production reactors so close together that an explosion in one could set off the others. A single plutonium production reactor, for all its buildup of radioactivity, is limited in the damage it can do. While it would be possible to contaminate thousands of square kilometers with plutonium and fission products, this would be a low-level contamination. It would be a long-term danger and not producing immediate injuries. A seismic event that would swallow Chelyabinsk-40 would have showed up on seismographs around the world, and a destructive landslide in the heavily wooded Urals seemed unlikely.

Medvedev reasoned that it must have been an underground storage tank filled with waste products from the plutonium processing, heating up in the confined space and causing a sudden, massive steam explosion from the water content of the solutions. It seemed almost plausible, but fresh, concentrated fission waste, straight out of a reactor that had been running at full power, only generates about 60 kilowatts of heat per ton. After a year of sitting quietly, the same

188 On the other hand, in 1958 there were more than 100 above-ground nuclear weapon tests in the world, with 71 detonations carried out by the United States alone. The suspended fission-product dust in the atmosphere was getting so dense, it was hard to tell that another bomb had been set off. The U-2 high-altitude photographs would have clearly shown effects of ground-level destruction, but if such evidence existed, would it have been released by the CIA?

waste is putting out 16 kilowatts of heat, and after ten years the rate has fallen to 2 kilowatts per ton. This is certainly enough heat to melt through a steel tank or even to blow the thing wide open, but it lacks the power concentration to cause the reported level of mayhem. Even if left-over plutonium were to make a supercritical mass in a waste tank, it would simply boil the water furiously until it modified its own configuration into subcriticality. As we have seen in all the criticality accidents, to be killed you must have been embracing the reactor. The reported power of the blast did not correspond to what would be contained in a steel storage tank.

Shortly after Medvedev's article appeared in the *New Scientist*, articles appeared in three British newspapers seeming to confirm his incredible story. The newspaper articles, all appearing on December 7, 1976, linked back to a letter to the editor in the *Jerusalem Post*, sent in by another ex-Soviet, Professor Leo Tumerman, former head of the Biophysics Laboratory at the Moscow Institute of Molecular Biology. Tumerman had written to strongly disagree with Medvedev's assertion that the supposed accident could have been the result of a reactor explosion. It was common knowledge in Russia, he claimed, that the catastrophe was the result of gross negligence on an industrial scale. He was not sure how, but the careless storage of radioactive wastes at Chelyabinsk-40 had resulted in massive destruction.

Tumerman had not been there at the time, but in an automobile trip in 1960 he had seen evidence of a disaster with his own eyes. He had been visiting his brother, an engineer, at the construction site of the Byeloyarsk power reactor, about 300 kilometers from Sverdlovsk in the Southern Ural Mountains. From there he had to drive about 180 kilometers to Miassovo, near Chelyabinsk-40, for a summer seminar on genetics. He reached the main highway heading south at about 5 A.M., and soon he passed a large declarative road sign. It was a warning to all drivers: DO NOT STOP FOR THE NEXT 30 KILOMETERS! DRIVE THROUGH AT MAXIMUM SPEED!

The next 30 kilometers of highway were quite strange. As far as the eye could see, there was nothing there. No cultivated fields or pastures. No herds of cows. No people. No birds. No insects splatting against the windshield. No towns or villages. No trees. There were only chimneys sticking up all over the place, with no houses connected to them.

Curious about what he had seen on the drive to Miassovo, he got an earful from some seminar participants. The whole countryside was hot from radiation contamination. It was caused by an explosion at either the plutonium production plant or a waste tank. Details were fuzzy, but thousands had been evacuated permanently, and their houses were burned down to prevent looters from hauling away contaminated objects and spreading the radiation farther than it was. Everyone called it the "Kyshtym Disaster." It actually occurred at Chelyabinsk-40, but Chelyabinsk-40 officially did not exist, and Kyshtym was the nearest town on the map. No further details were forthcoming, and experts were puzzled. Finally, in 1989, formerly secret files concerning the Kyshtym Disaster started finding their way out of the crumbling Union of Soviet Socialist Republics, and the unexpected truth began to crystallize. Nobody had speculated correctly.

Under the Stalin and later the Khrushchev governments in the Soviet Union, the safety of the environment and even workers was not exactly a primary concern. The six graphite plutonium-production reactors at Chelyabinsk-40 used open-loop water cooling, pumping water out of Lake Kyzyltash and dumping it back in. At first, liquid waste from the plutonium extraction plant was simply dumped into the River Ob and allowed to empty into the Arctic Ocean. In 1948, plutonium production was at a fever pitch, and there was no time to work out the details of making the process efficient. It was extremely important that a nuclear weapon be successfully tested before the official celebration of Stalin's seventieth birthday, which would be on December 18, 1948. They did not quite make it, but the RDS-1 plutonium-fueled bomb was successfully tested on August 29, 1949. The rest of the world was stunned by this development, thinking that the Soviet Union was farther behind than that.

In the crash program to produce fissile bomb material, a great deal of plutonium was wasted in the crude separation process. Production officials decided that instead of being dumped irretrievably into the river, the plutonium that had failed to precipitate out, remaining in the extraction solution, should be saved for future processing. A big underground tank farm was built in 1953 to hold processed fission waste. Round steel tanks were installed in banks of 20, sitting on one large concrete slab poured at the bottom of an excavation, 27 feet deep. Each bank was equipped with a heat exchanger, removing the heat buildup from fission-product decay using water pipes wrapped around the tanks. The tanks were then buried under a backfill of dirt. The tanks began immediately to fill with various waste solutions from the extraction plant, with no particular distinction among the vessels. The tanks contained all the undesirable fission products, including cobalt-60, strontium-90, and cesium-167, along with unseparated plutonium and uranium, with both acetate and nitrate solutions pumped into the same volume. One tank could hold probably 100 tons of waste product.

In 1956, a cooling-water pipe broke leading to one of the tanks. It would be a lot of work to dig up the tank, find the leak, and replace the pipe, so instead of going to all that trouble, the engineers in charge just turned off the water and forgot about it.

A year passed. Not having any coolant flow and being insulated from the harsh Siberian winter by the fill dirt, the tank retained heat from the fission-product decay. Temperature inside reached 660° Fahrenheit, hot enough to melt lead and cast bullets. Under this condition, the nitrate solutions degraded into ammonium nitrate, or fertilizer, mixed with acetates. The water all boiled away, and what was left was enough solidified ANFO explosive to blow up Sterling Hall several times, being heated to the detonation point and laced with dangerous nuclides.[189]

Sometime before 11:00 P.M. on Sunday, September 29, 1957, the bomb went off, throwing a column of black smoke and debris reaching a kilometer into the sky, accented with larger fragments burning orange-red. The 160-ton concrete lid on the tank tumbled upward into the night like a badly thrown discus, and the ground thump was

189 ANFO (Ammonium Nitrate/Fuel Oil) is a tertiary explosive, commonly used as a blasting agent, consisting of a mixture of an oxidizer, ammonium nitrate, plus a flammable organic compound. On August 24, 1970, one ton of ANFO loaded into a Ford Econoline van was parked in front of the physics building, Sterling Hall, at the University of Wisconsin-Madison. As a protest against the university's military research, the explosive mixture was detonated at 3:42 A.M., causing massive destruction. Parts of the van were found three blocks away on top of an eight-story building, and overall damage to the university campus was over $2.1 million. A radical anti-war group called the "New Year's Gang" claimed responsibility, and one member, Leo Burt, remains at large.

felt many miles away. Residents of Chelyabinsk rushed outside and looked at the lighted display to the northwest, as 20 million curies of radioactive dust spread out over everything sticking above ground. The high-level wind that night was blowing northeast, and a radioactive plume dusted the Earth in a tight line, about 300 kilometers long. This accident had not been a runaway explosion in an overworked Soviet production reactor. It was the world's first "dirty bomb," a powerful chemical explosive spreading radioactive nuclides having unusually high body burdens and guaranteed to cause havoc in the biosphere. The accidentally derived explosive in the tank was the equivalent of up to 100 tons of TNT, and there were probably 70 to 80 tons of radioactive waste thrown skyward.

It took a while for the government to rush into damage-control mode. A week later, the Chelyabinsk newspaper published a cheery story concerning the rare display of northern lights in the sky last Sunday, showing "intense red light, sometimes crossing into pale pink and pale blue glow." It "occupied a large portion of the southwest and northeast part of the sky." At the same time, massive evacuation measures were enforced, eventually emptying 22 villages along what would become known as the "East Urals Radioactive Trace," or the EURT. No explanation was given as to why everybody had to leave. Over the next two years, around 10,000 people were permanently relocated. The reason for storing nitrate solution and organic solution together in the same tank has not been revealed. The EURT is now disguised as the "East-Ural Nature Reserve," as an explanation for its prohibited access.

Although about 475,000 people were probably exposed to dangerous levels of radiation due to this incident, figures detailing radiation sickness and deaths are simply not available, even with the KGB files broken open and published. Refugees from the area reported that all hospitals within a hundred kilometers were inundated with people affected by the blast, coming in with burned skin, vomiting, hair loss, and every symptom of having survived an atomic bomb detonation. Hundreds of immediate deaths are commonly quoted, with thousands of sickened survivors. As has been noticed time after time in mass nuclear disasters outside the plant gates, an information blackout can turn a healthy population into a suffering mob just from the twisted psychology of fear and dread. Rumors can make people sicker than radiation exposure.

Studies of the effects of this disaster are extremely difficult, as records do not exist, and previous residents are hard to track down. A late study by the former Soviet Health Ministry cites 8,015 delayed deaths due to radiation effects in the area from 1962 to 1992, but on the other hand only 6,000 death certificates from all causes of death have been found. Add to that the possibility that just about everybody over 12 years old in the area smoked Turkish cigarettes, and cause of death is a toss-up between lung cancer and the effects of alcoholism. There are no hard records of immediate deaths due to the chemical explosion or acute radiation poisoning on site. Recent epidemiological studies suggest that 49 to 55 people along the EURT have died because of radiation-induced cancer, and at what is now the Mayak plant, 66 workers suffer from chronic radiation sickness dating back to 1957.

All this ranks the Kyshtym Disaster as possibly the worst, most senseless catastrophe in the history of nuclear power. Hopefully the conditions that caused it have subsided and this will never happen again. There would be more mischief in the Union of Soviet Socialist Republics as the world became more information-conscious, and we will have a hard look at it. But first, let us examine how all those nuclear weapons, cited as necessary for world peace, were handled with loving care by the Armed Forces.

Chapter 8

THE MILITARY ALMOST NEVER LOST A NUCLEAR WEAPON

"The information in it is pretty good. In some cases, a little too good."

—Tom Clancy, referring to Chuck Hansen's *US Nuclear Weapons*

IN 1993, I WAS AWARDED a contract to solve a problem for the Defense Nuclear Agency (DNA), a now-vanished branch of the federal government that was responsible for placement and security of thousands of nuclear weapons.[190] It tried to be inconspicuous to the point of invisibility, and its

190 DNA was the Defense Atomic Support Agency (DASA) from 1959 until 1971, when it became DNA. In 1996 it was changed to the Defense Special Weapons Agency (DSWA), and in 1998 it was combined with three other agencies to form the Defense Threat Reduction Agency (DTRA). Every time it changed, they had to change the seal and all the stationery. I liked the old DNA seal, the blue background of which was peppered with little mushroom clouds. The latest seal has the Eagle of State looking to his right, at the olive branch of peace, but in his left talon he holds an A-bomb instead of a bunch of arrows.

headquarters were away from the bustle of Washington, D.C., in a plain, unmarked building out on Telegraph Road in Virginia. It was, however, obviously a secret government building, because the entrance was completely blocked to any vehicular traffic, in addition to the fact that it did not have an actual numerical street address, as far as I could tell.

On my first meeting with my sponsor, he started the briefing with a stern lecture detailing the fact that the armed forces had never lost a nuclear device. Always one to assert superior knowledge when possible, I casually blurted out that Chuck Hansen had documented 32 instances of atomic or hydrogen bomb loss between 1950 and 1980.

This was the wrong place and the wrong time to bring this up. Showing a marvelous level of self-control, my sponsor launched into a spirited criticism of Chuck Hansen, his abilities as a historian, and any of the people who had divulged this information to him. I felt the skin peeling off my face.[191]

Did the Air Force ever lose an A-bomb, or did they just misplace a few of them for a short time? Did they ever drop anything that could be picked up by someone else and used against us? Is humanity going to perish because of poisonous plutonium spread that was snapped up by the wrong people after being somehow misplaced? Several examples will follow. You be the judge.

Chuck Hansen was wrong about one thing. He counted thirty-two "Broken Arrow" accidents.[192] There are now *sixty-five* documented

---

191 Chuck Hansen published *US Nuclear Weapons: The Secret History* in 1988, printed by Aerofax, Inc. and distributed to the trade by Orion Books. He collected information concerning nuclear weapons with 30 years of dogged use of the Freedom of Information Act. His collected papers are now housed in the National Security Archive at George Washington University. Hansen was 55 years old when he died of brain cancer in 2003.

192 A *Broken Arrow* is an accident involving a nuclear weapon, warhead, or components that does not cause World War III. A *Bent Spear* is an accident or a mistake made while transporting a nuclear device from one location to another. An *Empty Quiver* occurs when a functioning nuclear device is stolen or lost. A *Dull Sword* happens when you have a perfectly good nuclear device but some malfunction or damage to the equipment means that you cannot set it off. A *Faded Giant* is a malfunctioning military power reactor. A *NUCFLASH* is the unauthorized deployment of a nuclear device, such as launching an ICBM or flying away with a loaded strategic bomber when not under orders to do so. The prefix *Pinnacle* added to any of these terms makes it immediately reportable to the Chairman of the Joint Chiefs of Staff. A *Pinnacle NUCFLASH* is really bad. This terminology is not used globally, but it is detailed in DoD Directive 5230.16, "Nuclear Accident and Incident Public Affairs (PA) Guidance," and other high-level documents.

incidents in which nuclear weapons owned by the United States were lost, destroyed, or damaged between 1945 and 1989. These bombs and warheads, which contain hundreds of pounds of high explosive, have been abused in a wide range of unfortunate events. They have been accidentally dropped from high altitude, dropped from low altitude, crashed through the bomb bay doors while standing on the runway, tumbled off a fork lift, escaped from a chain hoist, and rolled off an aircraft carrier into the ocean. Bombs have been abandoned at the bottom of a test shaft, left buried in a crater, and lost in the mud off the coast of Georgia. Nuclear devices have been pounded with artillery of a foreign nature, struck by lightning, smashed to pieces, scorched, toasted, and burned beyond recognition. Incredibly, in all this mayhem, not a single nuclear weapon has gone off accidentally, anywhere in the world. If it had, the public would know about it. That type of accident would be almost impossible to conceal.

The danger of an accidental nuclear detonation was given full attention by design engineers before the first atomic bomb was dropped on Japan in 1945. War machines and materials have never been safe to handle or to stand near, and accidents happen all the time, whether in the chaos of battle or on a quiet Thursday morning. Chemical munitions can accidentally blow up entire towns, but the introduction of nuclear explosives kicked the danger level up by a factor of a million, and serious engineering went into ensuring that it would never happen.[193]

Any nuclear explosion begins with a chemical detonation, used to assemble or compress fissile material, and there is only so much you

---

193 On the morning of December 6, 1917, two ships collided in the Narrows of the Halifax, Nova Scotia harbor. The SS *Mont-Blanc* was fully loaded with explosives bound for Europe, and at 8:45 A.M. it tried to occupy the same space as the SS *Imo* from Norway, headed to New York to pick up supplies. *Mont-Blanc* promptly caught fire, and its crew abandoned ship, knowing that it was about to explode and rowing as fast as possible. The burning ship drifted gracefully into Pier 6 and disintegrated in a blinding fireball at 9:04 A.M. The chemical explosion was the equivalent of a 2.9-kiloton nuclear weapon. The entire community of Richmond was obliterated, 2,000 people were killed, 9,000 injured, and the shock was felt as far away as 220 miles in North Cape Breton. Buildings shook and shelves were rearranged 62 miles away. The harbor was scoured dry for an instant, then water rushed into the void and caused a tsunami that wiped out yet another community in Tuft's Cove. The following day, the remains of Halifax were buried under 16 inches of snow, complicating relief efforts.

can do to prevent it from going off accidentally. The explosive fission and nuclear fusion, however, are different. Conditions for the nuclear part to happen are subtle and precise, and prevention of an unplanned destruction of lives and property by nuclear means has been remarkably successful. In the early 1950s, the one-of-a-kind bomb dropped on Nagasaki was hastily modified to make it safe to handle. The round, plutonium ball at the center, or the "pit," in the original bomb had to be placed at the center of the shells of explosive material in the complex assembly procedure. When loaded into the aircraft bomb bay, there were few reasons why it should not explode. A couple of SAFE plug-ins had to be manually replaced with ARMED plug-ins to enable the electrons inside, and it had to actually drop free of the airplane, which would jerk out a set of enabling wires on top of the bomb casing. These were not considered adequate measures, and there was no way to fly a "safe" bomb in this configuration.

This problem was solved by opening a round hole in the front of the big ball of chemical explosive. There was a removable piece of explosive, which was solid and looked like dull brown plastic, made to fit in this hole. For normal transport or just cruising around with a device hanging in the bomb bay, the center of the bomb was left empty, with no pit installed. The pit was stored separately as a "capsule," to be installed on demand with 30 minutes of handiwork in the bomb bay. The flight crew member turned designated armorer would remove the cover on the nose or tail of the weapon, unhook the explosive segment, pull it out, and push the plutonium sphere into place. Replace the explosive segment, button up the hole, and she's ready to rock and roll.

The capsule was carried in an open, rectangular metal box made of aluminum tubing, called the "birdcage." It was designed to prevent anyone from storing two or more pits close together, as they were quite capable of becoming an out-of-control mini-reactor if within a foot of each other. The capsule was stuffed inside a metal tube at the center of the birdcage, closely resembling an old-fashioned soda-acid fire extinguisher, including the handling hoop, looking like a halo atop the canister.

The arming operation was not quite as easy as it sounds. The work area was not pressurized, and it was a cramped space. The special T-handle wrench was conveniently clipped to the wall of the bomb bay, but tools were easy to drop as hands froze in the high-altitude environment. It was a good first step. This procedure was improved soon by introduction of the Automated In-Flight Insertion (AIFI) mechanism. The capsule tube was mounted with the bomb in the bomb bay, with the explosive hole left open. Nothing that could happen to the airplane, from an electrical fire in the bomb bay to a vertical dive into the ground, could cause a nuclear explosion, because the pit could not be uniformly compressed without being at the center of the sealed explosive sphere. On a moment's notice, a crew member could arm the device by pushing a button on the bomb control panel. An electric-motor-driven screw pushed the capsule into the open hole in the bomb and closed it. In a minute, the red light came on indicating that the device was armed and ready to drop.

The first weapon using manual in-flight capsule insertion was the MK-4 Mod 0, made available on March 19, 1949, after two years of development. The last device to use manual insertion was the MK-7, used in the atomic depth charge named "Betty." It was finally retired in June of 1967 after 15 years on call as an anti-submarine weapon. The last bomb to use AIFI was probably the MK-39 mod 1, a big, ugly, gravity-dropped thermonuclear device. It was taken off the inventory in September 1965.

In every case of the airborne loss of one of these early bombs, the capsule was separated from the device, and in some instances it was not even in the airplane. In 1955 there was a radical change in bomb-core design, in which the traditional solid plutonium sphere was replaced with a thin, round shell of plutonium, collapsed by implosion onto a void. With this interesting design came a new paradigm for bomb safety. No longer would anything have to come apart and travel as two separate pieces, and the device would be a sealed, integral unit, proclaimed "one point safe."

In this new design, when the hollow shell of fissile material is explosively collapsed, the bomb does not have the hyper-criticality characteristic of a nuclear explosive. In fact, it reaches a point just short of simple criticality, in which as many neutrons are being produced by fission as are being lost. Explosive criticality requires enough fissile material concentrated to be more than two critical masses in the same spot. The only way to simulate hyper-criticality in the new bomb is to introduce a heavy burst of high-energy neutrons at the center, as if the mass were fissioning way out of control. This is accomplished using "boosted fission" after fission is forced to start using an electrical neutron initiator. To achieve boosted fission, a mixture of tritium and deuterium gases, held in a small pressure tank, is vented into the void at the center of the plutonium shell right before detonation.[194] The extreme temperature and pressure exerted on this gas mixture as the bomb implodes causes a percentage of the deuterium and the tritium, both heavy isotopes of hydrogen, to fuse into helium-4. Left over from each reaction is a neutron, clocking out at an impressive 14.1 MeV and very likely to cause fission in the collapsing metal ball. Just before the fusion starts, a sudden blast of neutrons comes forth from the initiator, an electrically driven deuteron accelerator, and sets off fission artificially, adding energy to the implosion event.

If the exact sequence of events, the implosion start, the initiator start, and the hydrogen fusion at the center, do not occur with perfect timing, then the device will not explode in a nuclear fashion. An accidental ignition of the chemical explosive used for the implosion caused by fire, shock, or an errant electrical signal will blow the bomb to pieces and it may even crush the fissile ball into a tiny kernel, but it will not cause a nuclear event. Furthermore, the fusion components in a thermonuclear device, the most powerful explosive ever made, require the explosion of an imploding fission bomb just to set them off. Without the fission device working, a hydrogen bomb is little more than dead weight.

A further advantage to this improved design is that the yield of the bomb can be dialed in from the control panel. A 5-megaton device can

194 There are at least five different boost tank designs. The most common one looks like a bratwurst—a small cylinder, rounded at both ends, with the electrically actuated release valve built into the center of the tank. One looks like a pipe elbow, and three designs look like squat cylinders with truncated conical ends, all with the valve attached at one end. In released documents they are never referred to as deuterium-tritium tanks, but only as tritium tanks or tritium cylinders. With only tritium in the tank, the fusion scheme would not work. There were cases in which tritium and deuterium were kept in separate cylinders, to be mixed upon injection. This allowed the tritium, which would deteriorate with age, to be replenished on a schedule.

be scaled back to a 1-megaton explosion just by turning a knob and pushing a YIELD SET button. This action sets the open interval on the gas-release valve, controlling the amount of deuterium-tritium mixture that is injected into the void at the center of the imploding sphere. More gas means more fusions in the explosion and thus more generated neutrons and a larger bomb yield. The energy release by the fusions that takes place in the center of the fission bomb does not add much to the explosion, but the dense cloud of high-speed neutrons produced by fusions determines how much of the fissile material burns before the explosive disassembly of the device. Full use of the boost gas also means that the fissile material, which can be plutonium, uranium, or both, is used more efficiently and less is needed than in the old-style bombs. The new bombs are lighter and more compact.

The neutron initiator is a vacuum tube, having an electrically heated filament anode coated with uranium deuteride. Downstream from this ion source is a cathode consisting of a thin metal plate coated with uranium tritide. A potential of 500,000 volts is established between the anode and the cathode, and deuterium ions (deuterons) boil from the uranium deuteride and are attracted to and vigorously accelerated by the cathode. Splattering into the cathode, the deuterons fuse occasionally with tritium held in the uranium, and a burst of neutrons develops. The neutron flux is amplified by the uranium content of the cathode, which tends to fission and release 2.4 neutrons for each received.[195]

The safety measures used in the nuclear bombs and warheads were well designed and effective, but there were still problems with the

195 This is an educated guess. The vacuum-tube neutron initiator is among those aspects of nuclear weapon design which remain classified SECRET. However, vacuum-tube electrical neutron generators are fairly common industrial items, and all work using similar principles. The technique for making neutrons electrically was first patented in Germany in 1938. For practical reasons, electrical initiators are not used in missile warheads, where a solid-state, explosive thing that looks like a roll of quarters rounded off at one end is employed instead. It uses the old-school polonium-beryllium neutron source scheme, with the two metals explosively mixed together. Sandia Labs in New Mexico has recently developed a new, solid-state neutron generator called the neutristor, based on integrated circuit technology. If used in nuclear weapons, the neutristor would advance bomb technology, developed in the vacuum-tube days of the 1950s, by at least 30 years. Part number for the neutron generator on the MK-28 bomb was MC-890A.

delivery systems.[196] The end of World War II was the beginning of a very exciting quarter of a century for the United States Air Force, particularly with the steep rise in strategic bomber capability, radical changes in strategy, and one very ambitious enemy, the Union of Soviet Socialist Republics. The new ultimate weapon would not be carried in formations of a thousand bombers, but by one lone aircraft. One city, one bomb. Starting with the propeller-driven B-50 Superfortress, an upgrade of the old B-29 from the last war, the Air Force would lunge forward, developing an ever-advancing line of exotic airplanes to carry the nuclear bombs. These aircraft would have to fly higher, farther, and faster than anything had before, and new boundaries would be drawn describing the limitations of men and machines. Each new bomber was built way out on the edge of what was possible, using new materials and design techniques, and none were forgiving of a slight piloting error, a switch left in the wrong position, or a gas cap left off. Plane crashes, explosions, and disappearances occurred with appalling frequency.

The first notable strategic bomber of the Cold War was the Convair B-36 Peacemaker, the "aluminum overcast," a 72,000-pound monster with a 230-foot wingspan. It was made airborne using six radial engines driving enormous propellers, swinging in circles 19 feet wide. Together, the engines developed 22,800 horsepower, but to top it off four General Electric J47-19 jet engines were added, two under each wing tip. Before takeoff, 336 spark plugs had to be changed, 600 gallons of oil were added to the engines, and a tanker truck had to top off the fuel after the B-36 had lumbered into takeoff position on the runway. For a while starting in 1949, it was the only airplane we had that could deliver atomic bombs to targets in the Soviet Union. It had a range of over 6,000 miles, and its bomb bay was the only one ever made

196 There was another problem with nuclear bomb safety to deal with: a bomber crew could go rogue and decide to deploy a bomb or warhead without an order to do so. In early strategic bombers, all the bombardier had to do was push some buttons, and away she goes. To fix this flaw in the system, the Permissive Action Link (PAL) was devised by the Sandia National Laboratories and implemented across the board by September 1962. The system was basically a mechanical combination lock, with the unlocking code known only at the executive level. The USAF Strategic Air Command got around this restriction for the Minuteman ICBM force by having the lock codes set to all zeros, a terrifying fact discovered in a 1977 shakedown.

that was big enough to carry the enormous MK-17 thermonuclear device.[197] It took a crew of 15 men to fly it, and amenities included a dining room, six bunk beds, and a wheeled trolley to ride from the tail to the nose. It was very prone to engine fires, due to the backward mounting of the engines, the fuel tanks in the wings would develop cracks and leak, and the vibration caused by firing the machine guns could disrupt the vacuum-tube electronics in the control system and send the big bird into an irrecoverable vertical dive. By the time it was retired from bomb-carrying service in 1955, B-36 airplanes had been involved in three crashes involving nuclear weapons.

In June 1951, the Air Force was introduced to its new, highly advanced jet-powered strategic bomber, the Boeing B-47 Stratojet. Everything about it was new, from its back-swept wings to its extensive electronic systems, and its many innovations helped Boeing to build its successful line of modern jet airliners. It was built like a fighter plane with six engines and a bomb bay, and it was flown by three very busy men: the pilot, the co-pilot, and the navigator/bombardier. It could almost break the sound barrier, and it remained on SAC alert, ready to fly with nuclear bombs against an enemy threat at a moment's notice, until 1965. But it still had a few bugs.

It liked to take off slow and land fast. Its hesitancy to leave the ground was solved by attaching groups of solid-fuel rockets to either side of the airframe, pointed slightly down. The 18,000 pounds of thrust from the rockets were a big help. It wanted to land at an exciting 207 miles per hour, and its design was so aerodynamically clean, it would rather continue flying and not touch down. This situation was helped by popping out a 32-foot drogue parachute from the tail for landing. This would cut the speed and shorten the landing roll without burning up the brakes. At optimum fuel economy altitude, about 37,000 feet, the airspeed had to be maintained precisely, with an

197 Although the B-36 could deliver the MK-17, there were doubts that it could make it back from the mission. The MK-17 was a high-yield weapon, 15 to 25 megatons, and the B-36 was a remarkably slow aircraft. With all the guns and nonessential equipment stripped out, it could probably reach 423 miles per hour, and at that speed it could not get out of the way of the fireball when the bomb went off, even if the bomb were dropped with a parachute. At the Operation Castle test in 1954, B-36s were flown at reasonable range to a 15-megaton explosion. They suffered extensive blast damage.

acceptable error of plus or minus three miles per hour. Four miles per hour too slow, and the plane would fall out of the sky. Four miles per hour too fast, and the wings came off. At lower altitudes it was easy to go faster than 489 miles per hour, but do so and the controls would reverse, going down when you say up.

The B-47 was a remarkable airplane and a forward leap in design, but it was involved in ten nuclear weapons incidents, the most interesting of which involved accidentally dropping bombs within the boundaries of the United States.

Learning a lot from having built the B-47, Boeing designed a new swept-wing bomber using eight jet engines mounted under the wings. The Boeing B-52 Stratofortress strategic bombers, introduced in February 1955, have been in service for an impressive 58 years, and they will probably be phased out around 2045. The grandchildren of people who flew the original batch of B-52s could be flying B-52s today. The last B-52H was built in 1962, and this last group of 85 planes still in service has been modified and improved several times. These bombers can go 650 miles per hour and climb to 50,000 feet with a range of 10,145 miles, and they have broken many flight records. They have flown around the world non-stop in 45 hours 19 minutes with in-flight refueling, and can fly from Japan to Spain with one load of fuel. A B-52 can land sideways in a heavy crosswind, using its in-board landing gear with coupled steering.

The flight crew can vary between six and nine, sitting in a fuselage that has work stations on two levels. A B-52B first dropped an MK-15 mod 2 thermonuclear device over Bikini Atoll on May 21, 1956, in the test code-named *Cherokee.*[198] A B-52 has never dropped a nuclear weapon in warfare, although it has bombed targets in several wars since 1955. B-52s put out the lights in Baghdad in Operation Desert Storm, 1991, with gravity bombs and launched one hundred air-launched cruise missiles into Iraq in Operation Iraqi Freedom, 2003.

---

198 The mod 2 on the MK-15 bomb was interesting. This modification gave the bomb a piezoelectric contact detonator, so that it would explode when it hit the ground. Most nuclear weapons were made to explode in the air, at least 1,000 feet up, so as to cover as much ground as possible with the shock wave while avoiding making dust. The contact explosion made the mod 2 useful against deeply buried targets. The experimental TX-15 version, code-named *Zombie*, delivered an impressive 1.68 megatons of explosive energy in the *Nectar* test of operation *Castle* on May 14, 1954. The mod 2 device carried a small drogue chute to slow its descent.

B-52s were involved in six nuclear weapon accidents. Common problems in the early versions of the plane were vertical stabilizers and wings breaking off in mid-flight. These structural deficiencies were identified, and all B-52s in service were eventually sent back to the Boeing factory for strengthening retrofits.

Next up was the B-58 Hustler, another wild experience in aircraft design from Convair of Fort Worth, Texas. Introduced to the Air Force on March 15, 1960, the B-58 could hasten along at Mach 2—twice the speed of sound. It looked like something out of science fiction, with its rakishly swept delta wings and its slender body, sharply pointed at both ends with a slight waist at the center, and four big General Electric J79 jet engines on pods under the wings. It was the first military aircraft in the inventory that was all transistorized, with computers monitoring and controlling everything. If the system detected a fault somewhere in the vast interplay of electronic, hydraulic, and human systems, a sexy recorded female voice, supplied by actress Joan Elms, would coo a warning into the communication system.

Everything in the airplane was from out on the frontier limb, including the material from which it was made. A special alloy of magnesium and radioactive thorium-232, "mag-thor" or HK-31, was used, giving the airframe superior strength and temperature endurance for supersonic flying. The thorium content gave the airplane a bigger radiation signature than the thermonuclear device it carried, but it was harmful only if ingested. There was no room in the fuselage, thin as a supermodel, for a bomb bay. The big, fat bomb was carried along with jet fuel inside the detachable auxiliary gas tank, slung underneath.[199] It was not easy to fly, and, despite all the electronic help, the three-man crew was kept busy. Altogether, the surfeit of innovation contributed to the accidental loss of 26 B-58s, over 22 percent of the 116 that were built. In ten years of flying around with bombs on

199 This cohabitation of fuel and bomb in the MB-1C pod did not last long. Predictably, there were unsolvable problems with fuel leaking into the bomb and out of the pod. This design was replaced with the TCP (two-component pod), in which either the fuel tank or the bomb could be dropped independently. The introduction of the 1-megaton MK-43 aerial bomb in April 1961 made this possible. The B-58 could carry four of these weapons along with the TCP. The MK-43 was interesting in that the wrenches, H745 and H1210, used to arm it were stored in a neat compartment recessed on the left side of the bomb.

board, only one nuclear-weapon incident occurred on a B-58, when in 1964 a fully loaded Hustler slid on ice at Bunker Hill AFB, Indiana, while trying to turn onto the runway. The landing gear hit a concrete electrical box, the plane caught fire, and the five nuclear weapons on board were destroyed.

With this combination of complex weapons, tricky flying machines, and men working under stress, packed down by a towering bureaucracy and a constant threat of war breaking out, it is a wonder that civilization survived. Somehow, the safety systems on both sides of the Cold War seem to have functioned well. But it was still not foolproof.

The era of nuclear-weapon accidents began on February 13, 1950, when B-36 no. 44-92075 was flying between Air Force bases Carswell in Texas and Eielson in Alaska on a simulated combat mission. The weapon on board was an MK-4. Its removable composite plutonium-uranium capsule, type 110, was replaced with a lead dummy for training purposes. Aside from the core, it was a complete MK-4, all 10,850 pounds and looking like a black teardrop with fins. The plane was an early, Model B example of the B-36, built before the jet engines were added.

The bomber had flown six hours when it started to ice up. It was at 12,000 feet, and the ice forming on the wings and propellers was so heavy, the crew could not climb out of the icing conditions. After a while, three engines caught fire. The Pratt & Whitney rotary engines were highly advanced for the time, but they were piston engines, mounted backwards on the wings. Usually, rotary engines were mounted with the carburetors in back, warmed by the hot air streaming through the cooling fins on the engine cylinders, but in this case the carburetors were in front, sucking in cold air with no advantage from engine heat. Ice formed on the air intakes, restricting the airflow and causing the fuel-air mixture to become mostly fuel. Raw aviation gas started to blow out the red-hot exhaust pipes, and fire was inevitable. With half the engines out, the lifetime of the airplane was severely limited.

The crew turned the bomber out over the Pacific Ocean, cranked open the bomb bay doors, and released the MK-4. It whistled off into the darkness, fell 5,000 feet, and made a bright flash as its airbust fuse detonated the implosion charge. Five seconds later, the shock wave rattled the bones of the big, crippled plane. The crew turned the aircraft around and made for land, steadily losing altitude. As soon as they saw Princess Royal Island below them, the crew bailed out. The copilot took one last look at the plane as he rolled over before his chute opened. He saw a brilliant blue-white stream of flame from an engine extending back to the tail. It appeared that the engine's magnesium heat exchanger had caught fire, and he guessed that the plane would self-destruct in a ball of flame any minute now as the fuel tanks ignited. Every crew member made it back alive.

Abandoned, the B-36 leveled out and turned due north. It flew inland for 200 miles, finally running out of gas and crashing on the remote, perpetually snowed-in Mount Kologet, British Columbia.[200] The MK-4 presumably blew itself to unrecoverable smithereens using the implosion explosives, without causing a full-nuclear event.

The next loss of an MK-4 nuclear weapon occurred two months later, when a B-29 carrying it took off at night from Kirtland Air Force Base in New Mexico. Taking off on Runway 8 at Kirtland was tricky, in that you had to make a right turn as soon as possible or you would run into the mountain at the Monzano Base Weapons Storage Area. The pilot, Captain John R. Martin, failed to execute the turn promptly and flew right into the hill three minutes after takeoff. The plane exploded and killed all thirteen people on board. The MK-4 was broken to pieces, with some of its explosive segments scattered and burned up in the gasoline fire. The capsule, stored separately in its birdcage, was recovered along with four spare detonator assemblies in carrying cases. The mission had been to ferry the bomb and some personnel to Walker Air Force Base in Roswell, New Mexico. For decades after the crash

200 You can find the crash site using Google Earth. Just look for Mount Kologet, and the crash site is indicated on the Wikipedia layer. A B-36 was a lot of metal, and although scroungers have carried off some interesting pieces, there is too much splattered all over the mountain to ever clean up completely. The wreckage was found and identified on September 3, 1953, by a team of Air Force investigators who hiked in on foot. Identification was confirmed by the number 511 found on the nosewheel door. The site was relocated accidentally by civilian surveyors in 1956, but they did not think to tell anyone about it. Finally, in 1997 one of them mentioned it to somebody and word got around. Both the United States and the Canadian Departments of Defense immediately launched expeditions just to look at it. The original discovery document, a single teletype page having the lat/long of the site, had long since disappeared into the filing system. Since the 1997 expedition, the location has been public knowledge.

you could still see pieces of the plane glittering in the sun on the west slope of the mountain if you knew where to look.

Three months later, a B-50, an upgrade of the aging B-29 bomber from World War II, was on a secret mission to transport an MK-4 nuclear weapon to a base in the United Kingdom. Just to be perfectly safe, the capsule was being transported on a separate flight. It was a clear day. After takeoff from Biggs Air Force Base in Texas, the plane was seen in a spiraling descent over Lebanon, Ohio, at about 2:54 P.M. local time. There were no parachutes visible. It stalled briefly at about 4,000 feet, then fell into a spin to the right and flew straight into the ground. The high-explosive components of the MK-4 went off on contact, and the bomb, the airplane, and the 15 men on board were obliterated. The cause of the crash was never determined. The separate plane carrying the nuclear capsule landed without incident.

A month later, on August 5, 1950, another B-29 transporting an MK-4 without nuclear capsule to Anderson Air Force Base, Guam, from California experienced two runaway propellers and landing gear that would not retract on takeoff. The pilot turned the plane around to try an emergency landing, but the left wing touched ground and the plane crashed, killing 12 crew members. Twenty minutes later, as the twisted wreckage burned from its full load of aviation gasoline, the atomic bomb's implosion explosive went off, and the shock wave, heard 30 miles away, wiped out a nearby trailer park, killing or injuring 180 military, civilians, and dependents. Among the fatalities in the airplane was Brigadier General Robert F. Travis. The Air Force base, Fairfield/ Suison, was renamed Travis Air Force Base to honor him.

The year 1950 was not good for MK-4 nuclear weapons.[201] Another one was ditched over the St. Lawrence River near St. Alexander-de-Kamouraska, Canada, from a B-50 on November 10. It was missing the nuclear capsule, but it still blew up into a million pieces when it hit the water. It was one of 15 MK-4s being air-transported from Canadian Force Base Goose Bay, Labrador, back to Arizona at the end of a

201 An odd incident, assumed to be an accident, occurred on March 10, 1956, somewhere over the Mediterranean Sea. A B-47 engaged in Operation Chrome Dome was ferrying two nuclear capsules to an overseas air base. It was scheduled for an in-flight refueling 14,000 feet over the Mediterranean. It never showed up. The airplane, its crew, and the two capsules simply vanished. Not a trace has ever been found.

six-week deployment beginning on August 26, 1950. The reason for having 15 nuclear weapons staged in Canada at that time is not clear.

Things were quiet until 1957, when on May 27 a late-model B-36J was ferrying a deactivated MK-17 from Biggs Air Force Base in Texas back home to New Mexico. The MK-17 was a thermonuclear weapon with a 15-megaton yield, made very complicated by its use of a cryogenic deuterium-tritium mixture for explosive fusion. It had been pressed into emergency service while a lighter bomb using solid-state lithium-deuteride was developed. The B-36J had lined up for the landing on Runway 26 at Kirtland and was coming in slow at 1,700 feet.

It was a written procedure at the time to pull out the locking pin on the bomb release before landing in a strategic bomber. It was designed to prevent any chance of an accidental drop while in flight, but if the landing started going bad, the base wanted you to jettison the bomb before you crashed and made a big crater in the runway. The locking pin would make a hasty decision to deploy impossible, so the captain always sent a man who was not needed for anything else to the bomb bay to disengage the pin and allow a drop if necessary. The captain, Major Donald F. Heran, sent a co-observer back to the bomb bay to pull it.

Disengaging the locking pin in the bomb bay was not easy. It was above the bomb, which was a hulking five feet in diameter, and there was not really any place to stand in the cramped space as the plane bounced around in the sun-baked turbulence at Kirtland. The co-observer found it by feel and jerked it out, but he lost his balance and grabbed for something. His hand found the bomb-release cable.

Instantly, the 42,000-pound MK-17 cleared the plane, taking the bomb bay doors with it. It hit the ground hard, detonating all of the primary high explosive, making a 25-foot-wide crater, and throwing shrapnel as far as a mile from the impact point. The largest piece found weighed 800 pounds, a half mile away.

Four more bombs were lost in quick succession. Two were dropped into the Atlantic Ocean, and two were burned up when B-47s blew out tires trying to take off, but on February 5, 1958, there occurred an odd incident. It happened off Tybee Island, near Savannah, Georgia.

Tybee has long been an interesting place, populated at various times by pirates, Spanish colonists, Yankee siege batteries, General George C. Marshall, tourists, hangers-on, and the world's first Days Inn. It has one of the last remaining 18th-century lighthouses which, when first built in 1736, was the highest man-made structure in America. It has been blown down and rebuilt several times, but the bottom 60 feet of the current building was erected in 1773. At 3:00 A.M. on Wednesday, February 5, 1958, Tybee Island was mostly asleep.

Major Howard Richardson, First Lieutenant Robert J. Lagerstrom, and Captain Leland W. Woolard left the ground at 2151 Zulu from Homestead Air Force Base, Florida, in a B-47 for a Unit Simulated Combat Mission (USCM), looping them up to Radford, Virginia, and back down to Homestead.[202] The bomber was loaded with an MK-15 mod 0 thermonuclear weapon, serial number 47782, having a dummy (lead) type 150 nuclear capsule installed. Burdened with the 7,600-pound dead weight, the B-47 made a decent approximation of a Soviet bomber trying to unload over the Savannah River Project (SRP), a secret plutonium production facility in Aiken, South Carolina. Code name for the flight was *Ivory 2*.[203]

A pair of F-86L Sabre interceptors, *Pug Silver Flight* and *Pug Gold 2*, were scrambled to find two B-47s and give them simulated grief.[204] *Ivory 2* did a simulated bombs-away over SRP at 0455 Zulu, then banked into a 200-degree turn and headed for his check-point at Charlotte, North Carolina. *Pug Gold 2*, one of the interceptors, caught up with *Ivory 2* as directed by the ground-based radar system, but the radar-return signal was weak and the position of the target was not well defined. The pilot, First Lieutenant Clarence Arville Stewart, keyed

202 "2151 Zulu" is armed forces lingo meaning 2151 Coordinated Universal Time, or the time at Greenwich, England, in 24-hour notation. The local time was five hours earlier, or 9:51 P.M.

203 *Ivory 2* was the second of two B-47s sent to simulate an attack. *Ivory 1* was spaced 4.5 minutes ahead of *Ivory 2*.

204 The F-86L was a special modification of the F-86D, equipped with electronic equipment linking it to the Semi-Automatic Ground Environment (SAGE) system. SAGE was a computer network of ground-based early warning and air-surveillance radars, built in the late 1950s to provide interceptor planes with real-time data for heading, speed, altitude, target bearing, and range of intrusive aircraft. L-modifications included an AN/ARR-39 datalink receiver and an AN/APX-25 identification radar. The directions from SAGE were uploaded into the interceptor's E-4 fire control system, automatically pointing the plane at the target, which in this case was a B-47. The SAGE system, which was vacuum-tube-based, was way ahead of its time. The purpose of this USCM, *Operation Southern Belle*, was to test SAGE and gain experience in its use.

his microphone and called "twenty seconds," indicating the impact time of his simulated rocket attack.[205] After 34 seconds of silence, his microphone keyed again without a voice signal. His F-86L had just collided with the B-47 as he ducked down in the cockpit, jammed the control stick forward, and accidentally hit the microphone switch. It was 0533 Zulu, or 12:33 A.M. With no remaining control of the aircraft, the pilot bailed out and watched his broken F-86L spiral out into the Atlantic Ocean.

The crew of *Ivory* 2 felt a severe yaw to the left and saw and felt a blinding explosion off the right wingtip. Number six engine was suddenly pointed up at 45 degrees and running out of control. Nothing would stop it until Richardson pulled the number six fire extinguisher and choked out the burners. The pilots started calling mayday while they executed a slow left turn and a descent. They noticed that the right wing tank was missing, and they jettisoned the left tank for the sake of symmetry, first noting no lights on the ground that would indicate civilization.

The heavily damaged bomber was given permission for an emergency landing at the closest field, Hunter Air Force Base in Savannah. They were in worse shape than they had thought, and the bomber came in too fast and too high to land. They had too much weight on board to make an emergency landing with the damaged wing. They decided, with permission, to take the MK-15 out over the water and lose it. They made a left turn to the east, out over Wassaw Sound off the shore of Tybee Island, and then released the bomb from 7,300 feet. There was no explosion when it hit the water 22 seconds later. Just a splash. The newly formulated implosion explosives, made insensitive to shock-detonation, performed correctly when it hit the water by not going off. The crippled B-47 and its crew made it back to Hunter and landed without catching fire, and Stewart and his parachute were recovered unscathed on the ground.

There remained a loose end. Somewhere out there was an MK-15 mod 0 thermonuclear weapon with a fake pit installed. The impact point was hard to pin down, as there was no usable radar scope

205 Rockets carried by the F-86L were the MK 4 FFAR, or the "Mighty Mouse," an unguided, inadequately spin-stabilized weapon with a 6-pound warhead. The only way you could hit something as small as a six-engine strategic bomber with a Mighty Mouse was to be very close to it, which is what caused this mid-air collision. Live testing of Mighty Mice in California caused massive brush fires and destroyed a lot of private property, as they would diverge from the aiming point and find their own targets.

recording made when the bomb was released, but it was near the coast, off Tybee Island. The U.S. Navy was given priority for the recovery operation, with the USAF Explosive Ordnance Detachment (EOD) acting as liaison at the scene. The EOD coordinated a search of the beaches for any bomb debris and an underwater visual search with hand-held sonar gear. The Navy brought in a submarine rescue ship, two minesweepers with high-resolution sonar, a Coast Guard cutter, a troop transport, a 1,000-ton barge, two 15-foot motorboats, a catamaran, two Higgins landing craft, and a helicopter. In the close confines of Wassaw Sound, the fact that no watercraft crashed into each other speaks well for the seamanship involved.

The search was called off on April 16, 1958, without any leads as to where the MK-15 could be. Since it was not a complete bomb with capsule, it was not worth further effort, and so the search was abandoned.[206] As far as the Air Force was concerned, the bomb had been swallowed by the mud at the bottom of the sound, and it was probably buried for good under 15 feet of silt by this point.

Nevertheless, the search for the missing MK-15 has continued sporadically over the past 50 years, mostly by amateur treasure hunters. In 2004, retired Air Force Colonel Derek Duke claimed to have narrowed the search down to a small area by towing a Geiger counter behind a motorboat and mapping the radioactivity over a patch of water. The Air Force came out to investigate and left disappointed.[207]

---

206 Assistant Secretary of Defense W. J. Howard stated in a 1966 congressional investigation that the Tybee Island bomb was a complete weapon, with the 150 capsule installed. The source of his statement is not clear. At the time of the loss, it was Air Force policy not to fly training or test missions simulating combat with the capsule on board, and the receipt signed by Major Richardson mentions no capsule.

207 There is an ongoing misconception about nuclear weapons. Many assume that such a device can be detected at a distance using a radiation-measuring instrument, such as a Geiger-Mueller counter. While it is true that uranium and plutonium give off gamma rays, these emissions are weak and are almost completely shielded from outside detection by the substantial metal bomb-case and the thick layers of chemical explosive that surround the nuclear components. The fusion materials in a thermonuclear weapon, lithium-6 and deuterium, are not radioactive at all. Even if the bomb had contained the plutonium capsule and if the case had corroded away, a couple of feet of water shielding would make it invisible to any type of radiation counter. It is true that the remaining bomb mechanism included a uranium tamper, which is radioactive, but the radiation in Wassaw Sound, often reported, is actually due to monazite sand on the bottom, composed of radioactive thorium oxide.

A little over a month after the bomb was jettisoned off Tybee, another B-47 was forced to drop their bomb. This time it was over Florence, South Carolina, in what became known as the Mars Bluff Incident. Unlike events happening out over the ocean, on Air Force property, or in remote mountain ranges, the public relations aspect of this Broken Arrow would prove challenging. The bomb was dropped on an American family, and the voting public would be smacked right in the face by the remote, abstract, and carefully hidden world of nuclear warfare.

It all started on March 11, 1958, 3:53 P.M. at Hunter Air Force Base, when a B-47E manned by Captain Koehler, the pilot, Woodruff, copilot, and Bruce Kulka, the bombardier, left the ground, headed for England on a training mission named *Operation Snow Flurry.* They had on board an MK-6 atomic bomb with a 30-kiloton yield. It was a practice exercise, so the nuclear capsule was not installed. In case of a sudden war breakout, the manually insertable capsule would be stored on board in its birdcage.[208]

Once they had climbed to 5,000 feet, Woodruff, sitting behind the pilot, rotated his seat 180 degrees so he was facing backward and pushed the lever to engage the locking pin on the bomb-release mechanism, which would prevent accidental release, as called for in the flight manual. The lever did not feel right, and the red light stayed on, indicating that the pin had failed to engage. For five minutes Woodruff tried wiggling the handle to make the pin drop in, banging on the handle with a palm, applying force with a knee, and criticizing it with abusive language, trying to shame it. Nothing worked. The not-engaged light remained on, indicating that the bomb was not locked into position and was free to be dropped on demand.

Captain Koehler asked Kulka, the bombardier, to go have a look at it and try to push the thing home. They were still climbing and had reached 15,000 feet. Kulka would have to enter the bomb bay, which

208 Most accounts of the Mars Bluff Incident refer to the bomb as an MK-6, although the bomb-type is blacked out on released government documents. This does not agree with *History of Strategic Air Command 1 January 1958–30 June 1958*, page 88, which states that the bombs assigned to Hunter AFB at that time were all MK-36 mod 1 thermonuclear weapons. The MK-36 weighed 17,600 pounds. Maybe the B-47 crew was given the much lighter MK-6 to save fuel on the round trip to England. By 1958 the MK-6 was an old design, and all had been upgraded to mod 6 with improved barometric and contact fuses.

was not pressurized, so they had to get this resolved before the air got too thin. The space was cramped, and the door to the bomb bay was small. Kulka had to take off his parachute and squeeze through, past the hulking, five-foot-wide bomb. He had no idea what a locking pin looked like or where it might have been in the complicated maze of levers and cables.

After 12 minutes of searching, Kulka decided correctly that it must be on top of the bomb, hidden from the casual observer. He tried to pull himself up off the bomb bay floor, where he could see over the bomb, groping for something to hang on to. He grabbed the emergency bomb release handle. Click, and down she went.

For a moment, the MK-6 seemed to hang there, resting along with Kulka on the bomb bay doors. It weighed as much as a Rolls Royce with six people in the back seat. The doors gave up, and the bomb exited downward. Kulka, finding his floor gone, flailed wildly for anything to hang on to. He grabbed a cloth bag, but it was not connected to anything. He felt himself, almost in slow motion, following the bomb into space. His hands found something solid, and he managed to pull himself up and out of the open doorway as he watched first the bomb and then the cloth bag get smaller and smaller, falling away.

On the ground underneath, it was 4:34 P.M. on a sunny Tuesday afternoon. Walter Gregg, a train conductor, his son, Walter, Jr., and Jr.'s cousin were making benches in their workshop. Gregg's wife was in the kitchen, and his three daughters were off in the yard somewhere. They had a playhouse next to the vegetable garden, but they were through pretending domestication for the day and were busy piddling elsewhere. They all heard the B-47 overhead. It sounded closer than it was.

All hell broke loose. The MK-6 centered the playhouse, the implosion charge ignited, and the garden went airborne. The blast and shock waves were tremendous, making an oblong crater, 75 feet long and 35 feet down at its deepest point. Huge chunks of earth, some weighing hundreds of pounds, were suddenly coming down all over the place, and in the thick fog of dust and debris Gregg could not see ten feet in front of his face as he and the boys staggered out the front of the workshop. Looking back at the woods, he could see the pine trees snapped off at the ground and laid out in a circle around the former garden. The house was shifted off its foundation, and the back wall was pierced by rocks and shrapnel, as if a huge shotgun had been discharged at it. The roof appeared to have melted onto the rafters. He could hear his family screaming all over the place before the air cleared and the reverberations stopped.

Captain Koehler and the crew felt the shock wave seem to lift their airplane several feet. Fearing the worst, they turned around and immediately saw the vertical column of smoke rising over Mars Bluff. They sent a specially coded digital message back to Hunter indicating that they had laid an egg. On the receiving end, they had never before seen such a message and did not know what to make of it. Getting no acknowledgement, Koehler radioed the tower at the Florence municipal airfield and persuaded the radio operator to place a collect call to Hunter Air Force Base and tell them that aircraft number 53-1876A had lost a device somewhere near here. The impact point would be easy to find.

Back on the ground, Gregg was certain that a jet plane had crashed on his property. Nobody was dead, but his wife had a cut on her head from flying plaster, he had a cut on his side, the cousin was bleeding internally and would have to get to a hospital, everybody was bruised all over from flying rocks, and neither of his automobiles would ever roll again. Neighbors came running, thinking that his propane tank had exploded. The family doctor took them in for the night after stitching up some cuts.

The bomber crew saw their careers melting before their eyes as they flew in circles around Mars Bluff. Evaluating the remaining fuel, Kulka suggested that they might do better to fly to Brazil. They were met on the airstrip at Hunter by security guards who deprived them of their side-arms before locking them in a room. It took a while to convince the Air Force that they had not dropped a bomb on South Carolina on purpose. The national and international press were all over it, descending on Mars Bluff like the black death. Newsreel cameras rolled, peering into the crater and showing airmen examining the ground with Geiger counters. After three days, the excitement died away. Eventually the Air Force was able to find 25 pounds of fragments identified as belonging to a MK-6. When it left the bomb bay, the thing had weighed 7,600 pounds. In August 1958, the Gregg family was paid $54,000 for their losses. They moved elsewhere.

Today, the crater is still there, although it is somewhat filled in and obscured with plants and trees. It is visible on Google Earth. Just look for Mars Bluff. The flight rules were changed immediately. After Mars Bluff, the locking pin was inserted by the bomb-loading crew while the plane was on the ground and remained in at all times, unless you were intending to drop the device. At great expense, all the existing nuclear weapons were upgraded to have reformulated chemical explosives that would not detonate on contact. The three crewmen were reassigned overseas and were never seen again.[209]

In the next few years, five nuclear drop-weapons were lost or destroyed in places that were non-residential, and thus escaped the same level of scrutiny as Mars Bluff. In 1960, a BOMARC IM-99B nuclear-tipped cruise missile exploded and burned at McGuire Air Force Base in New Jersey when its helium fuel-pressurizing tank blew up without warning, but there was little notice outside the base. The next unusually serious accident occurred just after midnight on Tuesday, January 24, 1961, in the tiny farming community of Faro, North Carolina. It is forever known as *The Goldsboro Incident*, and it was impossible to keep it quiet.

By 1961, the big, technically sophisticated Boeing B-52 had taken over as the Strategic Air Command's most prized long-range bomber and the twitchy, crash-prone B-47 took a back seat. The B-52 carried two MK-39 mod 2 thermonuclear weapons, individually strapped into the double bomb bay by what looked like a very heavy bicycle chain.

The MK-39 was an ugly-looking, blunt-nosed cylinder, nearly 12 feet long, painted olive drab with notations (part number, serial number, modification level, etc.) stenciled on the sides with yellow paint. On top was stenciled "DANGER," in red, "DO NOT ATTEMPT TO REMOVE ARMING RODS." The rear section of the bomb was slightly larger than the rotund body, containing four tightly packed parachutes.[210] Four very stubby aluminum fins were bolted to the rear section, looking like an engineering afterthought. It was designed to drop nose-down, with the parachutes out the back slowing its fall. The "frangible" nose section,

---

209 Attempts to find Bruce Kulka, known to his colleagues as the "Nuclear Navigator," dry up after his service in the Vietnam conflict. He moved to Thailand and stopped answering his mail.

210 The rear section was a combined desiccant pack and parachute tube, with the arming rod sockets on top. The four parachutes were deployed sequentially. First, a six-foot drogue chute deployed, and the drag from it pulled out a 28-foot ribbon drogue to stabilize the bomb and make sure it was pointed down. The third chute was a 68-foot octagonal canopy to decelerate the bomb, and the last chute was a 100-foot solid canopy to lower it gently to the ground.

painted yellow, was made of crushable aluminum honeycomb, intended to ensure a soft landing before it detonated. It weighed over 9,000 pounds, and would explode with the force of 3.8 megatons. To put that in perspective, that is more explosive power than the sum of what has been detonated in every war in the history of the world, including the two A-bombs dropped on Japan in World War II. These bombs were sealed and ready to go, with *no* removable capsules as had been the saving grace in so many previous accidents. They were protected from accidental detonation by a series of six steps that must occur before the device will explode. The newly implemented boosted fission feature prevented the plutonium-fueled fission stage from ever exploding because of fire or shock to the system.

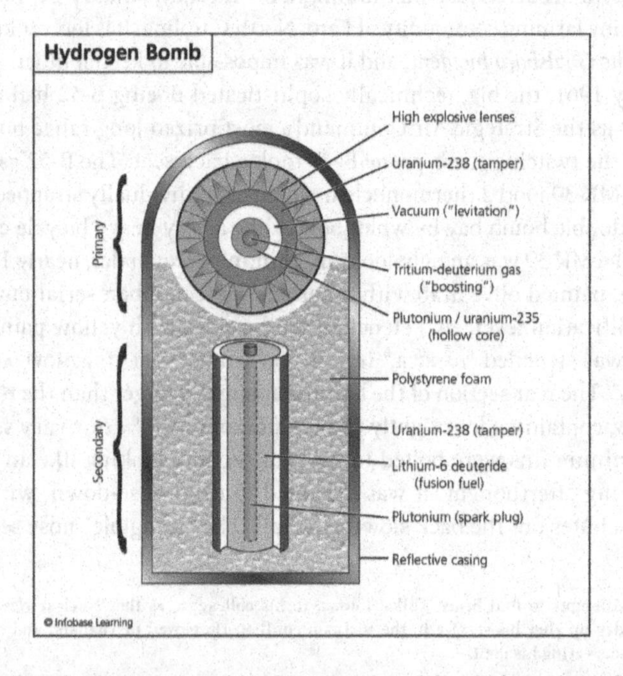

The hydrogen bomb is more powerful and more complicated than the fission devices built in World War II, although it uses an off-the-shelf atomic bomb as the primary explosion. Excess neutrons streaming from the fission detonation convert lithium in the secondary component to tritium, which then fuses with the deuterium under pressure from the x-ray shock wave caused by the disintegrating primary bomb. The fusion process releases energy in addition to that from the fission explosion. Tritium and deuterium are both isotopes of hydrogen.

As was the case with every nuclear weapon built, including the first ones, "safety rods" had to be pulled out of sockets on top of the bomb to make the explosion sequence begin. The safety rods, also called "Bisch rods," were thin metal wands stuck in holes on the bomb and hard-connected to the ceiling of the bomb bay. When the weapon exited the bomb bay, it pulled out the rods, threw a switch, and automatically started the arming steps. The weapon prepared itself for a full nuclear explosion on the way down.

The B-52G number 58-187 took off from Seymour Johnson Air Force Base in Goldsboro, North Carolina, at 10:52 A.M. on January 23.[211] It was a "Coverall" mission, in which the bomber's task was to fly around for 24 hours with two thermonuclear bombs, locked and loaded. It was the peak of Cold War tension, and the Strategic Air Command was keeping one third of their airplanes in the air at all times, ready to strike. Major Walter Tullock was the Command Pilot, Captain Richard Rardin was the Senior Pilot, and First Lieutenant Adam Mattocks was the Third Pilot. They would take turns flying the plane during the long mission. An additional five airmen rounded out the crew. The flight was code-named *Keep 19*.

After 10 hours and 33 minutes, the routine, boring mission began to unravel. Tullock was at the controls for the second mid-air refueling. The B-52 was connected to a tanker plane by a hollow boom, through which JP-4 fuel was being pumped rapidly while the two airplanes were flying together, 30,000 feet off the ground. The Boom Operator on the tanker, looking out his rear window while steering the refueling pipe, noted a stream of fuel, about six feet wide, exiting the right wing on the B-52, behind the number three engine. This was not good. The width of the stream rapidly increased to 15 feet. At the same time, the Flight Engineer on the bomber noted that the fuel in main tank number three went from full to empty in 90 seconds. That is about 44,000 pounds of fuel. Major Tullock decided to cut the mission short and proceed back to Seymour Johnson.

The tanker shifted to a position behind the bomber so that its crew could see where the fuel was leaking. It appeared to be coming out of

211 At the time, this airbase was unofficially referred to as "Seymour Johansen, the Swedish Air Base." I have no idea why.

the wing, between number three engine and the fuselage. There was a real chance of the leaking fuel catching fire, so engines number five and six were shut down. This was starting to look like a full-blown emergency. *Keep 19* was cleared to land at Seymour Johnson, but they were advised to make one orbit to make sure that the leaking fuel tank was empty. Tullock then lined up on Runway 26 and put the landing gear down. So far, so good.

Tullock cautiously lowered the flaps for landing. With this added stress on the wings, the crew heard disturbing cracking, thumping noises, and the plane started to barrel-roll. As the roll reached 90 degrees right, everyone heard the right wing explode, and the commander gave the bailout order over the interphone as the plane started breaking into large pieces. The electronics warfare officer and the gunner were not able to get out, and the radar navigator landed in a tree and died from a broken neck. The other five crewmen parachuted successfully.[212]

The two bombs also abandoned the stricken aircraft. As it fell, the bomber broke apart between the fore and aft bomb bays. The bomb in the aft bay twisted slightly clockwise and slid forward, leaving the plane nose-first as it slipped out of its chain. As it rolled out of the bomb bay, the arming rods were jerked out, and it began the detonation sequence with actuation of the Single Pulse Generator, MC-845. Next, the MC-834 Explosive Actuator fired, then the MC-543 Timer ran down and stopped. The MC-832 Differential Pressure Switch, detecting that the correct altitude had been reached, closed all contacts. Two more steps to go, and the bomb would make Goldsboro into a large inland bay. The MC-640 Low Voltage Thermal Battery was turned on and warmed up. Fortunately, the MC-772 Arm-Safe Switch

---

212 The command pilot and senior pilot in a B-52 are shot out the top of the plane when they eject. The third pilot, Mattocks, was on his own to find an open door and jump. He has gone down in history as the only man who ever successfully abandoned a B-52 through the opening left when the command pilot ejected upward. After he landed, Mattocks, the only crew member of African heritage, was driven the 12 miles to Seymour Johnson AFB by a farmer and his wife, who dumped him out at the gate. Mattocks, who seemed to have lost his military identification, was immediately arrested for having stolen a government-owned parachute, which he still carried, wadded up in his arms. (His treatment was not as bad as it sounds. Without an ID, nobody was allowed on a SAC base. The only way the guards could pass him through the gate was to charge him with a federal crime.)

had not been turned to ARM. That would have required the radar navigator to pull out a knob on his control panel using both hands, shearing off a copper retaining pin, and turn it to ARM. The bomb did not explode. It hit the ground nose-first, and its parachutes were found draped in a tree. The nose, which crushed as intended, was buried 18 inches deep in the soft ground.

The other weapon took the entire bomb rack with it as it fell away from the front half of the plane at about 7,000 feet. Somewhere on its way to the ground, the rack came loose and pulled out the arming rods, but the sequence in which the rods were pulled out was incorrect, so the parachutes did not deploy and no remaining steps in the deployment sequence were activated. The bomb did a free fall, nose down, reaching a terminal velocity of about 700 miles per hour. It landed in the middle of a plowed field and left a crater 15 feet in diameter and six feet deep. It made no explosion. Its MC-543 Timer had only run for 12.5 seconds when it hit the ground. The only way you could tell that a bomb had fallen from the sky was the crater it made.

The bomber itself rained down on the tiny farming community of Faro, North Carolina, over a swath two miles long, on a line starting at Big Daddy's Road and running northwest to southeast, 1.5 miles south of where Big Daddy's tees into Faro Road.[213] To the sparsely placed residents awakened by the house-shaking thumps of airplane sections hitting the ground, it was the end of the world. Flames were everywhere as the remaining jet fuel burned off.

Lieutenant Wilson, the navigator, landed in a swamp, feeling exhilarated at being alive. He stripped off his chute and started running toward civilization. He could see the porch light at a farmhouse in the distance. He jumped a fence, landed wrong, and broke an ankle, but still he hurried along. He made it up onto the porch, took a second to get his breath, and knocked. The farmer came to the door and opened it cautiously. Wilson appeared to be a homeless person with possible intention to rob. He was able to convince the man that he had, in fact,

213 Search for Faro, NC, with Google Earth. It labels the intersection of Big Daddy's and Faro as the spot, but this is incorrect. Go southwest from there on Big Daddy's for 1.5 miles. Off the northwest side of the road is a clump of trees in a plowed field. West of those trees 114 feet is where the secondary component of the bomb is still buried. It is at 35.492817° lat, -77.859307 lon.

bailed out of the airplane that was burning in the distance. The farmer got his wife out of bed, and she made coffee so that they could sit in the kitchen and hear all about it.

Major Tullock, the command pilot, a World War II veteran who had flown everything from B-29s over Japan to B-36s, had never actually parachuted before. He found it curiously relaxing, right up until the point when he got hung up in a tree. It was so dark, he could not see anything, but he thought that the ground was just out of reach of his feet. He released the parachute harness and did a 20-foot free fall into the Nahunta Swamp. It was not clear which way to go to get out of the cold water, so he wrapped himself up in his parachute and waited for daylight, shivering and occasionally finding himself face down in the water.

Captain Rardin, the senior pilot, landed in the trees, but he was able to extricate himself from the tangled chute harness and put both feet on the ground. He could see some lights, and he started walking across a large field. By the time he got to the road, his further progress was slowed by the convergence of "various and sundry dogs" demanding to know who he was and what he was doing here.[214]

By morning, the Air Force had taken over the town. Everyone was cleared out, the area was cordoned off, and nobody, particularly a stray journalist, was allowed in. Residents could not help but notice as they were removed that the Air Force was very interested in the crater in the field. They were told that it was caused by a seat that had fallen out of the plane.

The first order of business was to secure the MK-39 thermonuclear weapons, with their valuable cores. The one that had landed with its parachutes was easy. Specialists opened the access port on the side of the bomb, pinched off the tritium injection tube, and removed the tritium tank. Without that component, there was no way a nuclear explosion could occur.

The other bomb was a problem. It was buried somewhere down in a crater made in soft, swampy ground. At 1:30 P.M. on January 24, specialists from the explosive ordnance disposal squadron began digging in the crater. They had burrowed down eight feet, resorting to shovels by

214 Captain Rardin's concise report reads in its entirety: "I could see three or four chutes against the glow of the wreckage. The plane hit ten or twelve seconds after the bail out. I hit some trees. I had a fix on some lights and started walking. My biggest difficulty was the various and sundry dogs I encountered on the road."

the end. They found a piece of the nose section. The next day, they were down 12 feet, and they found the top of the parachute pack. It rained the next day, making the swampy ground even softer and less stable, but they were able to completely expose the rear section of the bomb. The arming rods had definitely been pulled out, starting the arming sequence. The internal structures of an MK-39 were mostly made of plastic, and they found broken pieces of it embedded in the mud. On the 27th, they were able to pull out the entire back end of the bomb, containing mostly parachutes, and they found the tritium cylinder. It was full. The bomb was never fully armed, and it would not have gone off.

By the 28th, they were down another 10 feet, using mechanical shovels, and the diameter of the hole had grown considerably. Water was filling up the crater, making further digging impossible, and 16 gasoline-driven pumps were used to drain it. At this depth, they could see that the primary, the fission-implosion bomb used to start the fusion process, had broken free of the bomb and the ball of chemical explosive was smashed to pieces. At the end of the day, they found the MC-772 Arm-Safe Switch. It was in the armed position. A question came up: Why had the MK-39, with its arming rods out and the switch in the armed position, not gone off?[215]

On they dug, with the walls of the crater now collapsing and mud sliding onto the workers. On the 30th, the hole was 22 feet deep, 50 feet wide, and 70 feet long, and it was becoming dangerous to dig. They found the plutonium shell of the primary, which was fairly intact, and more fragments of the high explosives, detonators, and tangled-up sections of electrical cable.

The secondary, the part that made it a high-yield hydrogen bomb and not just an A-bomb, had obviously broken free and shot through

215 This question was answered when the parts were sent back to Los Alamos for inspection, but that does not stop the recurring concerns that the bomb *could* have gone off. In reality, the Arm/Safe switch (MC-772) proved to be in neither armed nor safe condition when found. It was severely shocked in the impact, and the switch contacts were distorted. It looked as if it was in the armed condition, but it was not. The MC-772 would be set remotely to armed condition by a 28-volt pulse sent through two arming rods on top of the parachute housing. The arming signal originated at the radar navigator's console, on the AN/DCU-9A Weapons Monitor and Control Panel. You could tell just looking at the MK-39 bomb whether or not it was armed by glancing at a small glass window on the side. If there was a green S in the window, then it was safe. If there was a red A in the window, then it was armed and ready to go.

the front of the bomb like a bullet, burying itself deep into North Carolina. The secondary was a hollow cylinder, 14 inches in diameter and 34 inches long, made of uranium-238, filled with lithium-deuteride powder, with a plutonium-239 rod running down the center. It weighed a little less than 200 pounds, and, believe it or not, it was encased in molded Styrofoam, the stuff used to make disposable coffee cups.

On February 7, the team gave up digging at a depth of 42 feet. The hole was now 130 feet wide at the surface, and cave-ins and water leaking in were exacerbated by rainy weather. It was not worth the effort to find the lost secondary unit, and the hole was filled in. Later simulations indicated that it was probably 120 feet below the surface. It is still there, somewhere. The Air Force paid $1,000 for the easement for a circle 200 feet in diameter in the field and will permit no digging on the property. On July 2, 2012, a historic marker was unveiled on Main Street in Eureka, North Carolina, commemorating the Goldsboro Incident, which actually happened about three miles down the road, in Faro. It reads:

> NUCLEAR MISHAP. B-52 transporting two nuclear bombs crashed Jan. 1961. Widespread disaster averted: three crewmen died 3 mi. S.

The last survivor of the crash, Adam C. Mattocks, was on hand. The young inhabitants of Eureka dismiss the story, finding it about as believable as a local version of the Loch Ness sea monster. To them, it is just another ghost story.

These accounts of air disasters with nuclear weapons could go on for another three volumes and start to get repetitious, but from this small sampling a general trend begins to form. Here is one last, fascinating tale before we move on. It is a little-known accident, occurring in Greenland near Thule Air Base on January 21, 1968. Pronounced "Too-lee," Thule, located well north of the Arctic Circle at the intersection of three moving glaciers, has long held the prize, unofficially, as the most miserable air base on the planet Earth. Rather than list its deficiencies as a vacation spot, I offer the following anecdote:

Late on a dark winter's night, an Air Force general, high up in the Strategic Air Command, landed at Thule Air Base in his flying command-center/palace for a surprise inspection. As the plane taxied off the runway, the general asked his pilot to radio the tower and have equipment sent out to empty the latrine on board the plane. Its waste-holding tank was full.

The tower radioed back that this would be accomplished right away, but the aircraft could not proceed to the disembarkation point until the latrine service truck had done its business and disconnected. Understood. We shall wait.

The general, who was by disposition opposed to waiting for something to happen, sat fuming in the cabin as time ticked away. After a 30-minute wait, he was fit to be tied. He exited the plane through the rear door so that he could pace up and down on the tarmac and give the tardy airman a lecture about promptness. Finally, the honey wagon showed up, approaching in no particular hurry. After making absolutely sure that his parka was all zipped up, the airman ambled over to the back of the truck, unrolled the hose, eventually connected it to the waste port on the plane's fuselage, and stuck his hands into his parka.

"Airman!" the general screamed. "You are intolerably late, and your lack of enthusiasm is obvious! This will go down in my report, and you will regret your slovenly performance here!"

The airman, barely awake, squinted so as to see who was talking. "General," he began, "it's 27 degrees below zero. I'm pumping shit out of an airplane. I'm at Thule. What exactly are you going to do to me?"[216]

It is always cold at Thule, but 35,000 feet over Thule in the dead of winter, Sunday, January 21, 1968, it was even colder than usual. The crew in the downstairs compartment aboard the B-52G strategic bomber, "HOBO 28," were freezing, and they called upstairs begging for more heat.

Gone were the halcyon days of SAC, the Strategic Air Command, when they kept a dozen bombers in the air at all times in "Operation Chrome Dome," ready to counter-strike a Soviet threat at a minute's notice. There were too many accident opportunities, and it was terribly expensive even if nothing crashed into anything. On January 17, 1966, a B-52 and a KC-135 tanker plane had a fender-bender over Palomares, Spain, and managed to spill the contents of the bomb bay

216 This bit of Air Force lore was given me by Colonel Eric Conda Murdock, USAF retired. It might even be true.

over land and sea, this time in a foreign country.[217] It was the latest in a number of loud wakeup calls, and the number of planes to be in the air at once was cut back to three. By 1968, the Ballistic Missile Early Warning System (BMEWS) was up and running in Greenland, and there was enough planet-killing power available underground at the throw of a switch in missile silos or in submarines to give airborne weapons a back seat.

The Department of Defense questioned the need for any bombers at all. SAC, fearful of losing their defining mission, argued that the BMEWS radar sites, particularly the one at Thule monitoring the over-the-pole route for enemy missiles, could be taken out by a stealthy Soviet overland strike, and the data link back to the command center in Colorado would go dark. We would have no idea what was being sent in over the North Pole with that radar knocked out. The only way to ensure that all was well at BMEWS Thule was to fly over it in a bomber and make sure it was still there. It was a weak justification, but the DoD went along and gave SAC permission to keep exactly one plane in the air. It would fly a "butter-knife" pattern, a sort of figure-eight route starting at Baffin Bay, over Thule, then over the BMEWS station, and back down to Baffin Bay, over and over, with in-flight refueling.[218] If the connection to BMEWS Thule was ever lost, the B-52 could look down and tell whether it was just a power outage or it had been blown to Kingdom Come. The operation was named "Hard Head."

217 In the Palomares incident, a B-52 overran his tanker in a refueling operation over southern Spain. Both the bomber and the tanker broke up and crashed, killing seven of the 11 crewmen and scattering the four MK-38FI thermonuclear weapons. Two of the bombs were destroyed on impact by detonation of their primary explosives, one landed safely by parachute on a tomato farm, and one sank in the Mediterranean Sea. With great effort the sunken bomb was recovered, and 1,400 tons of soil and vegetation contaminated with fissile uranium and plutonium were dug up and shipped to the United States for controlled storage.

218 The long-range incoming threat detection equipment at BMEWS Thule at that time used AN/FPS-50 radars operating UHF at 425 MHz. The antennas, permanently aimed at Russia over the top of the world, were 165 feet tall and 400 feet wide, called "fences." How powerful was this setup? On October 5, 1960, the moon rose directly in front of the antennas, and the radar detected it as an extremely large missile coming straight at them. The long signal return, 2.5 seconds, clued the operators that it could not really be a missile, and World War III was averted.

The old reliable pair of MK-39 thermonuclear weapons carried in the B-52 bomb bay had been replaced with a four-round clip of sleek, slender MK-28FI bombs in case something of global significance happened while the bomber orbited over Thule. The newer bombs were only 22 inches in diameter but were 12 feet long. They looked exactly like cigar tubes, dull but attractive aluminum with perfectly hemispherical noses, and each weighed only 2,300 pounds.[219]

The Command Pilot on HOBO 28 was Captain John M. Haug, the Copilot was Captain Leonard Svitenko, and the Third Pilot was Major Alfred J. D'Amario. The crew was Major Frank F. Hopkins, Radar Navigator; Captain Curtis R. Griss, Jr., Navigator; Captain Richard E. Marx, Electronics Warfare Officer; and Staff Sergeant Calvin W. Snapp, Gunner. Their mission, *Junky 14*, started at 7:30 A.M. with a pre-takeoff briefing in the Bomb Squadron room at Plattsburgh Air Force Base in New York. Captain Haug and his crew loaded their equipment into HOBO 28 and started engines at 8:44 A.M.

The Third Pilot, D'Amario, had to sit downstairs in "the hole," an uncomfortable, fold-away jump seat bolted to the bulkhead of the lower crew compartment. The back rest was a thin cushion glued to a door at the back of the cabin. The sealed doorway led to the front landing gear compartment, and a door on the far side of that space led to the forward bomb bay. There was a seat belt and a headphone jack, but it was the only seat in the plane that would not eject from the aircraft on demand. Officially, the seat was designated the IN, for Instructor Navigator.

At D'Amario's feet was the "egg crate," the grid over the folded-up ladder for the aircraft's front door, a hatch underneath the ladder.

---

219 Also designated B28FI, the F means "fully fused" and the I means "carried internally." A fully fused MK-28 could be dropped in a free fall for an airburst, dropped with a retarding parachute for an airburst, or lowered to the ground gently for a delayed burst on the ground, called a "laydown drop." The actual bomb was a cylinder only three feet long. The entire back section of the casing held the parachutes, and the long nose was filled with balsa wood, intended to cushion the shock of landing nose-down. The intended yield of the bombs carried in HOBO 28 is unknown. It could have been anything from 70 kilotons to 1.45 megatons. The fissile material at the center of the bomb's secondary stage consisted of highly enriched U-235, and not the usual Pu-239. Four bombs were conveniently mounted in an MHU-14/C clip-in subassembly plus MHU-19/E bomb cradle. Loading this four-bomb cluster onto a B-52G was quick and easy using a special trailer with a hydraulic lift.

He looked into the backs of the ejection seats for the Navigator and Radar Navigator, and on his left and right were racks full of electronic equipment. At his left elbow was the urine canister, piped downstairs from the crew's toilet, to which the yellow ladder in front of him led.

There was no place to put extraneous things in a B-52G. D'Amario had gotten some extra cloth-covered foam rubber cushions to make his long stay in the hole more comfortable. Before takeoff, he stowed three of them under the jump seat, right on top of the "hot air spray tube," or what anywhere else would be called the "furnace vent." After takeoff, D'Amario found that the extra cushion he was sitting on did not do as much good as he had hoped, so he crammed it on top of the other three cushions, and in front of them he moved a metal box to put his feet on.

By 1:09 P.M. they were entering the orbit area over Greenland, where it was dark 24 hours a day. The pilots had just finished a successful in-flight refueling, and D'Amario went upstairs to spell Svitenko in the copilot's seat. Svitenko immediately noticed that it was freezing cold downstairs, and he called for more heat. The temperature control knob was already turned all the way up, so D'Amario turned on the emergency cabin heater system, taking hot air from the jet engine manifolds on the right side of the plane. It was still cold downstairs, but on the flight deck it immediately started getting uncomfortably hot. As the pilots puzzled over the temperature paradox, Marx called from downstairs that he smelled something burning. Rubber?

A brief search found smoke coming from underneath the IN seat. Marx unhooked a fire extinguisher from the wall, pulled the safety pin, and released its contents all over the seat. One fire extinguisher did not do anything, so he quickly found another and emptied it in the direction of the smoke. Svitenko pulled out the metal box in front of the cushions, and flames erupted into the downstairs compartment. Haug immediately called Thule, reporting an emergency on board and requesting an immediate landing. Marx opened the sextant port to try to vent the smoke out of the compartment, while Griss tried to smother the fire with an A-3 bag.[220] It was only growing worse. As the plane

220 A-3 is an "aviator's kit bag," a shapeless Air Force duffel still used to hold a crewman's spare clothing and accessories for a long mission. It has become a fashion accessory, which shows that anything is possible.

descended, at 3:32 P.M. D'Amario dumped the cabin pressure, trying to smother the flames by depriving them of oxygen. The smoke was so thick, the pilots could not see the instrument panel. Three minutes later, all electrical power failed, and Haug shouted the bailout order.

Haug could not see much of anything, with the only visible light being a red glow coming through the hatchway to the lower compartment, but he distinctly heard four ejection seats fire off in quick succession. He looked down at the end of the left armrest and grabbed the "magic handle," painted yellow with black stripes, with his left hand and rotated it up. The ejection trigger snapped into position, the hatch over his head unlatched and blew away, the inertial reels on his shoulder harness locked, and the control column with which he had been flying the plane spaced forward, out of the way. All the smoke vacuumed out of the cockpit, and for a second the sound of air rushing past the open hatchway was deafening. He squeezed the trigger. The next thing he knew, he felt as if he were floating. He could see lights on the ground.

D'Amario, unlike Haug, was not wearing a winter-weight flying suit with thermal underwear. He had the right undies underneath a summer-weight flying jumper, but as he saw the situation deteriorating he pulled on his snow parka. After he was blown out the top of the plane and the chute opened, he occupied himself with deploying his arctic survival kit.[221] He could see the lights on the west end of the runway at Thule.

Hopkins, the Radar Navigator, was surprised at the violent jerk of his ejection seat. The eject trigger set off an explosive charge, and the high-pressure gas generated in the explosion vented into a pneumatic cylinder at the back of his seat. The seat, mounted on rails, left the

221 When strapped into his seat, an airman was sitting atop his survival kit. It was two molded fiberglass halves, held together like a clamshell with aluminum hardware. When the seat fell away and the parachute opened, the kit was held against the airman's backside by the lap belt. If there was a chance that he was coming down over water, he was supposed to reach back and pull a handle. The kit would come apart at the seam, and the back half would fall away, attached to a 20-foot lanyard. At the end of the lanyard, a life raft would inflate, ready for him when he hit the water. After he was on the ground, another handle, almost impossible to manipulate with cold hands, would open the remaining half of the kit, containing food, water, a knife, an aluminized Mylar "blanket," a radio beacon, a book on how to survive, and, best of all, a tiny folding rifle, good for keeping rats away from your food providing they were slowed down by the cold. (In 1968, the rifle was probably an Ithaca M6, single-shot, over-under .22 long rifle and 410 shotgun with a palm-squeeze trigger that can be pulled wearing heavy mittens.)

plane very quickly. Somewhere in the event his helmet and his left glove were lost, and his left arm seemed to have gotten caught in the hatchway as his seat exited downward. He could not wave his arms and stop himself from oscillating back and forth under the parachute canopy. To the east he could see the lights at Thule. His arm was hurting, which thankfully meant that it was still attached to him.

Marx was not properly dressed, having only a light summer flying suit and thermal underwear. As soon as his seat cleared the plane, he became acutely aware of the cold air. He could see the base, but he was drifting in the wrong direction.

Snapp, the gunner, was not well dressed either, and it seemed to him that it took an awfully long time to hit the ground. He had lost his helmet and both gloves in the ejection. While floating, he tried to release his survival kit, but his hands were too cold.

Griss, the navigator, felt his arms flailing as his seat ejected downward, and he was afraid that he had broken both shoulders. He could not reach for his survival kit on the long way to the ground.

Svitenko, the one stuck in the IN seat without an ejection charge, tried to jump out the hole left when the navigator seat dropped out. His head hit the hatch-frame on the way out, and he never regained consciousness.

With nobody on board, the eight-engine bomber made a slow left turn, heading out over Wholstenholme Fjord but quickly losing altitude. Over the vast plain of thick sea-ice on the western shore of Greenland, the plane's left wingtip touched the surface and started to dig a trench. The nose slammed down on the ice as the bomber began to cartwheel, and it disintegrated in a bright yellow flash, spreading finely divided debris in a north-south line two and a half miles long. Hopkins was the only one who happened to be looking west when the plane hit the ground, and he witnessed it disintegrate in the darkness, leaving a large field of burning jet fuel.

Haug made an easy landing and tried desperately to open his survival kit. The survival kit was important in that it was always a secret, sealed box, said to contain many interesting things, including the coveted folding rifle, but it could never be opened and examined under normal conditions. It was a sort of prize given only to those who had lived through the extreme event of being shot out of the plane in an ejection seat, and Haug and the other survivors wanted to look inside. His hands were too cold. He finally abandoned it and his parachute and walked 600 yards to the comfort of a heated hangar on the base.

D'Amario yanked the quick-release lever on his harness as soon as he felt his feet touch ground, and it abruptly dumped him out onto his right side. He stood up, felt all over, and could find no injuries. He collected his survival kit and walked the 200 yards to a hangar.

Hopkins could not reach the survival kit handle with his hurting arm, so he just let it go. He started walking after he hit the ground, and soon he ran into Marx. When Marx hit the ground, the snow felt like concrete to him, and he was so cold, his numb hands could not open his survival kit, so he gave up and started walking. Seeing that Hopkins was injured, however, he went back for his kit and finally got it open, but sadly, there was nothing inside that could help Hopkins. They started walking together, and 30 minutes later they were picked up by a rescue helicopter.

After he landed, Snapp spent a frustrating time trying to open his kit as his hands got colder and colder. It seemed that there was a turning point for each of the men, at which the 53 degrees below zero chill factor or even physical injury began to take precedence over the desire to see inside the secret box. Unsuccessful, he dropped it and started walking. Two rescuers found him an hour and a half later and took him to the base hospital.

Griss was the first to eject and the last one to be found. The pain in his shoulder kept him from even trying to open his survival kit. He wrapped himself up in the parachute and lay down in the snow. The rescue team found him 22 hours later, frostbitten but alive.

The United States Air Force, together with the Royal Greenland Trade Department and some dog-sled teams, had found and rescued the surviving members of the crew of HOBO 28 under terrible conditions of cold and dark. It was a job very well done, but now there was a larger problem. Greenland belonged and still belongs to Denmark, and the Air Force was allowed to land, take off, and house airplanes on the ice at Thule, but the written conditions of this agreement stated clearly that no nuclear weapons were to be present. There were now four such devices, spread out over a large blackened area on the ice covering the fjord, about seven and a half miles west of Thule, that had fallen out of the plane as it crashed. To police the area of debris would not be an easy effort.

The first thing the Air Force did was to give the task a name: Project Crested Ice, or, unofficially, "Dr. Freezelove," based on the name of a Stanley Kubrick film from 1964, *Dr. Strangelove, or How I Learned to Stop Worrying and Love the Bomb.* All the Americans had seen the film. It was unfamiliar to the Greenlanders.

Weather conditions for the cleanup project were extreme, with temperatures as low as 76 degrees below zero and winds up to 89 miles per hour. Sunlight would not peek over the horizon until February 14. The hand-held radiation detectors which worked so well in a laboratory would fail as the batteries froze. Radiation contamination from nuclear weapons torn to pieces and burned was spread out over three square miles on the ice shelf over the fjord. The goal was to pick up every interesting-looking piece at the wreck site so that Eskimos would not collect them and take them home, subjecting themselves and their families to long-term alpha-radiation exposure. Unfortunately, the crash occurred in the middle of an election in Denmark, and the U.S. Embassy in Copenhagen was immediately notified that the Danish government wished to send help for the cleanup operation. There was no objection, as it was diplomatically correct to accept the assistance, and the Air Force needed all the hands it could muster to collect every small bit of debris off the ice. Danger to personnel from radiation exposure during the operation was, for all practical purposes, nonexistent and paled in comparison with the threat of frostbite. Any plutonium or uranium smoke originating in the crash had condensed onto the ice in the extreme cold, so it could not be breathed. Nobody's bare skin touched anything in the frozen environment, so there was no chance of absorbing alpha-emitting material during the mission.

The Danes turned out to be a bit more fussy than the Americans on the subject of disposal. The Americans wanted to dump the bomber and its cargo into the deep fjord, where it would be out of sight, out of mind, and gone forever where it was too cold and too deep for anybody or anything to ever pick it up. The Danes wanted every little piece, every dust particle, and every ton of contaminated ice to be packed up and shipped to the United States. The Air Force, wanting to keep its base in Greenland, agreed to this stipulation.

Camp Hunziger was erected at the crash site, consisting of many igloos, a heliport, several generator shacks, communications buildings, a large prefabricated building, two buildings mounted on skis, a decontamination trailer, living quarters, and a latrine. A straight line of 50 men would start walking over a search area, looking down and picking up anything that was not clean ice. Back at the camp, bomb experts would determine which blackened fragments were from a nuclear device, and these parts were separated out and loaded into steel drums for transport back to Pantex, where the weapons were originally assembled. Ice showing any radioactivity at all was loaded into tanks for shipment back to the Savannah River Project in South Carolina.

About 93 percent of the airplane and its cargo was accounted for. The remaining seven percent had probably fallen through the ice and was considered unrecoverable. Carefully reconstructing the bombs from all the tiny pieces, the specialists at Pantex were able to account for everything except the secondary unit, the cylinder made of uranium and lithium deuteride, from one of the weapons. An underwater search in the fjord was attempted using a Star III mini-sub, but nothing was ever found. After almost eight months of work, the operation was completed on September 13, 1968, when the last tank of contaminated ice was sealed and loaded onto a ship.

The Danish workers did not go anywhere near the crash site. They worked on the vehicles used in the cleanup and at the port where the steel tanks full of contaminated ice were loaded. Nineteen years later, about 200 of the workers were convinced that they had been exposed to radiation, and they took legal action against the United States. The action was successful only in forcing the release of many classified documents concerning the Thule accident and Operation Crested Ice, making this detailed account possible. The Danish government wound up paying 1,700 workers 50,000 kroner, about $8,300, each.

The airborne alert program came to an immediate stop at Strategic Air Command. As one observer pointed out, if HOBO 28 had crashed into the BMEWS site, we would have concluded that the Soviets had simultaneously destroyed our radar outpost and our look-down aircraft, and the Third World War would have commenced.

On the whole, the Air Force, Navy, and Army were good stewards with the thousands of nuclear weapons and warheads in inventory in the 20th century. They managed to never detonate one, even with planes crashing, ships colliding, and mistakes and errors so simple, it was impossible to predict what would happen next. A solid conclusion that one can draw from the representative subset of accidents here is that a technically advanced warplane is safe only when it is on the ground with the engines turned off.

The accidents that did happen were not technically nuclear in nature, and they certainly were not nuclear-power accidents, and yet they added to the distortion and the magnification of the public nuclear concept. When one of the tobacco farmers in Faro was advised to leave town while the specialists dealt with the thermonuclear weapon standing on its nose in his field, he and his wife stole a glance at it on the way out. The wife said she could feel it. The radiation was making the right side of her body burn. Psychologically, that was true, but it was not alpha, beta, or gamma rays streaming out of the weapon. It was difficult to detect the plutonium and uranium deep inside the bomb chassis using the most sensitive electronic instruments pressed against it, much less two hundred feet away with sensitive skin. It was not the radiation, but it was the very thought of radiation that made her skin react. It was the strictly imposed secrecy, the complete lack of knowledge of what was inside the bomb case that would let the imagination run wild and make a normal person break out into hives. The major edict in the instructions for cleaning up a lost nuclear weapon, the Air Force procedure known as *Moist Mop*, is "Don't scare the locals!" The effect seems just the opposite.

It will take decades or maybe centuries for the fear of deliberate nuclear destruction to subside, but that emotion is not even necessary to sustain the atomic dread. Commercial nuclear power, the benign type that makes electricity come out of your wall, would itself cause enough mayhem to keep the fire burning, now that the Cold War is over and continual transport of nuclear bombs via airplane is not as routine. The next event would happen in Dauphin County, Pennsylvania, on a fine March day in 1979, and it would become known to nuclear engineers everywhere as "TMI." As was the case of the aircraft accidents, the initiating fault would be so minor, so insignificant that it would be impossible to predict what was going to happen.

Chapter 9

THE CHINA SYNDROME PLAYS IN HARRISBURG AND PRIPYAT

> "The most important man in a nuclear submarine? That would be the inconspicuous seaman who goes all around the sub and drips oil in the bearings. You lose one bearing in something like a valve-actuating motor somewhere, and you can lose the whole boat."
>
> —Paul "Spider Fuzz" Field, former submariner and Research Technician at Georgia Tech

ONE DAY I WAS IN the E.I. Hatch Nuclear Power Plant near Baxley, Georgia, on a mission to install my life's work in the former operator's break room, which had been converted into an equipment bay. There was a list of new equipment mandated by the Nuclear Regulatory Commission after the Three Mile Island disaster, and my contribution was four ROLM MSE/14 minicomputers and associated hardware, taking up four racks for each of the two reactors. These machines would work hard, 24 hours a day, constantly updating a long list of data considered crucial to the safe operation of the two reactors at the plant and making it available to the operators on demand.

I was supremely confident that my MSE/14s would be the meanest, toughest pieces of hardware in the entire room. They were built to military specifications and were meant to run on the upper deck of a navy ship, pitching in the waves and taking fire in salt water spray and rocket exhaust.[222] I was so, so wrong. In the adjacent rack was an unidentifiable piece of electronic gizmo bolted in with very heavy screws. The front face was half-inch steel armor-plate, and the thing looked like it weighed about 500 pounds. My mil-spec equipment looked delicate in comparison. I was concerned to notice that the meters on the gizmo's face were smashed flat, and the extremely robust controls, built for use by a gorilla, had been crushed and sheared off.

I turned to my handler and asked nervously, "Uh, what happened to this thing?"

"Oh," he responded. "That. Well, the installer complained that the pipefitters were in his way. My suggestion is, don't upset the pipefitters."

I had noticed all the plumbers with their bending machines strung out all over the yard, working slowly and carefully to install piping upgrades and new tubing runs all over the place, including in the ceiling of the new equipment bay. Quickly I learned working at the plant that the pipes, valves, and pumps had a much higher coefficient of importance than any electronic gadget in the facility. There were reasons for this hierarchy.

The great meltdown accident at the Three Mile Island Unit 2 in Pennsylvania would seem to have been the end of the Exuberance Period in atomic energy, but, of course, it was not. The excitement and mystique of nuclear power had pretty much faded out years before then, as the cold realities of loan interest and wavering public power demand put a lid on it. It was too bad, because the adolescence phase is interesting even for

222 The MSE/14 was the mil-spec equivalent of the Data General Eclipse S/140, a 16-bit industrial control computer. The military designation was AN/UYK-64(V) ("yuck sixty-four") or 1666B. They were used in the Ground-Launched Cruise Missile and Sea-Launched Cruise Missile ("glick'em—slick'em") military programs as the launch-sequencers, but the GLCM (BGM-109C Gryphon) went away with the INF Treaty in 1988. SLCMs (BGM-109 Tomahawks) were last used on March 22, 2011, against targets in Libya.

technology, and nuclear power technology had been declared mature while it was still wearing short pants. There remained unresolved problems, some of which were known and some of which would snap into clarity with a couple of hard jolts.

By the late seventies, a world standard for commercial nuclear power reactors had been loosely established by what utilities chose to buy. It was the light-water-moderated and -cooled reactor, mainly the pressurized-water-reactor concept that Admiral Rickover had developed with spectacular results for his nuclear submarine program. Liquid-metal-cooled breeders and all oddball reactor types, such as molten-salt, gas-cooled, and pebble-bed designs, were largely abandoned and suffered a lack of development funds. About 80 percent of the reactors being built or run in the United States were PWRs made by Westinghouse, Combustion Engineering (CE), or Babcock and Wilcox (B&W), with the remainder being the simpler boiling-water reactors built by General Electric (GE).

An unresolved technical detail was the Emergency Core Cooling System (ECCS), a collection of devices used to prevent a fuel meltdown in case of an accidental breakage of the primary cooling loop. In the spring of 1972, the AEC held a series of hearings to address concerns that insufficient attention had been paid to this unlikely but potentially disastrous type of accident. Interim ECCS designs were being sold to utilities, and these were extremely complicated add-ons, consisting of multiple auxiliary water-injection systems with a great deal of plumbing. These systems were reminiscent of "five-mile-per-hour bumpers," clumsy-looking things added to cars because a small tap on the front of the vehicle could cause a lot of expensive damage. A single pipe break could bring down an entire generating plant, and the insurers were justifiably frightened. These auxiliary coolant systems required electricity to operate valves remotely, run pumps, and provide power for the control room, so backup generators had to be installed as well, ensuring that there would be power for the ECCS even if the turbine had stopped and there was no power available on the utility network. The ECCS, even in its possibly inadequate form, ran up the cost of a nuclear power plant. The AEC hearings ran for a year and a half, and a few improvements were mandated.

The technical questions about the use of an ECCS in an emergency are very simple: if a major pipe breaks open and the reactor core is denied water for cooling, the ECCS is supposed to make up the lost coolant by throwing water in from an alternate source. What is to keep the ECCS water from leaving the core through that same hole?

In a water-cooled reactor, the fuel is made of little uranium oxide pellets, lined up in thin metal tubes, and the tubes are kept upright and apart by light sheet-metal spacers, designed so as to add as little non-productive metal to the inside of the reactor core as possible. If the fuel were denied coolant long enough, perhaps minutes, this fragile metal structure would start to sag and bend, disrupting the normal down-up flow of water as it is added by the ECCS. With too much disruption, the metal would melt and collapse into a heap at the bottom of the reactor vessel. How does the auxiliary cooling water get to the hot fuel in the middle of the heap without spacers to keep flow-channels open? There was no concern about heaped fuel going critical, heating up by uncontrolled fission, and burning through the bottom of the nine-inch-thick steel vessel. Being denied coolant also meant denial of moderator, and the three-percent-enriched commercial reactor fuel was incapable of forming a critical mass without interstitial water. The heat from the recently fissioned fuel, however, was enough to cause an irreversible reactor wipeout, with the internal structure reduced to a chaotic mass of melted parts.

There were no power-reactor disasters back then to study and contemplate. The nearest thing we had to working data was from computer simulations of theoretical accidents and some experiments with the Semi-Scale simulation at the NRTS in Idaho. Neither source could possibly point out everything that could happen in a billion-watt power plant, but in the 1970s confidence in the inherent safety of the pressurized water reactor was high.

There were some dangerous problems with the system in general, and with Babcock & Wilcox reactors in particular. The primary fault was in the training of reactor operators. The Navy was supplying reactor operators to the nuclear-power business the same way the Air Force was supplying airline pilots to the air transportation industry. A young man who had been rigorously trained in Rickover's Navy to run a submarine reactor with a few years under water could retire early and snag a fine job in a nuclear generating station. He was considered to be at the top of the game, having run the reactor on one of Rickover's flawlessly performing boats with military discipline and polish.

It saved the power company the cost of having to train an operator from scratch, and veterans from the submariner or nuclear aircraft carrier service were always welcomed.

It seemed a good policy, but there were fundamental problems. Those attack submarine reactors used in the first years of the nuclear navy were tiny, almost toy-like, producing only 12 megawatts to run a sub at full speed.[223] Small reactors have small problems, and the mega-disaster capabilities of an extremely complex billion-watt power reactor were unknown to any submarine veteran. The submarine reactor was run by two men, sitting at a console about as complex as the dashboard of a twin-engine airplane. A power-plant console is completely different. It sits in a room the size of a basketball gymnasium, and it takes several men to run it, all standing up. There are 1,100 dials, gauges, and indicator lights, 600 alarm panels, as well as hundreds of recorders, switches, and circuit breakers. That is just the front of the main panel, towering over everything in a wrap-around, U-shaped configuration, as wide as the room and seven feet tall. In back of the main panel is the larger secondary panel, containing all the indicators and dials for which there was no room in front, and there is little reasoning in the positioning of anything. Finding the immediate status of some important subsystem in the plant can involve remembering where and on which panels the various bits of information may be located. The slightest problem is brought to the attention of the operating staff by an alarm sounding off and blinking a light behind a square plastic tile having the fault identifier printed on it. At any one time, there could be 50 alarm tiles lit up from minor problems and needing attention. Off to the side in a B&W plant in the 1970s was a small computer, keeping track of all the alarm conditions and printing them on a continuous roll of paper. As the automatic control system in the plant detected a fault, the computer identified it on the print-out with the time of day at which it occurred. The printer was a pin-matrix unit, running at a sedate 300 baud.

223 The reactor used in most pre-ballistic missile subs was the S2W, built by Westinghouse, and all operators were trained on the S1W at the NRTS, running at about 5 megawatts. The first nuclear aircraft carrier, the USS *Enterprise*, which was basically a floating air-base, used a great deal more power than a streamlined attack submarine, but it did not use a scaled-up power plant. It used eight little reactors instead of one big one. This design was brilliant, and it solved many potential problems.

Their training in the Navy had not prepared the operators for this level of available information sitting atop an enormous amount of raw power.

What the Navy had pounded into these men was an absolute need to "not let the pressurizer go solid." But what exactly did that mean? Nuclear accident investigators started to notice this curious phrase coming up in most operator debriefings as soon as power reactors started having accidents. Its exact meaning would become important as the conditions that caused the problems at TMI gradually lined up and self-organized into a disaster.

In a PWR, the reactor coolant/moderator is liquid water, forced to circulate by electrical pumps in two continuous loops. Water is heated to several hundred degrees in the reactor core, and this energy is used to make steam by circulating it through two steam generators. A steam generator is a vertically mounted cylinder, about 75 feet tall, and it works like an old-fashioned steam boiler, using heated water rather than fire to boil water into vapor. The primary water, cooled by its trip through the steam generator, is pumped back into the reactor vessel to be reheated.

The reactor vessel is a thick, forged-carbon-steel pot, cylindrical and about 39 feet high, with a stainless steel liner to prevent corrosion. To maintain the water in the vessel in a liquid state, it must always be at very high pressure, else it would boil and turn to steam. The only way to maintain the fission process in a PWR reactor vessel, which is small compared to other designs, is to make sure that the moderator, the water, is constantly at maximum density, or liquid state.

This high-pressure condition is maintained by the pressurizer, which is basically a large, 42-foot electric water heater connected into the top of the reactor. The pressure in the reactor vessel is automatically monitored and kept at the correct level by either turning on the heater coil in the pressurizer to increase the pressure or spraying cool water into it to decrease it. The entire primary coolant system, including two steam generators, four main coolant pumps, the pressurizer, and all the pipes, is kept completely filled with water, with no bubbles or voids. There are no bubbles in the system except for the pressurizer, which always has a void sitting at the top of its water column. The pressurizer is constantly kept about 80 percent full.

The reason for this discrepancy involves a second role for the pressurizer. Not only does it keep the pressure high within the reactor, it also acts as a shock absorber. Any sudden jolt in the water running around in the primary cooling system, such as a valve slamming open or shut or a pump starting or stopping, causes a shock wave to travel through the incompressible coolant. The water cannot be broken by a shock, but the metal pipes and cylindrical structures in the system are not flexible, and a "water hammer" can take apart a cooling system instantly. This problem is solved by giving the system a section that can be compressed, the bubble of steam atop the water in the pressurizer. If it is kept big enough to absorb the hammer, then this void can prevent any harm to the precious plumbing by compressing and absorbing the transient pulse of the shock wave.

If you "let the pressurizer go solid," it means that you have mismanaged the water level in it and allowed the shock-absorbing bubble to disappear—making the pressurizer become a "solid" block of water. This was the absolute worst thing that could happen in a cramped submarine power plant, and operators were trained to avoid it at any cost. It was not a bad lesson to bring to the power plant, but in the increased-power realm, worse things could happen. Much worse.

A remaining, nagging problem with nuclear reactors in general is the decay heat of fission. Each fission event releases an enormous amount of energy, 210 MeV, but only 187 MeV is immediately available. The remaining 23 MeV is released gradually, as fission fragments radioactively decay in a cascade of sub-events over the next few billion years.[224] The rate of energy release is exponential, which is engineer parlance meaning that at first the rate falls like a lead brick on your foot, but then it slows to a dead crawl.

The issue with decay heat is that it is quite easy to instantly shut the fission process down and stop the reactor, but there is always a coast-down period in which the machine is still making power at a greatly reduced and falling percentage. If the power before shutdown is not too great, then there is no problem with the coast-down, even if all reactor cooling systems are not working. The fuel will still be hot, but reactors are built to withstand overheating. An S2W reactor on an attack submarine built in

224 Only 200 MeV are eventually recoverable. Of the 210 MeV released by fission, 10 MeV is in the form of escaping neutrinos from beta decay of fission products. Neutrinos have such a minuscule interaction cross section with matter, they can go clean through the Earth and out the other side without disturbing anything.

the 1960s made 12 megawatts when running at full speed. Immediately after an emergency shutdown, that reactor was still producing 6.5% of the 12 megawatts, or 780 kilowatts. That is not enough power to melt anything in the fuel matrix, which is made of high-temperature zirconium alloy and uranium oxide.

A typical PWR, on the other hand, can produce 1,216 megawatts of electricity. The efficiency of the steam-to-electricity power conversion process in a PWR plant is about 32%, which means that the reactor is producing 3,800 megawatts of heat to make that 1,216 megawatts of electricity.[225] Upon a sudden shutdown, a PWR is still making 247 megawatts of heat; and in the confines of a reactor vessel, that is enough power to melt solid rock. It is therefore important to keep the cooling system running after the reactor has been stopped cold. If the coolant pumps are shut down for some reason or the pipes are blocked, then the ECCS takes over, spraying cool water into the vessel to soak up the heat that is still being produced.

The temperature in the fuel falls rapidly as the fission products decay away, and after one hour the power has dropped to 57 megawatts, which the system is probably able to withstand without external systems taking away the heat, but that one hour after shutdown is extremely critical. The fuel melts at an extremely high temperature, over 5,000° Fahrenheit, and the zirconium fuel tubes and structures come apart at over 3,000° Fahrenheit, but this level of temperature is achievable in an uncooled reactor core running at a hundred megawatts. After a day of sitting in shutdown mode, the high-power PWR is still making 15.2 megawatts, or more than enough to run a submarine at flank speed. After a week, the reactor has cooled down to only 7.6 megawatts. The only way around this problem with nuclear fission is to ensure a shutdown cooling system, particularly

225 Much argument is made by anti-nuclear factions concerning the power conversion efficiency of only 32%. No great effort is made to efficiently run the steam turbines in a nuclear plant, and 68% of the power is lost in the cooling towers. The thermal pollution into the environment is therefore unusually large. A coal-fired power plant is typically 40% efficient, but to achieve this, a great deal of effort is required using a lot of fancy, expensive, and crash-worthy plumbing. This difference is because of the expense of mining coal and transporting it to the power-plant site, and the environmental damage done by burning it. The amount of coal used must be minimized at any cost. Fuel cost and its transportation are no problem at a nuclear plant, and the steam system is made as simple as possible for the sake of reliability.

in that first hour after shutdown, using redundant, multiple devices. If one auxiliary reactor cooling device fails, then there are still other ways of cooling the fuel held in reserve. This is the application for which the ECCS was designed and installed on all power reactors.

Given these minor systemic flaws, the nuclear establishment complex of manufacturers, customers, and regulatory bureaucracy was confident that installed power plants were safe against the worst possible accident, a catastrophic steam explosion throwing fission products into the atmosphere. The thought was, if we built power plants to withstand the worst accident, then the resulting physical strength and over-engineered systems will prevent any minor accident.

Everything in the nuclear power world seemed safe and running smoothly right up until March 22, 1975, when things began to unravel at the Browns Ferry Nuclear Power Plant on the Tennessee River near Decatur, Alabama. At the time, Browns Ferry was the largest nuclear plant in the world, having three General Electric BWR reactors capable of generating 3.3 billion watts of electricity. It was owned and operated by the Tennessee Valley Authority, a government program created by congressional charter in 1933 under the Franklin D. Roosevelt administration.

It was 12:20 P.M., and Units 1 and 2 were running at 100 percent power, while Unit 3 was in the last phases of construction.[226] As a rule, rooms in a nuclear plant are airtight so that a negative pressure can be maintained in the reactor building using a very large blower. This prevents any radiation leakage that could get into the air outside the plant and spread into the surrounding territory, and every room must be airtight to prevent leak points. This rule applied to the spreading room, a large chamber underneath the control room and adjacent to the reactor building, used simply as a space in which electrical signal and control cables can meet and crisscross in an orderly way. In this room thousands of cables were neatly arranged and labeled on trays and in open conduits. This room had been the last one sealed, because cables from Unit 3 were still being installed, and any instrumentation change in Units 1 and 2 required an unsealing of the room.

---

226 Unit 3 came online on August 18, 1976, and is licensed to operate until July 2, 2036. While under license to operate, it has achieved an impressive capacity factor of 99 percent.

Temporary seals around cables entering one of the four walls around the room were accomplished using a self-foaming polyurethane compound in an aerosol can, occasionally referred to as "great stuff." When workers had to hack away at the seals to install a new cable, it was easy to then spray in some great stuff and watch it expand, seal the opening, and harden. Unfortunately, the hardened foam, consisting of extremely thin plastic bubbles, has an enormous surface area, and it burns like gasoline.

A technician tested his latest sealing job using a proven method: he lit a candle and held it up to the new foam. The seal was imperfect, and the flame was sucked into a small hole by the negative pressure in the spreading room. The foam caught fire, of course. Efforts to extinguish the blaze by beating it with a flashlight were unsuccessful. As the situation quickly became desperate, two men tried to smother the flames using rags, but the flames were spreading into places where a rag could not reach.

Ten minutes later, at 12:30 P.M., someone had dragged up a carbon dioxide fire extinguisher, and they emptied it into the fire in the spreading room. It looked like it had gone out, but one minute later it flamed up again, and this time it had crossed the concrete wall through a small hole and was now in the reactor building. A worker ran up to the guard at his post in the entrance to the reactor building and asked to have his fire extinguisher, remarking that there was a fire below. Honestly, it would not seem as if there was anything to burn in a nuke plant. Everything is concrete and steel, and there is enough water on-site to fill a lake. What burns? Paint? Thousands of pages of operating and procedures manuals? Obviously, the plastic foam plus tons of plastic wiring insulation can make quite a bonfire. The Public Safety Officer sitting nearby picked up his phone and called the control room. "The building's on fire," he began. It was 12:35 P.M., and the fire alarm started going off as the announcement came over the public address system. The fire was spreading down the cable trays, about 20 feet off the floor, stopping just short of coming through the sealed penetrations in the ceiling and into the control room. Smoke was accumulating in the rooms below, making it hard to see or breathe. Both reactors were still running at 100 percent power, oblivious to the developing problem.

At 12:40 P.M., the evacuation alarm sounded in the spreading room, just as the ECCS alarm panel for Unit 1 began to light up with irrational indications of problems. The plant operator, seeing that everybody was out of the spreading room, pulled the handle to actuate the room's carbon dioxide flooding system. Nothing happened. It had been de-energized because workers were in the room. He found where it had been shut off, turned it back on, and pulled the handle again. Whoosh. The room filled with misty carbon dioxide, but still the fire burned. Another employee grabbed the handle. "You didn't do it right. Let me show you." Another loud rush, and the room clouded up, but the fire did not care. It turned out that the ventilation system was still running, blowing fresh air into the room and assuming that men were still in there, working.

The alarm indicators in the Unit 1 control room were acting crazy, indicating problems that did not exist, and at 12:51 an operator pushed the scram button with his palm. Unit 1 dropped off the power grid as the turbine coasted down. Nine minutes later, the operating crew began to lose control of Unit 2 as the fire spread to its cable trays. All systems in Unit 2 began reverting to their fail-safe conditions, and the ECCS system came on by itself. At 1:03 P.M., the cables to the main steam isolation valves burned through, and remote control of the cooling systems failed.

An assistant shift engineer took command of the fire brigade, as they passed carbon dioxide and dry chemical extinguishers hand over hand into the highly congested maze of cable racks in the spreading room and discharged the flood system a third time. At 1:10 P.M., the assistant shift engineer decided to call the Athens Fire Department and beg assistance.

Twenty minutes later, the lights went out in the reactor building. The power feed had burned up.

At 1:45, the fire department arrived, took a quick evaluation of the fire, and suggested that the plant's electrical wiring in the instrumentation and control systems be soaked with water as soon as we can get a hose in there. The recommendation was not immediately followed.

By 5:30 P.M. it was becoming clear that they had done everything possible with fire extinguishers and they would have to dump water on the wiring. It was always risky to put water on electrical circuits. Water conducts electricity, and the damage that could be caused by random short circuits in the complex instrumentation and control wiring was unpredictable. By this time, they were out of choices. Somewhere around 6:30 P.M., remote control of the pressure-relief valves in Unit 1 was lost. At 7:20 P.M., water was finally released into the cable trays, and ten minutes later, the fire was extinguished. It had burned for seven hours and ten minutes, and it had done a great deal of damage to the Browns Ferry plant and to the confidence level of nuclear engineering. A single candle flame had brought down two operating reactors and destroyed the electronic process-monitoring and control systems.

The reactor and steam systems were left in perfect order, but the control systems had been put out of action. Although flames in the cable spreading trays were considered unlikely, there was a large tank of carbon dioxide, the flooding system, installed with piping for the sole purpose of putting out a fire in the room, just in case. There was comfort in knowing that the interlocks that kept the fire from being smothered were there to keep workers from being smothered. No unusual radiation was released into the environment. It was a severe industrial accident, and pleasantly unbelievable that no one had been harmed.[227]

There were changes in nuclear power-plant codes and standards implemented after the Browns Ferry fire, from the use of silicon sealant instead of plastic foam to the rapid recharging of respirators. Unit 1 was down for a year while its wiring system was rebuilt, this time using non-flammable covers on the cables.[228]

In the fall of 1977, Cleveland Electric and Toledo Edison were proud owners of a new Babcock & Wilcox model 177FA pressurized-water-reactor power plant, built to generate 889 megawatts of electricity and

227 This was hardly the first or the only fire in nuclear-plant cable trays. There were two cable fires at San Onofre Unit 1 in 1968, one at Nine Mile Point Unit 1 during startup testing, and a cable tray fire at Indian Point Unit 2 that started in a wooden scaffolding. There were 11 cable-tray fires in nuclear plants before Browns Ferry in the United States alone, and foreign cable fires are numerous. On May 6, 1975, a fire very similar to the Browns Ferry incident occurred at the *Deutsches Electronen Synchrotron* (DESY) near Hamburg, Germany.

228 There was some further bad luck at Browns Ferry. The operating license was pulled in March 1985 because of operational problems, and the entire plant was idled for over a year. It turns out that all nuclear plants do not have the iconic concrete, convection-driven cooling towers looming over the site. Browns Ferry instead has six smaller, forced-draft cooling towers, with water flowing over slats made of redwood. With the reactors down, there was no cooling water on the slats, and they dried out in the Alabama sun. On May 10, 1986, an attempt to start up the electrical fans ignited the dry wood in a tower, which was four stories tall, 30 yards wide, and 100 yards long, and the entire structure burned to the ground. It is the only case of a nuclear-plant cooling tower being destroyed by fire.

located in Oak Harbor, Ohio.[229] The plant is named Davis-Besse, and a PWR is not a small machine. The reactor pressure vessel alone is 700 tons of steel with walls nine inches thick. It contains 100 tons of uranium fuel in 36,816 rods. It makes scalding hot water, which feeds two steam generators, each 73 feet high and weighing 400 tons. On September 24, the plant was six months old, and they were still testing it, running at low power just to see if something would break. The reactor was at nine percent power.

All was quiet and peaceful in the control room when the floor gave a shudder. There was a distant rumble, seeming to come from the turbine building, and the operating staff on duty assumed a collective "what-the-hell-was-*that*?" look. The long U-shaped console lighted up with trouble indicators, and alarms started going off. Six operators scanned the meters and alarm panels, seeking to quickly evaluate the status of the system. Shift Supervisor Mike Derivan, trained as an engine-room supervisor in the nuclear Navy, looked first at the level of water in the pressurizer. It was shooting up rapidly, shrinking the steam bubble at the top. He instinctively reached for the red scram button and pushed it. The control rods slammed into the reactor core and stopped the fission process.

The coolant pumps for the number two steam generator seemed to have stopped for some unknown reason, and that had caused the pressure to rise in the reactor and force water up into the pressurizer. With one steam generator out of commission, the reactor was suddenly making too much heat. That much made sense. The operations crew now watched, perplexed, as the pressure in the reactor dropped by several hundred pounds in less than a minute. It was not clear what was going on.

The reactor, noticing that no operator was moving to prevent a meltdown, then decided to fend for itself, automatically turning on the ECCS. This first component of the ECCS was the High Pressure Injection system (HPI), spraying cool water into the reactor vessel at an aggressive pressure of 1,900 pounds per square inch.

Derivan, still locked on the pressurizer water level, decided that there was no problem with the size of the steam bubble now, and he manually overrode the automatic system and shut down the ECCS. Inexplicably, the

229 B&W reactors were named by the number of fuel assemblies (FA) in the core. In this case, it was 177. Each fuel assembly holds 208 fuel pins, which are zirconium tubes filled with cylindrical pellets of uranium oxide, each 12 feet long.

water level began again to rise in the pressurizer, indicating an increase in reactor vessel pressure, even though the reactor was shut down and cooled by the water injection. The pressure should have been dropping.

The staff was now completely confused, and somebody suggested that they cut off the other two coolant pumps. Coolant pumps, after all, generate some heat on their own, just by stirring the water, and perhaps that extra energy was causing the pressure to rise. Grasping at straws, they stopped the pumps.

The water level in the pressurizer shot up and off scale. More alarms started blaring, and by this time hundreds of trouble lights were blinking all over the console. Number two steam generator boiled dry, and a particularly insistent alarm indicated that the air pressure in the reactor containment building, which should be below atmospheric pressure, was now abnormally high. Was there a break in the pipes? Was steam escaping into the building? This was getting very serious.

Derivan ran behind the console to look at the containment building pressure gauge. It was at three pounds per square inch above normal and rising. Finally, he understood what was going on. The Pilot-Operated Relief Valve (PORV) atop the pressurizer had blown open and failed to reclose. It was designed to open automatically if the pressure in the primary coolant loop reached 2,200 pounds per square inch and allow the steam to blow off into a holding tank in the containment building. The containment building held the reactor vessel and the steam generators, or everything in the potentially radioactive primary cooling system, and was a secondary safety against radiation escape into the surrounding environment. The relief valve prevented damage to the plumbing in the primary cooling system when the pressure in the reactor spiked too high. The act of relieving the pressure would cause it to drop, and the PORV was supposed to close again when it fell below 1,800 pounds per square inch. If the valve failed to close, then the pressurized water reactor was no longer pressurized, with all its energy free to escape into the air in the building.

A normal steam relief valve is a simple affair, consisting of a steel spring holding down the cap on a hole in the highest point of a boiler system. In this case, the hole was a rather large four square inches, and it took 3.5 tons of force to keep it closed. That would be a very, very large steel spring, awkwardly heavy, and to mount that on top of the pressurizer would be asking for trouble from vibration effects. Instead, on a nuclear reactor of 1970s vintage, the force used to close the relief valve was supplied by the steam pressure underneath it. A "pilot tube" conducted the steam through a control box, wired electrically back to the control room, and fed a cylinder and piston connected to the valve cap.[230] The concept was elegant, very lightweight, appealing to engineers, mechanically complex, expensive, and notoriously subject to random failure.

"Shut the block valve!" Derivan yelled.

In case the PORV was stuck open, a second, simple valve, similar to the one used to turn on the water in a sink, could be operated remotely from the control room. It was named the block valve. An operator reached for the switch handle, gave it a quarter turn, the steam leakage stopped, and 20 minutes of hellish confusion ended. After 26 minutes of settling down, everything was back to normal.

The Nuclear Regulatory Commission and a number of top engineers from B&W were all over this incident. It was unnerving and very serious, because if the reactor had been running at a higher power, such as 50 percent, the entire core could have melted. With this level of operational chaos, a pure disaster was possible. The government and manufacturer representatives investigated in depth.

The problem was indeed the PORV, but it was not the fault of the PORV. A pen-chart recording showed that the valve had rapidly slammed open and then shut nine times, beating itself to pieces and leaving the steam line gaping open. The fault was back in the control room, in a rack of relays behind the control panels. An unnamed worker had found a bad relay in a circuit he was repairing. The same type of plug-in relay was used all over the system, and he needed one. He found a perfect replacement unit for his circuit in the PORV panel, so he unplugged it and used it. The PORV was used only for

230 The PORV was able to keep enough force on the valve to keep it closed without using a spring even though the same steam pressure that was trying to open the valve was used to keep it closed. This was possible using a hydraulic trick. The surface area of the piston used to hold down the valve cap was larger than the opening of the pipe that the valve was closing. This ensured that the force from the piston was always bigger than the force opposing it from the steam pipe. This was a fine concept, but unfortunately the mechanical precision required was more than could be supported in the rough-and-tumble world on top of a power reactor.

emergencies, and it would probably never be needed, he must have reasoned. The PORV, missing a logic element, went berserk when called to action by the primary steam-pressure sensor.[231]

That explained the problem, but what explanation was there from the operating staff for having shut down the ECCS? The HPI pumps had been manually killed only four minutes into the incident, at least 16 minutes before they had any clue as to what was happening. The ECCS was designed to keep the reactor from overheating and melting out the reactor core. To turn it off looked like sabotage.

The answer to the question was both simple and disturbing: they shut off the HPI to keep the pressurizer from going solid. This glaring problem with operator training, to undo this component of the Navy training, was discussed at length, but not to the point where operating power plants were notified of this finding, and the analysis of the frightful Davis-Besse incident got lost in the bureaucratic tangles at the NRC and at B&W. None of the other seven owners of B&W reactors were told about the dangerous confusion that can result when the PORV sticks open.

At about the same time, in the fall of 1977, Carl Michelson, an engineer working for the TVA in Knoxville, Tennessee, was studying the reactor building layout of the B&W model 177FA, when he noticed something.

In the training diagrams, in nuclear engineering textbooks, and in any diagram of a PWR primary cooling system simplified to the point where you can tell what is going on, the pressurizer is shown on top of the reactor vessel, usually connected to one of the hot pipes coming out near the top of the vessel. That is where the heated water comes out of the reactor and is piped to one of at least two steam generators. The pressurizer is the highest point in the system, so the steam bubble that is allowed to exist is always trapped in the uppermost part of the pressurizer tank. Because it is the highest point in the system, the water level in the pressurizer is used to evaluate the

231 The PORV used in the Davis-Besse reactor was a Crosby model HPV-SN. It was repaired and put back into service, but it failed again on 5/15/78 (broken valve stem) and 10/26/79 (pilot valve and main disk leaking). The Crosby valve was not used in subsequent B&W installations of 177FA reactors. The missing unit in the rack was specifically the "seal-in relay."

water level in the reactor, which is vitally important. If the water level ever falls below the top of the reactor fuel, which is blazing hot even with the reactor shut down due to the delayed heat production, then the internal structure of the reactor is going to melt.

Instead of putting expensive instrumentation on the reactor vessel to monitor the water level, the operators are taught just to look at the water level in the pressurizer. If there is any water at all in the pressurizer, then the reactor vessel must be completely full, and there is nothing to worry about. Just worry about the water in the pressurizer, and everything will be all right.

But in the B&W layout, the pressurizer is 43 feet tall, or about 10 feet taller than the reactor. There is not enough room below the fueling floor in the containment building for the pressurizer to be on top of the reactor. If it were, it would stick up out of the floor, so they had to lower it. In its position next to the number two steam generator, its inlet pipe had to be looped underneath a coolant pump line. The loop of pipe looks just like a sink drain trap, used to keep sewer gas from backing up into the sink. Michelson realized that no matter what condition might prevail in the reactor vessel, the water level in the pressurizer would never go below the trapping loop. The pressurizer would always be 20 percent full, even if the reactor was boiled dry. The operating crew in a 177FA control room actually had no idea of the water level covering the hot reactor fuel, and this struck him as dangerous. His finding caused a lot of commotion in the TVA, the NRC, and at B&W, but it never escaped the tangle and was never passed down to the operators at the eight reactors that B&W had built. There was a disaster, set up by a combination of policy and engineering, and it was waiting to happen.

On November 29, 60 days after the relief-valve fiasco, Davis-Besse experienced another emergency shutdown. The cause was traced to a wrongly wired patch panel in a control-room computer. In the middle of trying to figure out what was wrong and correct the problem, the operations staff, apparently acting on pure, ingrained instinct, again turned off the ECCS. The incident investigators found this action difficult to comprehend. Why did the operating staff at a nuclear power plant tend to disable the emergency core cooling system during an emergency?

These bits of information would have been useful at Three Mile Island, Pennsylvania, where a new B&W 177FA had been running "hot, straight, and normal" for almost three months. The first reactor unit built there, TMI-1, was down for refueling, and TMI-2, the new reactor, was running at 97 percent full power, putting 873 megawatts on the power grid for the owner, Metropolitan Edison of Pennsylvania. The two B&W reactors were built on a three-mile-long sandbar in the middle of the Susquehanna River, just south of Harrisburg, the state capital.

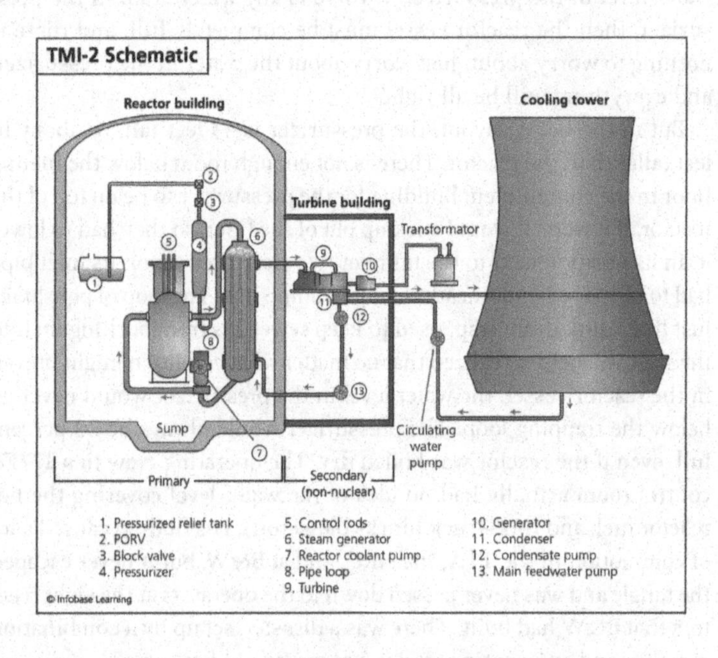

The TMI-1 reactor system was typical of pressurized-water reactors used all over the world, but some oddities of the B&W design contributed to a disastrous breakdown. The complete lack of water-level instrumentation in the reactor vessel was a big problem that the Nuclear Regulatory Commission would make illegal after the accident.

The TMI-2 reactor had been originally contracted for the Oyster Creek Nuclear Generating Station in New Jersey, supplementing a BWR brought online by General Electric back in 1969, but the Jersey craft-labor corruption was starting to get out of hand. Jim Neely, the negotiator for Jersey Central Power and Light, had been dealing with the mob ever since a worker made a point by dropping a wrench into the gearbox of the crane hoist while they were lifting the 700-ton reactor vessel into place for Unit 1 at the plant. That was alarmingly close to a disaster. Now a mob representative wanted one percent of the construction budget for the new B&W unit to ensure peace among the workers. That would be $7 million for one individual, and he was only the first in line. It was just not worth it to build a nuclear plant in New Jersey anymore, and Neely gave up. He amended the license application, and gladly transferred the contract to Met Ed, Pennsylvania. Maybe they would have more luck with it. TMI-2 was constructed without incident and began delivering power on December 30, 1978.

On March 16, 1979, a movie, *The China Syndrome*, opened in theaters nationwide. It was a fanciful cautionary tale about a potential nuclear power accident that could spread deadly radiation covering an area "the size of Pennsylvania."[232]

It was March 28, 1979, heading toward 4:00 A.M., and the Shift Supervisor on duty was Bill Zewe. Zewe was 33 years old and had learned the nuclear business in the Navy. The previous shift had left him a problem. The steam that runs the turbine is turned back into water by the condenser beneath the turbine deck, dumping the excess heat to the twin, iconic cooling towers out back. The towers are each 30 stories tall, and they cool a million gallons of water per minute, making white, fluffy clouds rise into the air above.

The water out of the condenser must be "polished" before it returns to the delicate, expensive steam generators, removing anything that may have dissolved in it as it cycled through the pipes, valves, steam generator, pumps, and condenser. A bit of rust, for example, could have been picked up along the way, but the water is made sparkling pure as it is pumped through a line of eight 2,500-gallon tanks in the basement of the turbine building en route to the steam generators.

232 The film starred Jane Fonda as a television reporter, Michael Douglas as her camera operator, and Jack Lemmon as the shift supervisor at a fictitious nuclear power plant in California. It was inspired by the Browns Ferry fire, which had occurred two years earlier, and it was a fair assessment of the growing public angst over nuclear power. Although there was a great deal of built-up worry and end-of-the-world dread in the movie, nothing melted, there was no radiation escape, and the reactor safety systems basically acted as they were supposed to. The only harm to come to any of the characters in the movie was by sniper rifle and motor vehicle. It was thus a parody of a typical commercial nuclear plant accident.

Each tank is filled with tiny balls of purifying resin, and they must be flushed out and replaced as they become contaminated or loaded with gunk out of the condensed water. Unfortunately, the resin beads tend to mash down and stick together, reminiscent of the problem that blew up on the Atomic Man back at Hanford, and the back-flushing system installed by B&W was underperforming. In Tank 7, the resin was stuck tight and was not moving. Zewe left two men working on the problem and climbed the eight flights of stairs to the control room. He asked Fred Schiemann, the foreman, a Navy man, to go down there and encourage them. As he left the control room, Schiemann gently reminded Zewe that the PORV was leaking, which was nothing new. In the 177FA design, B&W had replaced the troublesome Crosby PORV with a Dresser 31533VX30. In terms of reliability, it was proving no better than the Crosby. There was nothing particularly fatal about a little bit of leakage out the top of the pressurizer, but it was a pain, having to readjust everything as steam slowly escaped the primary coolant system and blew off into the containment building. It was just an irritant, and nothing more. It would be on the list of things to be corrected during the first refueling shutdown, which was not scheduled for two more years.

The men who had built this plant were an industrious, creative lot, and when they found that the resin beads could not be flushed using the factory-designed system, they added a compressed air line from the general-purpose compressed-air system in the plant. The air pipes were about the size of a garden hose. You could just open a valve, and the air bubbles would stir up the beads in the tanks and break them loose from sticking. There was not quite enough air in the system, so they cross-connected it to the instrument compressed-air system, which could then be used to open and close valves remotely, using switches in the control room. But ceasing to manually check those valves would be a problem, as somebody on the previous shift had air-flushed the tanks, but had forgotten to close the air valve. The one-way check valve in the air line was leaking, so for the past 10 hours, pressure from the 5,000 tons of water per hour running through the tanks had forced water up the instrument air line, almost to the point where it would cut off the air going to the valves on top of the eight tanks, which would slam them all shut at once and stop the flow through the steam system. There was an electrical backup system that would prevent such an improbable, almost impossible catastrophe, but it had not been wired up. The valves were supposed to be left open while the steam was running, but you could call up the control room and ask an operator to close the inlet valve on one of the eight tanks. Tanks could thus be cleaned out one at a time as the plant ran at full power.

Schiemann, down at the resin tanks, tried to assess the situation. They had not been able to dislodge the beads, and they had tried everything. A water flush, compressed air turned up all the way, and even steam had been unleashed on Tank 7. Schiemann climbed on top of the enormous water pipe so he could watch the sight glass and see the level of water in the tank. It was hard to see in the dim light. It was 3:58 A.M., and suddenly there was an awful quiet in the normally rumbling water pipe. Uh-oh. The water had backed up in the air pipe just enough to close all the valves atop the resin tanks.

He could feel it under his feet, a water hammer, caused by the sudden perturbation in the steam system, coming down the pipe, hot and fast, like a ballistic missile. He leaped free, just as the pipe jumped out of its mounts and ripped out the valve controls along the walls. Scalding hot water blew out into the room as the pump at the end of the pipe flew apart.

Back in the control room, every alarm tile on panel number 15 came lit up at once, and the warble-horns started going off. The turbine, sensing that it was not going to get any more steam, threw itself off line, and the reactor followed eight seconds later with an automatic scram, ramming all the neutron-absorbing controls deep into the core. The main safety valves in the secondary loop opened and blew the excess steam skyward. It sounded like the building was coming apart. The floor in the control room trembled, as the four main feed-water pumps shut down. Pressure in the reactor vessel, now denied its two primary cooling loops, rose sharply, and in three seconds the PORV opened automatically, blowing extremely hot water and steam into the drain tank on the containment building floor. On the control console a red light came on, indicating that the PORV had received an OPEN signal. Ten seconds later, a green light came on, indicating that the PORV had received the CLOSE signal. The sharp spike in the primary loop pressure had quickly dropped below 1,800 pounds per square inch, so there was no longer a need for an opening in the normally closed cooling system.

The senior men, Zewe, Faust, and the operator Ed Frederick, had seen this before, and knew it was a turbine trip. Regardless of the blinking lights and the pulsating horn blowing in their ears, it was nothing to get panicky about, and all the systems were acting correctly.

This feeling of tense calm lasted about two minutes, when the two high-pressure injection (HPI) pumps, a main part of the ECCS systems, switched on automatically. Now this was something new to Zewe, Faust, and Frederick. Why did the reactor think it needed emergency cooling? The temperature in the reactor was too high for this lockdown situation, and the pressure was too low. A minute later, Schiemann made it to the control room, gasping for breath after having sprinted up the staircase.

At 4.5 minutes after the turbine shutdown, Schiemann had been watching the water level rise in the pressurizer, and he ordered that one HPI pump be turned off, and throttle back the other one. The last thing he wanted was for the pressurizer to "go solid," and with the pressure this low, the HPI was capable of filling it up. Still the water level rose, and meanwhile the two steam generators had boiled dry, making solidity in the pressurizer a very real possibility.

It had been eight minutes since the trip, and to the horror of the operating staff, the pressurizer was rapidly going solid. Frederick turned off the second HPI pump, thinking that the flow of water into the system was flooding the pressurizer. Still, the temperature in the system rose while the pressure kept falling. It made no sense. Everybody could see that something was wrong, but they did not know what.

Zewe had a hunch. He asked an operator to read him the temperature of the PORV outlet. If it was unnaturally high, it would mean that the green instrument light was wrong.[233] If the PORV had not been closed, steam was escaping from the top of the pressurizer, and that would explain the low pressure. The operator shouted the value back to him: 228 degrees. That was not an unreasonable temperature. The

233 The green light was correct. The PORV had been signaled to close. But that did not mean that the valve had closed, only that the request had been made. In this way, the instrumentation in the B&W control room was not adequate. There should have been another set of red/green lights to show the physical state of the valve, opened or closed.

valve had, after all, been leaking since January, and that was a little bit of steam getting past the valve cap. Unfortunately, the operator had read the wrong temperature readout. The outlet temperature of the PORV was actually 283 degrees, and the entire primary coolant inventory was draining out through it. The valve was jammed wide open, and the water was boiling out of the reactor core, forcing water up in the pressurizer and out the top. At that moment, when the temperature readout was off by 55 degrees, the TMI-2 power plant was lost, and a half-billion-dollar investment flushed down the drain. At the low pressure allowed by the open valve, the reactor could boil dry.

The critical time is that first hour, when the energy rate from the decaying fission products, freshly made in a core that was running at nearly full power, is falling rapidly from 247 megawatts down to 57 megawatts. If you can just keep water covering the fuel for that first hour, then everything else will work out fine. It does not have to be cool water or clean water, and it does not have to cover anything but the naked fuel pins, but if any fuel is left without water to conduct the heat away, it is going to start glowing cherry-red and melt down the supporting structures. It happens with merciless dispatch. With its gas-tight metal covering melted away, the uranium oxide and any soluble fission products embedded in it are free to dissolve in whatever water or steam is left in the reactor vessel, and this becomes a perfect vehicle for the highly radioactive, newly created elements to escape the normal confines of the tightly sealed PWR primary cooling system. It goes right out the jammed PORV, with the steam, into the drain tank. Fortunately, all of the fission products are solids, and even if the drain tank is opened they tend to stay inside the building, stuck to a wall or some expensive piece of equipment as the water evaporates.

All, that is, except the iodine-131 and xenon-133. They are gaseous. Iodine is not too bad, because it will corrode any metal in the building and bond to it, and there is a lot of metal in the building for it to cling to and thus not escape. Xenon-133, however, is guaranteed to escape into the outside world, as it will never bond with anything. It has a half-life of 5.24 days, undergoing beta-minus and gamma decay.

After 15 minutes of taking water from the PORV outlet, the drain tank was completely full, but there was still a lot of primary coolant left. The cover on top of the tank ruptured, as it was meant to in an overfill emergency, and the water cascaded down the sides of the tank, across the floor, and into the sump ditch at the lowest point in the building. After a while, the sump was full to the top, and the pumps came on automatically, designed to transfer the runoff into a big tank somewhere else in the building. The pumps, however, were connected wrong. They started pumping the coolant, which eventually would be made radioactive by having dissolved fission products out of the red-hot fuel, into the auxiliary building. It was shared by the two reactors, TMI-1 and TMI-2, and it was not a sealed structure.

After an hour, 32,000 gallons of water had left the cooling system. The main coolant pumps, now pushing steam around, started shaking violently. The operators could feel it through the floor. After 14 minutes, they could stand it no longer and shut off two of the four pumps. The two remaining pumps felt like they were going to explode, so after another 27 minutes, they shut them down. There was now no known cooling system operating in a reactor that had been running nearly full blast less than two hours ago, and the staff had no idea what was happening. The level of water in the pressurizer, which was solid, indicated that the reactor vessel was still completely full.

The fire alarm went off in the containment building. Frederick canceled the siren, and then it went off again. This time, it was the control room fire alarm. But they were in the control room, and a quick glance proved that there was no fire here. Zewe walked around to the back of the console to have a look at the less important gauges. Here he found that the pressure in the containment was going up. What was going on? What was making the air pressure in the reactor building climb? Was something amiss in the primary cooling system? The phone rang. It was Terry Dougherty, former nuclear submarine machinist's mate, calling to say that the sump pumps in the containment had switched on. As he was talking, Dougherty noticed that the hand-frisker in the hallway, a Geiger counter that checked workers' hands for radioactivity at the doorway, was sounding its radiation-limit alarm. Its meter read 5,000 counts per minute, which was entirely abnormal.

Brian Mehler, the Met Ed Shift Supervisor for Three Mile Island, having been roused out of bed at 5:00 A.M. by a problem at the plant, finally arrived and was appalled at the conditions indicated by the instruments. The operators were all clustered around the pressurizer instruments, fretting about the high water level. Mehler turned to Schiemann. "Shut the block valve on top of the pressurizer," he said, thus effectively shutting the barn door after all the horses had escaped. It would have been, of course, the correct action, but it was too late, two hours and eighteen minutes after the shutdown. Now, shutting the blocking valve simply closed off the only outlet for heat that the reactor had, which was the evaporation of the coolant. With that last, noble gesture, the melting began in earnest. It took about eight minutes for the top of the core to collapse.[234]

The radiation instruments monitoring the reactor core began to take off, as if it were trying to restart itself. Zewe called for a coolant analysis. If for some reason the boric acid concentration, normally high in a new uranium core, was brought down by dilution from the emergency water that had been injected, perhaps the reactor could go critical with all the control rods fully in? A power reactor was designed to have as much excess reactivity as was safe, to allow a long time between refuelings, and in a PWR some boron in the coolant was there to counteract the reactivity. As the fuel burned up, the excess reactivity would go away and the boron would be chemically removed from the coolant.

Before he was able to make any conclusions, there came another shrill call from Dougherty in the aux building. Something had filled up the sump in the building, and it was now overflowing and going down the floor drain. Just then, the radiation alarm went off in the aux building. There were now 50 people standing in the control room, simultaneously gripped by the sound of the radiation alarm in the containment building. Zewe picked up the intercom microphone and

234 Unfortunately, at this moment the confusion factor in the control room was at its peak. George Kunder, the plant engineer, was on the phone in the back of the room with Leland Rogers, the B&W rep on site. He was familiar with the Davis-Besse incident and was an expert on the 177FA reactor systems. He listened to the symptoms and then asked Kunder, "Is the block valve closed at the top of the pressurizer?" Kunder replied yes. It had been closed moments before. Unfortunately, Rogers did not ask "How long has it been closed?" If only he had known that the pressurizer drain line had been open for over two hours, he would have instantly known what the problem was, and he would have told them to re-open it until they could get enough water into the system to re-start the coolant pumps. With his partial information, he did not realize that 250,000 pounds of coolant had blown out of the system, and all subsequent diagnoses were wrong.

announced a Site Emergency, indicating a possibly uncontrolled release of radioactivity. TMI-2 was going down. If it had been a submarine, everyone would have drowned.

Thirty minutes later, Gary Miller, the TMI Station Manager, became aware of the situation and declared a General Emergency, and at 7:02 A.M. Zewe called the Pennsylvania Emergency Management Agency. Captain Dave, Traffic Reporter for a top-40 radio station in Harrisburg, WKBO, picked up an odd State Police conversation on the CB radio in his car. They were babbling about an emergency at the plant. He called it in to the news director, Mike Pintek, who rang up the Three Mile Island Nuclear Generating Station. The switchboard operator, not knowing what to do with someone wanting to know if the plant was going to explode, switched him to the TMI-2 control room. Pintek connected with the reactor operator who was closest to the phone and got a sense of frantic chaos. The story aired at 8:25 A.M., and the cat, so to speak, was out of the bag.[235]

At this point, although it was not fully realized, the TMI-2 power plant was a total loss. There were several things tried to bring the reactor back to some normal shutdown condition, but all failed. An entirely new set of goals had to be set. The fission products must be kept inside the primary cooling loop and not be allowed to escape into the area surrounding the plant. The public must be informed of the problem and any developments on a timely basis, but not in overly technical terms, causing panic and a mass stampede to get out of Harrisburg. The state and federal emergency services must monitor the landscape for a possible radiation plume and be prepared to relocate anyone under threat of harmful radiation exposure.

By 9:00 A.M. the radiation counter in the ceiling of the containment building was reading 6,000 rads per hour, indicating that not only had the coolant escaped through the overfilled drain tank, but that the

235 After his short conversation with a reactor operator, Pintek called Blain Fabian, Communications Director at Med Ed, and eventually got a prepared statement for his radio announcement: "The nuclear reactor at Three Mile Island Unit Two was shut down as prescribed when a malfunction related to a feed-water pump occurred about 4:00 A.M. Wednesday [March 28]. The entire unit was systematically shut down and will be out of service for about a week while equipment is checked and repairs made." The announcement proved to be wildly optimistic.

fuel pins were no longer containing the fuel.[236] Hot fuel had lost its zirconium cladding and had dissolved in the steam, sending fission products out of the primary loop. The sealed containment building, made of concrete and steel five feet thick, was a solid blockage between the wet, steamy radioactive waste and the outside world.[237] It held throughout the danger period of the accident, and still stands today.

The uranium-oxide fuel in the reactor, laid bare of any effective cooling, reached temperatures as high as 5,000 degrees Fahrenheit. Normal temperature with the reactor running at full power was 600 degrees Fahrenheit. Nothing approaching this had ever happened in a full-sized, billion-watt power reactor, and the core temperature was way outside the range of the control instrumentation.

The operating crew, supplemented now by a mass of experts from the factory and the NRC, could only guess what was happening. By the time the NRC showed up, at 10:30 A.M., radiation had started to leak into the control room, and everybody had to wear a respirator to keep from breathing radioactive dust.

At the elevated temperature, the zirconium alloy fuel pins and supporting structure not only melted, they reacted chemically with the steam left in the reactor vessel, making zirconium oxide. This chemical action stripped the oxygen out of the water, making hydrogen gas. At first, the hydrogen escaped with the steam and floated to the top of the containment structure, but when the block valve was closed, it was sealed tightly in the reactor vessel. There were things wrong with the B&W reactor, but vessel integrity was not one of them. With the block valve closed, nothing could escape the reactor. The hydrogen, imprisoned in the vessel, floated to the top, formed a bubble, and exerted gas pressure on the structure. The bubble grew

---

236 The radiation dose rate was correctly measured in rads and not rem, because there was not a human being in the containment building. A rad is simply the radiation dose that causes 0.01 joule of energy to be absorbed per kilogram of matter. To be expressed in rem or sieverts, the dose must be multiplied by a factor that depends upon the type of radiation and its effect on an average-sized man.

237 Commercial PWR power stations at the time were built with containment structures that could withstand a direct hit from a jet airliner, which was the worst accident that an engineer could think of. This was long before the 9-11-2001 terrorist attack in New York City, in which two airliners were used to take down two skyscrapers.

to a highly compressed 1,000 cubic feet, and by Thursday, March 29, it was 20,000 cubic feet.

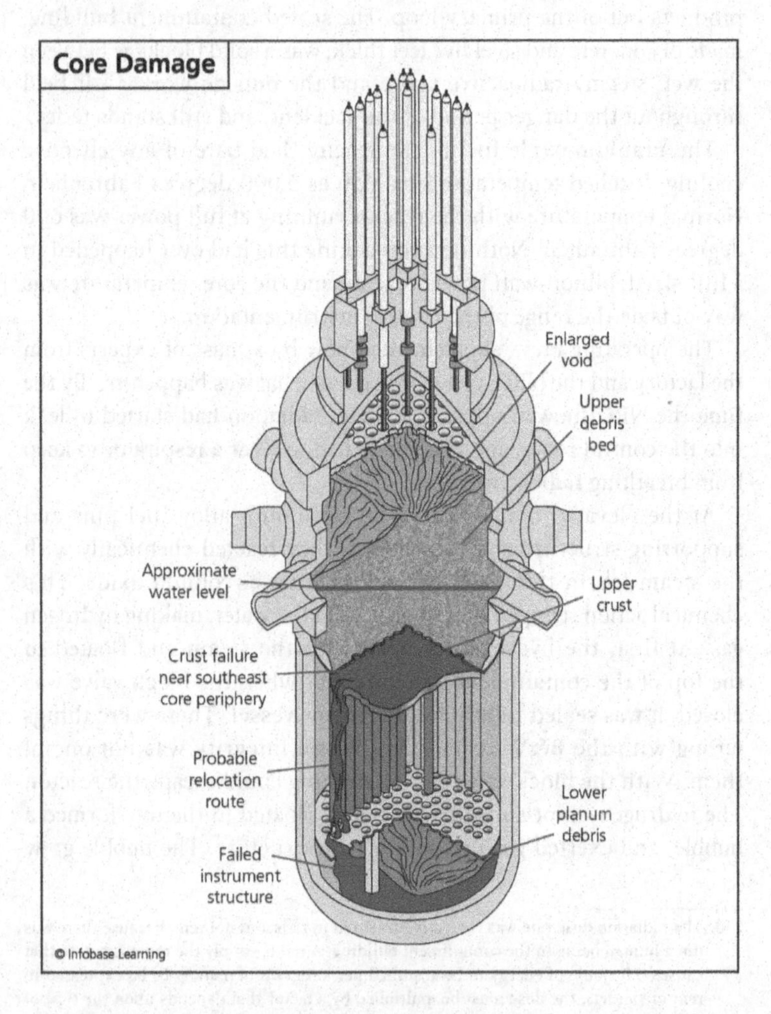

The fuel, the controls, and the oxidized zirconium structural elements in the reactor core melted together into one hard ceramic pool in the bottom of the steel reactor vessel. The cooling system and the vessel never failed, as was widely feared, but the reactor was not salvageable. Special boring tools had to be invented to remove the melted insides of the reactor and bury the radioactive debris.

This situation caused a great deal of anxiety in the control room. There was fear that the hydrogen pressure could either break open the 9-inch-thick, stainless-steel-lined reactor vessel, or it could explode, or both.[238] Furthermore, the fuel, now molten and dripping into the bottom of the vessel, could melt through it. Either scenario would contaminate the area downwind, as far as 10 miles. On Friday evening, March 30, at 8:23 P.M., the Associated Press had gotten wind of the worries about hydrogen exploding in the reactor, and they issued an urgent advisory to the public.[239] Two thirds of the people around Harrisburg who heard this announcement interpreted it as a warning of an impending massive nuclear explosion, a "hydrogen bomb," and 42,000 left town as quickly as was possible. The next day on the TV show *Saturday Night Live*, the venerable comedy team of Bob & Ray announced a contest to name a new capital of Pennsylvania. By Sunday, 135,000 people, or 20 percent of everybody who lived within 20 miles of the plant, had voluntarily evacuated.

At the same time, the various emergency holding tanks in the containment building and the auxiliary building were reaching maximum capacity. By Thursday night, March 29, the low-level waste-water tank, containing minimally contaminated water from the toilets, drains, showers, and laundry had reached its capacity of 40,000 gallons. The plant workers did what they always did under this condition. They opened a valve and let it drain slowly into the

---

238 The level of concern about the strength of the reactor vessel and a hypothetical hydrogen explosion shows the extent of technical hysteria in the control room. To explode, hydrogen must have a similar volume of free oxygen, and there was zero oxygen in the vessel. If there had been, then the hot zirconium would have scavenged it away. Moreover, every water-cooled power reactor has passive hydrogen recombiners built into the primary cooling loop. A catalyst, usually platinum, is used to recombine hydrogen and oxygen back into water after it has been broken down in the severe radiation environment of controlled fission. The reactor vessel for the B&W 177FA was designed to handle 2,500 pounds per square inch of internal pressure, and the hydrogen pressure never approached this value.

239 There was a hydrogen explosion at TMI-2, but nobody knew about it until much later, when data recordings were analyzed. Hydrogen entrapped at the concave ceiling of the containment building was set off by a spark and combined explosively with the air in the building. In the control room it was felt as a big thump, but nobody was certain what it meant. The containment building, which is quite robust, was not affected. I have mixed feelings about AP's announcement. I am all for telling the public what is going on, as this very book attests, but to let them leap to the conclusion that it would be a thermonuclear explosion was criminal.

Susquehanna. There was nothing illegal or even unusual about dumping the water tank, except under these frantic conditions of alert and anxiety. When the Governor of Pennsylvania, Richard Thornburgh, got wind of this in Harrisburg, he hit the ceiling and forbade any further release of anything radioactive. That was unfortunate, because anything that could have been disposed of that had a small enough radiation load to be safe for disposal, would have to wind up on the floor of the auxiliary building, and this made things more complicated than they had to be.

There was no governor's mandate that could stop the other radiation release, which was the gaseous fission products. The iodine-131 mainly bonded to the inside of the containment building, and although at the peak concentration there were 64 million curies of iodine in the reactor core, the amount that escaped was barely detectable. On the other hand, a great deal of xenon escaped, and that was 13 million curies.[240] That is a lot of radiation, and if it were any other fission product, there would have been long-term evacuation and contamination cleanup in an area ten miles long and a mile wide in an east-northeasterly direction from the plant, or in the direction that the wind was blowing. Xenon, however, is different. Its body burden is practically zero, because human metabolism has no use for a noble element that cannot chemically bond with anything, and so it is not able to bond to our biological bodies and cause us harm. It just floats in the atmosphere, decaying into non-radioactive cesium-133 over 52 days. One can breathe it in but will most likely breathe it back out without experiencing a radiation release in the lungs.

Although it is not generally known, all nuclear power stations make radioactive xenon nuclides while they are generating power, and it eventually goes up in the atmosphere by way of the otherwise inexplicable "smokestack" on site. Gaseous fission products find

240 For the sake of simplicity, I have limited this account to Xe-133. Actually, fission makes 18 xenon nuclides, from Xe-129 to Xe-146. Most have half-lives of seconds, and they never make it out of the fuel. Xe-135 is sometimes measured in reactor off-gas, but it has a half-life of only 9.1 hours. Three of these nuclides are not even radioactive. Xe-133 is probably the culprit in all radiation measurements outside the TMI power plant.

their way into the primary cooling loop, and they are drawn off in the water makeup system, located in the auxiliary building. Scavenged gases are compressed and stored in the "decay tank." When the decay tank gets full, a worker turns on a valve, and up the stack it goes. This is routine. The instant it scrammed, TMI-2 stopped making radioactive xenon. The zirconium fuel pins try to keep any gas from getting away, and a lot of it decays in the fuel without escaping, but xenon is good at finding its way into the coolant. The only reason that TMI-2 released a big slug of xenon was that the fuel pins had disintegrated, so the normal hindrances were gone.

The decay tank was purged at 8:00 A.M. on Friday, March 30. A helicopter directly over the stack measured a 1.2-rem-per-hour dose rate at 130 feet over the plant, and the reading immediately tailed off as the gas dissipated. It eventually made a narrow but diluted plume, 16 miles long. All off-site radiation measurements, peaking at about 0.007 rem per hour, were probably due to the xenon gas. In general, nuclear workers are allowed to absorb 5 rem per year, and civilians are allowed 0.5 rem per year. A population group, such as the citizens of Harrisburg, is allowed a collective 0.170 rem per year. Standing at the fence around the Three Mile Island plant for a year, an individual would have received 0.005 rem.

On April 7, 1979, at 2:03 P.M., Three Mile Island Unit 2 achieved cold shutdown. TMI-2 would never again generate any electricity. In the history of the world, it had been the worst industrial disaster in which not one person was harmed. Over the next 20 years, there were certainly cancers among some people who were downwind of the plant, as happens in any group of people over time, but it was difficult to correlate these illnesses with any aspect of the reactor meltdown at Three Mile Island.[241] The most popular T-shirt slogan was "I survived Three Mile Island . . . I think."

---

241 Immediately after the incident was declared over, hundreds of people in the area reported a metallic taste in the mouth, nausea, skin rashes, and hair loss, the symptoms of having stood atop a reactor as its power spiked way above design limits. There were rumors of cancer, sudden infant deaths, and stillbirths. These reports did not correlate with any measured radioactivity. The only thing that was able to escape the containment building was some radioactive gas, and it was gone within hours. Radiation-induced cancer has a latent period of about 20 years. The TMI-2 accident did cause illness, but it was most likely psychological and not radiological.

Many changes in nuclear power training, control-room instrumentation, and pressure-relief valves came down from the NRC in the following years. "PORV" now means "Power Operated Relief Valve," and not "Pilot Operated Relief Valve." Dresser Industries, maker of the PORV in TMI-2, put a full-page ad in the *New York Times*, with Dr. Edward Teller claiming that he was not afraid of nuclear power, but he was terrified of Jane Fonda.

The thorough cleanup operation, costing $1 billion, was completed in 1993. $18 million of the cost was contributed by the government of Japan, with the provision that we include Japanese workers to have experience in a nuclear power cleanup. The final report concluded that 35 to 40 percent of the fuel had melted, while 70 percent of the core structure had collapsed. A surprise to everyone was that there was never a chance of melted uranium oxide burning through the bottom of the reactor vessel. The meltdown had, in fact, formed an insulating layer of ceramic material, a durable mixture of zirconium and uranium oxides, at the bottom of the vessel, and it was impervious to extreme temperature.

TMI-1, the other reactor sitting next to TMI-2, has been quietly generating power and making money for its owner, the Exelon Corporation of Chicago, ever since 1985, when it was allowed to resume operation. Its operating license runs out in 2034. B&W never received another order for a full-sized civilian power reactor. They are now developing a small modular power reactor called mPower.

Could TMI-2 have been cleaned up, refurbished, and restarted? Economically, no. The entire inside of the containment building and every tank, pipe, valve, and piece of equipment inside was hosed down with radioactive fission products having a complex, ever-changing array of half-lives and radiation types, actively breaking down for thousands of years, and it had soaked into the fairly new concrete. It would have been cheaper to have bulldozed the plant into the ground and started from scratch, if only it had been legal to do so.

In the years afterward, there was an eerie quiet in the world of nuclear power. It was as though the worst had happened. Nature and probability seemed to have nothing else up their sleeves, and all was still.[242] Then, early in the morning of April 26, 1986, all hell broke loose in an ancient town in Ukraine named Chernobyl.

In 1986, Ukraine was a close member of the Russia-based Union of Soviet Socialist Republics (USSR), a large conglomeration of geographically connected countries making up Eastern Europe. The government, Communism, was a 20th-century invention being beta-tested, and there was a big ongoing contest with "the West," which was basically Western Europe and the United States, to find which experimental government system, soviet communism or a democratic republic, could develop the stronger, more dominant economic system. The USSR was still in its pre-war mode, implemented by Communist Party Head Joseph Stalin, to win the competition by having the larger population percentage of engineers, technicians, and scientists, thus advancing more rapidly in a world where technology seemed important. The West was not quite as tightly organized but was giving the USSR a lot of heat. To meet its goal of economic domination and modernization, the USSR saw fit to construct big, powerful nuclear power plants as quickly as possible, while building a vast inventory of nuclear drop-weapons and warheads.

Both goals, electricity and bombs, are met simultaneously using the RBMK reactor concept, a design that was original to the Soviet Union and not, as were some other mechanical motifs, a copy of Western machinery.[243] The RBMK uses blocks of solid graphite as the neutron moderator and water as the coolant, boiling in metal tubes running vertically through the reactor core. It therefore suffers from the worst characteristic of two reactor concepts, the possibili-

---

242 It was relatively still, at least. In September 1982, the No. 1 RBMK reactor at the Chernobyl Nuclear Power Plant experienced a partial core meltdown. Such minor incidents were not brought to the attention of the greater world of nuclear power, as the Soviet government was very careful not to release any information unless they had no other choice.

243 RBMK (Реактор Большой Мощности Канальный) means "high-power channel-type reactor." Although it was considered obsolete even in Russia by the 1970s, there were RBMK power plants under construction as late as 1993. There were 26 RBMK reactors planned, eight were cancelled in the middle of construction, and 12 are still operating.

ties of a graphite fire plus a steam explosion in the same machine, and it thus wins the prize for the most dangerous method for making power using fission. The advantage of it is the fact that it can be used both for power production and for plutonium-239 conversion. The neutron-energy spectrum produced by the use of graphite plus the fact that it can use natural uranium as fuel made it optimum for plutonium production, and the fuel assemblies can be swapped out while the plant is running at full power.[244] Timely, selective removal of the fuel, as opposed to changing it out during a refueling shutdown, means that the disadvantageous production of plutonium-240 can be minimized. It was designed in the 1950s, when a commercial power reactor in the United States made 60 megawatts of electricity. An RBMK was designed to make a gigantic 1,500 megawatts of electricity, under the belief that overwhelming size would be a factor in winning the global economic contest.

There were some serious design flaws. The reactor core is big—a graphite cylinder 46 feet in diameter by 23 feet high. Each fuel assembly is 12 feet long, and the machine that automatically pulls one out and exchanges it for another requires a space 114 feet high over the top of the reactor. There was no practical way to construct a sealed containment building over this tall machine, so the world is protected from fission products in the reactor by a single barrier, a round, concrete lid, eight feet thick, held by gravity in the reactor room floor. A sheet-metal roof keeps rain off the equipment.

Western reactors use the "scram" system to rush all the controls into a reactor and shut it down as quickly as possible. It takes about three seconds to complete the scram on a General Electric BWR power reactor, from the instant of hitting the big red button to having the controls top out in the reactor core. The equivalent Soviet system is the AZ, or "Rapid Emergency Defense." Push the big red AZ button, and it takes 20 seconds for the control rods to be

244 Natural uranium fuel was a goal, but very low, two-percent enriched fuel was usually used in RBMK power plants. This was still cheaper than the higher enrichment used in the PWR and BWR reactors which were so popular in the West. After the Chernobyl disaster, the RBMK fuel enrichment was increased to 2.4 percent to compensate for revisions in the control-rod design.

completely in.[245] A lot can happen in 20 seconds, but that is not the worst characteristic of a Soviet-style scram.

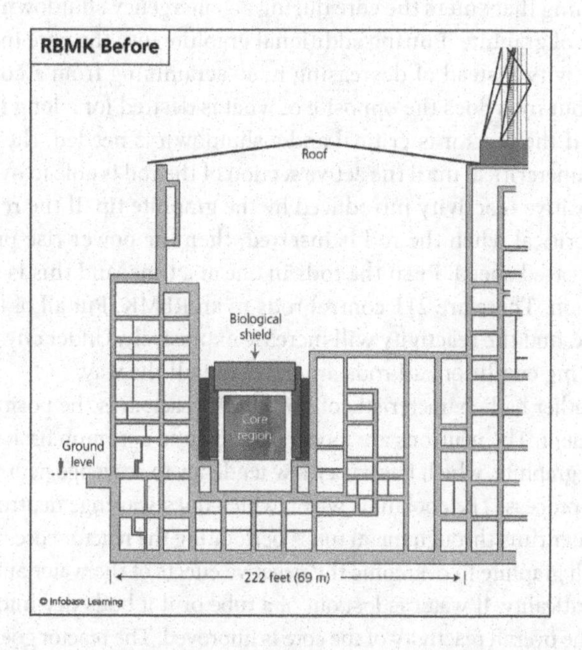

The RBMK reactor at Chernobyl-4 was housed in a very large concrete building, but the only thing between the top of the core and the sky above was the "five-kopek piece," a large concrete disc.

The control rods are as long as the reactor is high, but the active region in the middle of a control rod is only 16.4 feet long. The rest is just hollow tubes, at the top and bottom of the boron neutron-absorber section. At the bottom of each control rod is a rounded tip made of pure graphite, designed to act as a lubricant that will ensure smooth

245 There is not one AZ button for an RBMK; there are five. By choosing a low-numbered AZ button, an operator can gently shut down the reactor with a minimum amount of inserted negative reactivity without shocking the system too badly. Hitting button AZ-5 throws everything in at once, giving a rapid, absolute shutdown. Unlike most Western reactors, the control rods are not hastened into the core using hydraulic cylinders, but are driven in by the electric servo motors that are normally used to position them, running at maximum speed.

running of the rod in its metal tube under conditions that would ruin ordinary grease. If a control rod is withdrawn all the way, then the first thing that enters the core during an emergency shutdown is a big chunk of graphite. Putting additional graphite into the core increases the activity instead of decreasing it, so scramming from a condition of all-out rods does the opposite of what is desired for a long five seconds. If the reactor is critical and a shutdown is needed, the reactor goes supercritical until the active section of the rod is able to overcome the positive reactivity introduced by the graphite tip. If the reactor is supercritical when the rod is inserted, then the power rise proceeds at increased speed. Push the rods in one at a time, and this is only an irritation. There are 211 control rods in an RBMK. Put all of them in at once, and the reactivity will increase explosively. Under any normal operating condition, all rods are never out all the way.

Another bad characteristic of the RBMK reactors is the positive void coefficient. The neutrons are moderated down to optimum fission speed by the graphite, which has a very low tendency to scavenge neutrons out of the process. The coolant is water, which does scavenge neutrons, and the water runs through metal tubes perforating the reactor core. There is enough graphite to overcome the negative effects of the water and maintain criticality. If water is lost out of a tube or if it boils dry into steam, then the overall reactivity of the core is improved. The reactor goes supercritical, and the power level starts to rise exponentially. This is not good. In a PWR or a BWR, which use water for both coolant and moderator, if water is lost or turned to steam, the reactor shuts down instantly. In an RBMK, the power goes up with similar enthusiasm.

The first 3.23 feet of each control rod is a hollow metal tube, and it displaces the water out of the guide pipe as it is pushed in.[246] From an all-out position, the control rod would first introduce graphite to the core, and then empty the cooling water out of the guide.

The water inlet tubes for the reactor all come up from the bottom, which is the logical way to do it, but fear of leakage from these tubes caused the engineers to design a large gallery under the reactor, sealed tightly and filled with water. The tubes run through the water tank, and this will dilute any fission products that happen to escape the fuel,

246 This is no longer the case. After the smoke had cleared from the Chernobyl disaster, the control rods in all RBMK reactors were ripped out and replaced with an improved design.

get into the cooling system, and leak out through cracks or failed welds. This seemed like a good idea, but if the core ever melts it will fall directly into the water underneath and flash it into steam. Being tightly sealed with no way out makes rapidly derived steam into a bomb, sitting right under the core. The force of such a blast would be directed upward, making short work of the thin walls and ceiling above the reactor and pushing the big traveling crane above the refueling machine skyward. The mechanical engineers who designed the plant were good at anticipating bad welds but gave insufficient thought to what happens when a reactor runs away.

Chernobyl is an ancient town in the Byelorussian-Ukrainian woodlands on the banks of the Pripyat River, where the land is a featureless steppe. It is at least 1,000 years old, and was most memorable for providing Prince Svyatoslav the First with a particularly spirited, almost feral bride around the year 963. In the 1970s, work began on a cluster of six RBMK-1000 reactors, built on a flat spot 11 miles northwest of the town. A new, modern village, Pripyat, rose up 1.9 miles west of the sprawling plant, just outside the safety zone. Its population grew quickly to 50,000 people. Most employment was at the power plants or in jobs supporting the families living in the area. There were schools, multi-story apartments, a park with a Ferris wheel, a bookstore, recreation center, library, and every trapping of civilization. Reactor No. 1 was completed and came online in 1977, followed by No. 2 in 1978, No. 3 in 1981, and No. 4 in 1983. No. 5 and 6 were still under construction in 1986. Central Planning back in Moscow was dreaming of a 20-reactor complex.

In April 1986, reactor No. 4 still had 75 percent of its original fuel load and was looking at a refueling in the near future, meaning that its core was saturated with a nearly full load of radioactive fission products in its 200 tons of uranium oxide. Nikolai Maksimovich Fomin was Chief Engineer at the station, Viktor Petrovich Bryukhanov was the Plant Director, Taras Grigoryevich Plokhy was Chief of the Turbine Unit, Anatoly Stepanovich Dyatlov was Deputy Chief Engineer in charge of operations for No. 3 and No. 4 reactors, and Leonid Toptunov was the Senior Reactor Control Engineer.[247] The foreman in charge of

247 Fomin rose rapidly through the leadership posts and was Deputy Chief Engineer for Assembly and Operations when he knocked Plokhy out of the position as Chief Engineer. Although the Ministry of Energy did not support the decision to promote Fomin to this post, he had the solid endorsement of the Central Committee of the Communist Party, Nuclear Division, and that was all that was needed.

the reactor section of the plant was Valery Ivanovich Perevozchenko and Aleksandr Fyodorovich Akimov was the Shift Foreman. None of these men and nobody in the entire power plant had a clear understanding of the nuclear end of the power plant. They were experts in turbines, wiring, and mechanical engineering specialties, but had no training or experience that would lead to a comprehension of graphite reactor dynamics. Dyatlov, a physicist, was the most unusually slow-witted and argumentative of the lot, and as one who had an inkling of nuclear experience, he was in charge. He had worked briefly on very small experimental naval reactors, and this may have given him a distorted view of how a power reactor should behave.

Reactor No. 4 was scheduled to run a safety experiment on April 25, 1986. The Nuclear Safety Committee had long been concerned that if an RBMK reactor were to shut down for emergency reasons while producing power, there would be a delay between having lost power from the generator and starting the diesel engines that ran the backup generators. In this time lapse, the coolant pumps would not be running, and this could cause damage to the plant from a sudden heat buildup. There were theories that one of the eight heavy turbine rotors would have enough momentum to keep its generator turning for a few minutes, supplying just enough power to keep the pumps running, but a previous test at Chernobyl No. 4 had been disappointing. The generator fields had been modified to reduce the drag, and now the committee wanted the test to be run again. The reactor and all the plant systems were to be shut down suddenly, as if a major breakdown had occurred, after which the performance of one turbo-generator would be observed over several minutes.

Gennady Petrovich Metlenko was in charge of the electrical aspects of the trial, and on April 11 he had a special control-panel switch installed, called the "MPA." The letters in Russian stood for "Maximum Design-Basis Accident," or the worst thing that could possibly happen. Actuate the MPA switch, and the worst happens instantly. It shuts off the turbine, disables the ECCS, turns off all the pumps, blocks the diesel generators from starting, and basically kills everything that keeps the reactor running smoothly, bypassing the automatic controls. This was a monumentally bad idea.

The experimental sequence started at 1:00 P.M. on April 25, right on time, when Dyatlov ordered the reactor power reduced. No. 4

had been running at maximum power. Five minutes later, the No. 7 turbine was kicked off the power grid and the station's power needs were switched to turbine No. 8. The reactor was now running at a little over half power. At 2:00 P.M., the ECCS was disconnected. One thing they thought they did not want during the experiment was 12,360 cubic feet of cold water gushing into a red-hot reactor from the ECCS, thinking that it would warp something. Just then, a call came in from the electrical load dispatcher in Kiev, requesting that they delay the experiment a while longer. Demand for electricity in the area seemed to be peaking that afternoon.

At 11:10 P.M. they were able to resume the power-down. At midnight was the shift-change, and the Shift Foreman Yuri Tregub was replaced by Akimov. Toptunov replaced the Senior Reactor Control Engineer. The goal was to level out at 1,500 megawatts, but they had disabled the local automatic control system (LAL) for the test, and Toptunov was having trouble keeping the flux profile balanced as the power dropped. There were too many neutrons on one side of the reactor and too few on the other, and things were getting out of hand as the operators juggled the controls.

A graphite pile is a ponderous beast, and controlling it with no automatic assistance is like driving a concrete truck on the Monte Carlo racing circuit. All actions must be performed slowly, or it will turn over in a curve. The reactor power slid through 1,500 megawatts and kept going, down to 30 megawatts, very quickly. Toptunov had now steered Chernobyl No. 4 reactor into the dreaded "iodine valley," from which there is no easy return.

What is an "iodine valley"? When a nuclear reactor produces power by fission, one of the many fission products is iodine-135. It is radioactive, and does a beta-minus decay into xenon-135 with a half-life of 9.10 hours. This is perfectly natural. The energy-releasing beta decay of iodine-135 is a small component of the delayed energy from fission. The product of this decay, Xe-135, is unique in that it has a monstrous thermal neutron-absorption cross section of 2.6 million barns. That is a reaction killer. If the Xe-135 builds up from I-135 decay, then it will snatch so many neutrons out of the normal fission transactions, it will shut down the reactor. Fortunately, one Xe-135 atom can only capture one neutron, activating into Xe-136, which is stable with a very low, 0.26 barn, capture cross section. Xe-135 also undergoes a beta-minus decay, becoming stable cesium-135, also with a low tendency to capture neutrons. A reactor running at high power both makes Xe-135, indirectly, and destroys it by providing excess neutrons to be captured. The neutron population reaches an equilibrium of I-135 production and Xe-135 destruction, and the reactor is able to remain in a critical condition and produce power at a steady rate.

If the power is reduced, then the equilibrium is disturbed. At the lower power level, there are fewer fissions per second and fewer neutrons produced per second, but the level of Xe-135 from the previous, higher power level remains. Recall, the production of Xe-135 from an I-135 breakdown is slow. It takes 6.57 hours for half of it to turn into xenon, so the production rate of Xe-135 continues on at the previously established level. There is now more Xe-135 than the reactor can knock down with surplus neutrons, because its power level has been reduced. More neutrons are captured than are produced, and the power level plummets. If the reactor has enough reactivity in reserve, held in check by the control rods, then the power can be stabilized at the desired level, and the now-excessive level of Xe-135 presence can be "burned off." After a few hours, the control rods are slowly returned to their previous positions, holding the excess reactivity in reserve. If there is insufficient reserve reactivity, then the power level drops to zero. The reactor is stuck in the iodine valley, and it will take about 45 hours for enough Xe-135 to have beta-decayed away to allow the reactor to be restarted.

Dyatlov was enraged. He paced up and down the control panel, berating the operators, cursing, spitting, threatening, and waving his arms. He demanded that the power be brought back up to 1,500 megawatts, where it was supposed to be for the test. The operators, Toptunov and Akimov, refused on grounds that it was against the rules to do so, even if they were not sure why.

Dyatlov turned on Toptunov. "You lying idiot! If you don't increase power, Tregub will!"

Tregub, the Shift Foreman from the previous shift, was officially off the clock, but he had stayed around just to see the test. He tried to stay out of it.

Toptunov, in fear of losing his job, started pulling rods. By the time he had wrestled it back to 200 megawatts, 205 of the 211 control rods were all the way out. In this unusual condition, there was danger of an emergency shutdown causing prompt supercriticality and a resulting steam explosion. At 1:22:30 A.M., a read-out from the operations computer advised that the reserve reactivity was too low for controlling the reactor, and it should be shut down immediately. Dyatlov was not worried. "Another two or three minutes, and it will be all over. Get moving, boys!"

At 1:23:04 A.M., Igor Kershenbaum, the Senior Turbine Control Engineer, closed the throttle on the No. 8 turbine to begin the test, just as an operator pushed the MPA button in the control room. With the power demand from the turbine stopped, the water in the still-hot reactor began to boil with fury, and the power level shot up as the water left the cooling channels dry. The reactor was now supercritical. The operators watched in horror as the power level rose rapidly, out of control. After observing it for an agonizing 36 seconds, Akimov shouted, "I'm activating the emergency power reduction system!" and he punched the red button for AZ-5, throwing everything in at once.

The reactor was now prompt supercritical at 1:23:40 A.M. The controls jammed after moving only 6.5 feet, versus their usual range of 23 feet. The guide-pipes had twisted and warped as the reactor began to melt. In seven seconds, the reactor jumped to a power level of 30 billion watts and started to disintegrate.[248]

---

248 Four nuclear power plants were not the only high-tech equipment headquartered near Chernobyl. There was also the Chernobyl-2 over-the-horizon radar station, code-named DUGAR-3 and known to ham radio operators in the West as "the Russian Woodpecker." It was a 10-megawatt radio transmitter feeding the world's largest directional high-frequency antenna, and its signals, which sounded like a woodpecker hammering on a tree, disrupted amateur radio communications on the 20-meter band beginning in 1976. The antenna was 150 meters tall and 500 meters wide, and this masterpiece of mechanical engineering weighs about 14,000 tons. The NATO code name was "Steelyard." It was looking for intercontinental ballistic missiles fired at Russia from the United States. Suddenly, at 1:23:40 A.M. local time, the woodpecker went silent and never jammed the 20-meter band again. When reactor No. 4 went wild, it disturbed the power feed to the transmitter, and the woodpecker went dark. The antenna is now in the radiation exclusion zone, and the station cannot be manned. See it on Google Earth by searching on the name "Russian Woodpecker."

At that moment, Valery Ivanovich Perevozchenko, foreman in charge of the reactor section, happened to be standing on a balcony looking down at the lid on the reactor, 45 feet below. This lid was affectionately called the *pyatachok*, or the "five-kopek piece."[249] It was 49 feet in diameter, and on top of it were 2,000 fuel bundle covers. Each cover was a heavy cube, weighing 770 pounds. There was a deep rumble, and the entire building began to shake. The cubes started dancing and then leaping into the air. It looked and sounded like a popcorn popper heated with an oxyacetylene torch. The cracking and popping noises were so loud, Perevozchenko could not hear himself scream as he took the stairway down, burning the skin off his palms with friction on the handrails, taking four steps at a time, and descending with the equivalent speed of falling down a well 130 feet deep. He hit the ground at level "+10" and sprinted 229 feet down corridors and through the safety lock to the control room, where he burst through the door shouting the ultimate nuclear understatement, "There's something wrong!"[250]

A loud, percussive noise rocked the building, and the operators bounced off the walls. The pressure-relief valves on the steam separators had all opened with a collective bang.[251] A second later, they all broke free, blasted through the roof, and whistled away into the calm night air. All four separators, each weighing 130 tons, tore out of the floor at level +24 and followed the relief valves, trying desperately to get away. The steam lines into the reactor came loose, and the

---

249 In the Old Days of Tsarist Russia, the five-kopek was a comically huge brass coin, made large so that one could use it to buy bread from a street vendor without removing a mitten. One kopek was 1/100 of a ruble, roughly the equivalent of a U.S. penny.

250 Level +10 is ten meters above ground. The turbine hall sump is at level -5.2 and the reactor building floor is at level +35.5.

251 The steam-relief valve was invented by the French physicist Denis Papin in 1679, making possible the subsequent invention of the pressure cooker, or "Papin's digester." The old valves used weights or steel springs to hold back the steam pressure, and these units would open gradually, with a groaning or wailing noise. By the time nuclear-powered steam plants were being built, the steam-relief valve had evolved into a mechanism that would open with digital precision amounting to a controlled explosion. There was something to be said for the old ways.

big tanks of water underneath the reactor instantly boiled to steam under the blast of neutrons radiating from the center of the reactor core. The white-hot zirconium fuel cladding quickly scavenged the oxygen out of every molecule of water it could find. Where there had been water a second ago, there was now superheated steam mixed with hydrogen gas.

The 500-ton cap on the reactor, the five-kopek piece, lifted off as the steam exploded and the middle of the reactor turned into a cloud of radioactive aerosol, taking out the 250-ton refueling machine, the 50-ton crane in the ceiling above it, and the roof of the building. Red-hot chunks of fuel from the periphery of the reactor core fell on the roof of the turbine hall, which was waterproofed with a generous coating of tar, and set it on fire. Endowed with a blast of fresh air through the top of the reactor building, the hydrogen went off with an earth-trembling roar, and the vaporized part of the power plant was lifted to 36,000 feet in the air, contaminating any commercial airliners within 100 miles. Spinning chunks of red-hot debris started falling on Chernobyl reactor No. 3, crashing into the roof and into the ventilator stack.

There had been 1,700 tons of graphite in the reactor back when it was working, but that had been reduced to a crater-shaped remnant of 800 tons. It started burning, first a dull red and then orange, lighting up the scene with an eerie glow. The reactor lid had risen, flipped like a pancake, and come back down into what was left of the core, cocked at a steep angle. The smoke was thick and black and rose vertically in what was charitably described as a "flower-shape." Half of the 100 tons of uranium was gone. In perspective, the atomic bomb dropped on Hiroshima vaporized 139 pounds of uranium infused with two pounds of fission products collected in one second of extreme power production. The explosion at Chernobyl No. 4 evaporated a 50-ton mixture of uranium, oxygen, and about 800 pounds of fission products produced over the previous three years of power production.

Two men, Protosov, a maintenance worker, and Pustovoit, who was the "odd-job" man at the plant, were night-fishing on the bank of the coolant run-off pond, right where the plant outflow occurs, 1.25 miles from the plant. The fish really liked the warm water, and it was a clear, starry night. It seemed like the middle of summer, and the fish were cooperating.

They turned to look when they heard two rumbling explosions, seeming to come from inside the plant. Then a third explosion reduced the top of the building to flaming splinters, and they watched with mild interest as steel beams and large concrete chunks spun overhead. The turbine hall burst into flames and illumined the enormous column of black smoke. They turned back to their fishing rods. If they got excited every time something around here exploded or burned to the ground, they would never get any fishing done. "They'll have that out in no time," opined Pustovoit. Whenever a steam relief valve popped off, which seemed quite often, it sounded like a Tupolev Tu-95 strategic bomber had crashed into the side of a building, and fires consuming switch yards or fuel depots were not rare at the Chernobyl plant. The men sat there and fished until morning, noting that the fish were becoming sluggish. The fishermen each absorbed 400 roentgens of mixed radiation, and they started feeling extremely ill, right where they were sitting.[252]

The nonstop vomiting was utterly exhausting. Their skins turned dark brown, as if they had been locked in a tanning bed for too long. They had no idea what had happened to them, but they staggered into the medical center and were quickly sent to Moscow for special treatment. Both survived, and Pustovoit became a celebrity of sorts in Europe, living proof that ignorance hurts.

Back in the control room, the earthquake-like shocks had crumbled the steel-reinforced concrete walls and floor, and light fixtures were hanging down by the remains of power wiring from the ceiling. The instruments and controls were all dead, and the only light was from some battery-powered emergency lights and the electrical arcs from broken power cables. All hand-held, battery-powered radiation detectors available were intended to find tiny contaminations around the plant on the micro-roentgen level. None seemed to be working, as their meters

252 The Soviets did not bother to put their radiation dose estimates into human terms using rem instead of R units or to use the metric SI system. The degree of radiation exposure, however, is starkly evident. One can survive a 400 R exposure, but it really stings. Firemen on the roof of the turbine hall were getting 20,000 R/hr, which in an hour will kill an adult male 20 times over.

would crash against the stops at the highest indicated level and just hang there. There was no instrumentation available that could read the 30,000 roentgens per hour radiating out of the reactor hole or the 5,000 roentgens per hour streaming from every chunk of fuel or graphite that peppered the site. Akimov, Toptunov, and especially Dyatlov were absolutely certain that the reactor was intact, and all it needed was for water to be pumped into it to bring it down to a proper temperature and establish the cold shutdown condition. Dyatlov relayed word back to Moscow to this effect. There was nothing to worry about. There was a fire on the roof on the turbine building, and their erroneous conclusion was that a 29,000-gallon emergency water tank in the main reactor hall had exploded. They would have the plant back in operation in a few weeks.

Akimov believed this fantasy as well, but he was bothered by the fact that the control-rod locations, as indicated by the clock-like Selsyn dials on the wall, seemed stuck only partway in to the reactor. He summoned two young trainees, Aleksandr Kudryavtsev and Victor Proskuryakov, and told them to run up to the reactor hall and find out why the controls were stuck. "Jump up and down on them, if you have to." After an arduous trip through a maze of twisted steel, an elevator blown from its shaft, and concrete chunks the size of automobiles that had once been part of the reactor support structure, the two finally made it to where there had once been a reactor. It was simply not there anymore. There were no controls to free up, just a gaping hole. They looked down into the glowing crater, amazed at the sight.

They were aware of strange sensations as they climbed their way back to the control room. The air seemed incredibly fresh, as if a spring rain had just ended, and they could feel the tingling way down in their insides. By the time they made it back, they were brown with "nuclear tans." Akimov and Dyatlov ridiculed their finding, accusing them of being so clueless, they could not find a reactor the size of a grain elevator. They must have been on the wrong floor and completely missed it.[253]

253 Kudryavtsev and Proskuryakov were immediately sent to the infirmary at reactor No. 3, and from there to Moscow for special treatment of extreme radiation sickness. They, and almost everyone else at reactor No. 4 who had survived the explosions, died in agony in a few weeks.

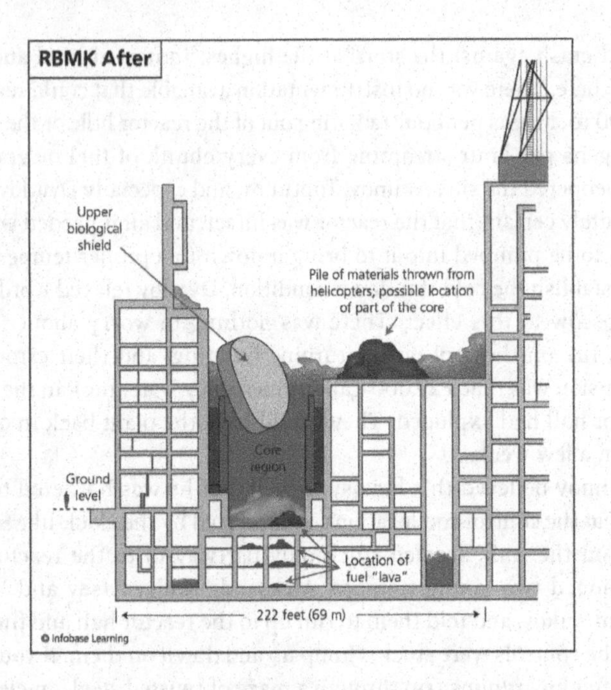

After the series of explosions at Chernobyl-4, the refueling floor at the plant was not recognizable. Hundreds of helicopter fly-overs had dumped sand into the building in an attempt to stop the graphite fire and shield the environment from the fission-product radiation.

Outside the plant building, the extent of the damage was more obvious. Flames on the turbine hall roof were shooting up above the vent stack, which was 600 feet high. The firemen had arrived within minutes, and they were aware of the gravity of a fire in the turbine hall. As is the case with all high-capacity generators, these were cooled by hydrogen gas pumped through the hollow copper field windings in the stators. There was a great deal of hydrogen gas stored in pressurized cylinders in the building, and with a fire raging they were likely to go off as bombs. It was hard to see how the condition of the plant could be much worse, but they were firemen and they were performing their jobs without contemplating the futility of it. The roof was collapsing onto the turbine floor, where broken pipes were spewing flaming oil over the linoleum, and a broken condensate pump was spewing water contaminated with fission products. Although in the stress of firefighting they could not feel it, the firemen on the roof were in a 30,000 roentgen-per-hour radiation field, and none of them would survive the radiation dosage.

Even as late as 1986, some governments would wish to keep the details and magnitudes of radioactive releases secret from the general, highly excitable local population and certainly from the external world. In this case, the Chernobyl reactor No. 4 was a military asset, and its loss of function was understood to be a state secret. At this early stage of the crisis, the Soviet Union leadership did not realize that the accident would eventually envelop the entire European continent, spreading radioactivity and bad feelings from Norway to Turkey, from Ireland to the Slavic Republic. At the time, the entire concept of the Union of Soviet Socialist Republics and its iron grip on Eastern Europe was in the midst of change, a softening and a relaxation, under a new and forward-looking General Secretary of the Communist Party, Mikhail Sergeyevich Gorbachev. As dawn broke on the morning of April 26, 1986, not even Gorbachev knew that half of the reactor core at No. 4 was airborne, and the question of whether or not anyone should be warned—let alone evacuated—had not come up.

The first inkling of a problem beyond the official border of the Soviet Union was picked up by a border patrol agent on the southeast border of Finland. Finland and the Soviet Union were in an unusual geopolitical relationship. One facet was a bilateral trade agreement, and Finland, coming up in the world, had also purchased a nuclear power plant from its new partner. It was a pair of new VVER-440 reactors, the Soviet version of the Westinghouse PWR, and not an RBMK. Part of the deal was that Finland had to return all the spent fuel to Russia for plutonium extraction.

Upon receiving the new power plant, the Finlanders were distressed by the poorly designed process-monitoring and control equipment in the plant, and so they redesigned and built their own replacement with some Western help. They realized that if all the Soviet reactors were built this way, eventually there would be a problem. Quietly, the government of Finland constructed a sophisticated radiation-indicator net covering the entire country.[254]

---

254 Countries that have nuclear power plants are by now equipped with networks of fixed radiation detectors feeding into central monitoring stations using Internet connectivity. The radiation-monitoring network in the United States, named RadNet, is operated by the Environmental Protection Agency. If you have been on top of a building lately, you may have noticed an innocuous metal box standing on a short pedestal. That is the radiation detector and data transmitter. In Japan, the similar system is named SPEEDI.

Early in the morning of April 26, the border agent noticed that the radiation instrument on his wall suddenly jumped to the red side. He picked up the phone and dialed the state department emergency number. His alarm made its way up to the highest floor, slowed by a government employees' strike, but finally informing the Prime Minister, Kalevi Sorsa. Being a man of unusual intelligence, he decided that no news medium was to be informed, as it would irritate the Soviet Union and only cause trouble.[255] The long radioactive cloud passed overhead in Finland, stalling and depositing dust in the northern territory. Entire herds of reindeer had to be buried, and people were advised not to pick mushrooms and berries. So began the dusting of western Europe, not just with some xenon and iodine, but with examples of every fission product across a wide spectrum of potential danger, human body burden, and half-lives.[256]

Meanwhile, in the light of the new day, the damage at reactor No. 4 was painfully obvious. The brave firefighters had to admit that the old rumors about graphite fires were true. Pouring water on a serious graphite conflagration only makes it worse. The Soviet government, now fully briefed and on board, sent legions of heavy-lift helicopters to Chernobyl carrying 4,000 tons of sand, clay, lead, and boric acid, everything they could think of that would quench the still-raging fire, stop the rising column of radioactive smoke, and discourage further criticality in a semi-reactor now devoid of neutron controls. The effort eventually worked at putting out the fire, but it also put an insulating blanket over a source of heat. Before long, the heat buildup melted everything below the covering. Concrete, steel, graphite remnants, and

---

255 Is this account of the first radiation fall from Chernobyl believable? Maybe. It was given to me by a very reliable source, but the monitoring station was about 760 miles away from the No. 4 reactor. The vanguard of the radioactive dust release would have to have traveled at 100 miles per hour to get there early on April 26. It is not impossible, as the radiation cloud very quickly reached 36,000 feet, which would put it at jet-stream altitude, where the wind blows at 100 miles per hour. The jet stream usually blows east to west, and Finland is north-northwest of Chernobyl, but the jet stream is capable of north-south looping. Officially, Finland discovered a radioactive dust-drop at exactly the same time it was noticed in Sweden, thus absolving Finland of any independent whistle-blowing. Later, the stream turned east-west and laid down a swath of radioactive dust nearly 1,000 miles long across the southern Soviet Union.

256 As of 2009, radioactivity in Finland mushrooms was still being detected.

remaining fuel melted into a flowing lava, dripping and spilling into the underground basement of the power plant. The helicopters and the men who flew them were heavily contaminated from the smoke after several trips over the crater. The helicopters had to be abandoned at a landing site, and the men had to go to Moscow for radiation-poisoning treatment. The accident was still a state secret, and no alerts were issued to anyone who was not in the immediate area.

The entire town of Pripyat had to be evacuated, population 50,000. Buses lined up for as far as the eye could see, and residents were told to bring just a toothbrush and a change of clothes. "It will only be for a few days," they lied, "so don't pack up your belongings." Everyone had to be relocated, never to see Pripyat again. In all, 135,000 people were evacuated from the area, and it had to be fenced off.

The next morning, early on April 27, nuclear engineer Cliff Robinson was walking down the hallway to his office at the Forsmark Nuclear Power Station, a triple BWR plant near Uppsala, Sweden. He had eaten breakfast in the break room and was now reporting to work. There were radiation detectors all over the plant, used to make certain that no worker would accidentally walk out of the building with any radioactive contamination on his clothes. They were pre-set to low count-rate levels, just above the background level of radiation. One went off just as Robinson passed it in the hall. Startled, he stopped, lifted a foot, and scanned the bottom of it with a hand-held "hockey-puck" probe. The counter went wild.

His first thought was that World War III had broken out and someone had dropped an atomic bomb nearby. His second thought was that something had broken loose in the plant. He called his boss, who picked up the phone to alert the Swedish reactor safety authorities, who demanded that the source of the radiation leak be found immediately. A thorough search found that it was not coming from anywhere in the plant. It had come in on Robinson's shoes. By 12:00 Universal Coordinated Time, they had determined that the source was a nuclear reactor, somewhere in western Russia. Finland agreed.

At that moment, the world became officially aware of the disaster at Chernobyl, and the Soviet Union started the long slide to oblivion. This accident was so large in scope, so nerve-rattling for everyone downwind, there was no hiding it or denying that anything had happened.

Of all the firemen, emergency response workers, and power-plant personnel at Chernobyl, 127 people suffered from acute radiation poisoning, and in the first three months, 31 died. Eventually 54 deaths would be directly attributed to radiation exposure due to the destruction of the reactor, but the even larger tragedy was the spread of the pulverized contents of the reactor core. The uranium fuel was not the problem. Having 50 tons of uranium distributed over most of Europe would not noticeably change the background radiation danger or even the uranium content of the landscape. Most minerals in the ground contain trace amounts of uranium, and it tends to be evenly distributed in the Earth's crust. Most of the exposed uranium has been washed into the ocean long ago. The transuranic nuclides, such as plutonium-239, are unnatural and dangerous, but they are a small percentage of the fuel mass.

The long-lived fission products are not the huge problem that one might expect. Much is made of the fact that fission products will be around for hundreds of thousands of years, and having them sprinkled over Europe seems disastrous. But a long half-life means low reactivity. Molybdenum-100 is a fission product that lasts a long time, but with a half-life of seven million-trillion years, it is barely alive from a radiological perspective and is hardly a threat to living things.

Nor are the short half-life fission products a problem. They are vigorously radioactive, but only for a short time. The big scare is from iodine-131, which decays with beta-minus/gamma radiation, and it is sought by the human body for use in the thyroid gland. However, with a half-life of only 8.023 days, its concentration in the environment decreases rapidly, and it is essentially gone after 80 days.

The big problem is in the middle ground, between the long-lived and the short-lived radioactive species. Strontium-90, for example, is chemically similar to calcium, and the human body seeks it for bone replacement. It has a half-life of 28.8 years, making it active enough to be very dangerous with a lifetime of several generations. The same is true of cesium-137, with a half-life of 30.07 years.

The total human impact of the Chernobyl disaster on human lives in the next 100 years will be difficult to measure. As many as 4,000 cancer cases due to fission-product uptake are predicted.

Even though the power-plant complex was badly damaged and littered with debris from reactor No. 4, the electrical needs of Ukraine were such that reactors No. 1, 2, and 3 were kept running. No. 3 was the last to be shut down permanently, on December 15, 2000. A sarcophagus made of concrete 660 feet thick was poured to cover up the remains of reactor No. 4. The radiation exclusion zone around the complex is a nature preserve, now populated with animals that had lived there before mankind cleared the land and built structures on it. In the deepest recesses of the destroyed reactor in a field of heavy mixed radiation, a new form of black fungus has found it a nice place to live. It seems to love the lack of competition for the space.

It seems unfortunate, but nothing was learned from the Chernobyl disaster. It did not, for example, lead to a better understanding of reactor accidents or an improvement in reactor design. At the time of the catastrophe, the graphite boiling-water reactor design was already history in the West, and its obvious mechanical flaws and lack of operational knowledge were two trains heading to collide on a common track, and the isolation that the Soviet engineers operated within was in large part to blame for this fundamental flaw in the plant's design.

The causes of the Chernobyl-4 disaster were complex, reaching deep into the social, political, and economic structure of the Soviet Union, and it was most likely a component of the complete collapse of this government four years later. It could theoretically have been avoided, but that would have required a complete rebuild of Soviet attitudes toward building the future, fiercely competing with the rest of the world, and placing value on the individual citizen. Such changes would begin to come about, but not before the Chernobyl reactor's self-destruction had deeply wounded the industrial fabric of the Union of Soviet Socialist Republics and embarrassed it to the rest of the European continent and beyond.

From what the Western nuclear engineering community could tell, the big Soviet graphite reactors were of questionable design and likely to give trouble. The Soviets had their own way of accomplishing engineering goals, and they were not necessarily open to contemplate negative Western opinions. Conversely, we were not interested in their opinions of our nuclear power systems.

The next disaster, though, would involve American engineering, and there would be no excuse for it. Events started tumbling toward disaster at 2:46 on a Friday afternoon, off the northeast coast of Japan.

Chapter 10

TRAGEDY AT FUKUSHIMA DAIICHI

> "HEAT SINK: a small metallic device attached to your CPU that, like the cooling tower at a nuclear-power plant, is the only device standing between safe, reliable system operation and a total core meltdown."
>
> —Howard Johnson, Ph.D., in *Manager's Guide to Digital Design*

It would be difficult to think of a worse place to build a nuclear power plant. Perfectly safe nuclear generating stations have been built in hazardous locations from Antarctica to the Greenland glacier, but putting one on the beach in Japan, looking out over the ocean, right there on the Pacific Ring of Fire, seems ill-advised. The Ring of Fire, encircling the entire Pacific basin, including the California coast, is under constant threat of earthquakes, tsunami waves, and volcanoes from the westward tectonic shifting of the North and South American continents.

The nuclear engineers, mechanical engineers, civil engineers, electrical engineers, and seismic specialists in Japan are well trained and experienced, and they know how to optimize a building project. That is why, for the 54 nuclear reactors that were recently generating 40 percent of the electrical power in Japan, there is not one cooling tower. Why build natural-draft, fog-making, hyperboloid cooling towers, 600 feet tall, when you can use the Pacific Ocean as the ultimate heat sink for your generating plant? It saves a lot of space, which is precious in Japan, a lot of concrete, construction time, and money.

Every nuclear plant in Japan is built on the coast. Those in the southwest of the main island of Japan, Honshu, where the electricity alternates at 60 hertz, stare into the Sea of Japan. Those plants in the northeast of Japan, where the electricity alternates at 50 hertz, overlook the subduction zone where the Pacific tectonic plate runs underneath the Okhotsk tectonic plate.[257] As the North American continent glides westward on the slippery, molten rock on which it sits, the Pacific Ocean gets smaller and smaller, the Atlantic Ocean gets bigger and bigger, and the Pacific Plate subducts under Japan at a blistering 3.6 inches per year. It is not a smooth movement. The Pacific Plate sticks to the Okhotsk as it tries to fold under, and it can hang, not moving, for a thousand years as the tension builds up. Finally, the situation between the plates is stressed to the breaking point, and it just lets go, all of a sudden, in the rocky depths miles below the bottom of the ocean off the coast of Japan.

The shock wave produced by this abrupt movement travels fast through the rock, and it hits the Japanese islands with more force than any atomic bomb could produce. Things sticking up out of the ground

---

257 In the early 20th century, Japan wanted to step up to electrical power, just like everyone else in the developed world, but there were at least two modes of alternating current in use. Europe had chosen an alternating frequency of 50 hertz, and America had chosen 60 hertz. Japan, still recovering from centuries of non-centralized government using the warlord model, found that the south end had chosen to buy generating equipment from Westinghouse, and it all ran on 60 hertz. The north end, on the other hand, had bought equipment from Siemens A.G., and it ran at 50 hertz. The two transmission modes are incompatible, and the north-south divide of electrical current exists to this day. In a broad emergency, this discontinuity makes it difficult to ship electricity from an undamaged southern end to the devastated northern end of Japan. A small number of frequency-converter stations exist on the boundary, but their capacity is overwhelmed in a disaster.

are sheared off as the Earth underneath suddenly shifts position by measurable feet. Destruction is widespread as the entire island shakes.

Moving more slowly, the second shock wave moves as an enormous ripple in the ocean water, beginning over the point undersea where the plates slipped and moving in a circle of growing radius. As the ripple nears the shore, the water gets shallow, and the shock wave, distributed evenly in the liquid, gets funneled down and concentrated. The amount of energy is still there, but there is less and less fluid to hold it. The ripple becomes a monster wave, and it hits the shore as a wall of water, the tsunami, coming very fast. It finishes off anything that was not knocked down by the initial shock wave through the ground and inundates inland territory.

That is why there are better places to build a nuclear power plant than on the beach in Japan, which would seem evident from the remains of what was once a roughly semicircular pattern of ancient, inscribed rocks, planted upright in the ground a few miles from the shoreline in northeastern Japan. These monoliths, probably 600 years old, are all graven with the same message, written in a long-forgotten Asian dialect. Scholars pooled resources for centuries, trying to decipher the message, seeing success some 30 years ago. Very roughly translated, the inscription reads: "Don't even think about building anything between here and the ocean." A thin layer of water-borne silt underneath the topsoil, ending about where the stones are placed, attests to the fact that a tsunami wave once washed this far inland.[258]

The secondary cooling in a Japanese reactor is breathtakingly simple. A concrete pipe extends for a couple of hundred feet out into the ocean from a pump stand on the beach. A screen keeps curious sea life from being drawn into the cooling process. Water is sucked in by the pump, circulated through the condenser in the floor under the steam turbine, and introduced back into the ocean at a point a few

258 There have been many recorded earthquakes originating off shore that caused tsunami destruction in Japan. The one referred to on the "tsunami stones" is probably the Jogan Sanriki earthquake, which occurred at the Pacific/Okhotsk subduction zone on July 9, 869 (the 25th day of the 5th month, 11th year of Jogan). Its magnitude is estimated at 8.6. The existing stones may be replacements for original warnings that eventually became unreadable due to advanced age.

hundred feet farther up the coast. A cooling system could not be any less expensive to build.

The quest for nuclear power in Japan began in 1951. The centralized state-run power company, established for national wartime mobilization in the last Great War, was formally dissolved, and nine small power companies were formed. One was the Tokyo Electric Power Company, Incorporated, now known as TEPCO, formed on May 1, 1951. Its territory was the upper eastern section of Honshu, the main island, operating on 60 hertz current.

There were never a lot of burnable resources in Japan, so generating power using steam was challenging. Liquid natural gas and fuel oil had to be imported, and this was not only polluting the air, it was extremely costly. Despite resistance from atomic bomb survivors, Japan would have to "make a deal with the devil" and embrace nuclear fission if it was to compete in the industrial world. Construction on the first plant, a British Magnox graphite pile, was begun by the Japan Atomic Power Company (JAPC) on March 1, 1961. They put it at Tokai, halfway down the east coast, and it had no cooling towers. The race to provide Japan with nuclear power was on.

TEPCO began its part of the nuclear expansion with the Fukushima I Power plant. Construction began on a single General Electric BWR/3 reactor with a Mark I containment on July 25, 1967. It would be the first of six boiling-water reactors built on the same property in the small towns of Okuma and Futaba in the Fukushima Prefecture, and it would become one of the world's biggest nuclear power facilities. With everything running simultaneously, it could produce 4.7 billion watts of electricity. A second plant, Fukushima II, was built a few miles down the coast, and, by the turn of the 21st century, two more GE BWRs were planned for Fukushima I. Unit 1, the first completed reactor in the new plant, was switched into the power grid on March 26, 1971.

At the time, General Electric was competing vigorously with Westinghouse for the hearts and minds of nuclear power-plant consumers, and the goal was to make the most economical plant with the least accident potential. Westinghouse was pushing their upscaled version of Rickover's flawless submarine engine, and GE was hooked on Untermyer's dead-simple BWR. To optimize these designs for the civilian market, there were several engineering modifications to make.

The Westinghouse reactor is comparatively small, and neutron flux is controlled by a concentration of boric acid in the coolant. Secondary protection from fission-product escape into the environment is a thick concrete building constructed around the reactor. As was demonstrated at Three Mile Island in 1979, the building is strong enough to withstand a hydrogen explosion *inside*.

The GE reactor had to be big and tall, about 60 feet high, because the engineers had simplified the structure to include the steam separators and dryers in the top of the reactor vessel. To use external separators would involve a lot of complicated plumbing, and an engineering goal for these civilian reactors was to minimize the pipes.[259] The separators and dryers kept water droplets, which could quickly destroy a steam turbine, out of the steam pipes. The resulting integrated reactor/separator vessel, made of six-inch steel, was so tall, there was no way to construct a sealed, reinforced concrete building around it thick enough to contain a possible steam explosion. There would be a stout, rectangular structure covering the reactor, the refueling pool, and the traveling crane above, but it could not be specified as pressure-tight.

Not having a separate building as the secondary containment structure meant that something else would have to encapsulate the reactor vessel, so a large pressure-resistant container with a bolt-down lid, made of inch-thick steel, was designed to be built around the reactor. This "dry well" is shaped like an inverted light bulb, being a vertical cylinder surrounding the reactor with a spherical bulge at the bottom. The dry well is supported off the floor of the reactor building by a large concrete pillar.

In a Westinghouse reactor, a steam explosion due to a major pipe rupture is controlled by letting the steam expand into the building that houses it without letting the steam breach the walls. The steam has enough room to dissipate its explosive energy in the process and condense into water, which runs down the inside walls and into the sump. With this arrangement, no fission products dissolved in the steam are able to contaminate the countryside. Under extreme conditions,

259 The GE BWR/1, designed in 1955 and installed at the Dresden Generating Station in Illinois, used an external steam separator/dryer. The Soviet RBMK reactors, such as Reactor no. 4 at Chernobyl, were boiling-water reactors using multiple external steam separators.

overpressure from steam trips a relief valve and it goes up the vent stack. In this worst case, radioactive steam is released, but the building remains intact and able to contain further contamination.

The dry well is not of sufficient size to contain such a steam explosion. To deal with this maximum accident, there are eight large pipes connected to the spherical bulge and pointed down. Circling the reactor at the bottom of the dry well is the "wet well," which is a large torus or a doughnut-shaped steel tank, 140 feet in outside diameter, into which the connecting pipes terminate. The doughnut-tube is 30 feet in diameter, or slightly larger than the pressure hull of an early nuclear submarine.

This wet well is half filled with water, 15 feet deep, and it is held off the floor by steel supports. In the event of a sudden high-pressure steam release, the hot gas rushes down the eight pipes and blows out under water in the torus. This tames the steam by cooling it. It condenses into water, which is then pumped back into the reactor vessel to keep the fuel from melting. The GE invention, the Mark I containment structure, is thus able to do in a relatively small space what a Westinghouse reactor needs an entire building to accomplish.

It seemed an exceedingly clever workaround for the problem of otherwise needing an impossibly large reactor building, but the General Electric Mark I containment turned out to be the most controversial engineered object in the 20th century. Disagreements of an academic nature concerning the design may have been flying around at GE for a few years, but the problems became semi-public at a "reactor meeting" in Karlsruhe, Germany, on April 10, 1973, when a paper was presented by O. Voigt and E. Koch, "Incident in the Wuergassen Nuclear Power Plant."

Wuergassen was the first commercial reactor built in Germany, on the River Wesser between Bad Karlshafen and Beverungen. It was an early BWR/3 with a Mark I containment. The Germans had spent four years building the plant, and they connected it into the power distribution system in December 1971.[260] They were aware of rumblings of

260 Technically, the Wuergassen plant was not a GE, as it was built by AEG. AEG, however, did not design the reactor itself, but paid for a license from GE to build their latest model BWR. If there was something wrong with the Mark I, it was not the fault of AEG.

possible discontent concerning the Mark I containment coming from many directions, and they decided to put the speculations to rest by performing a simple test. Responsible engineers at GE claimed that the Mark I "did not take into account the dynamic loads that could be experienced with a loss of coolant." It was great design on paper, and it looked wonderful standing still in the basement of the reactor building, but would the steel torus stand up to live steam being suddenly blown down into it, or would it experience transient forces for which it was not prepared? The Germans ran the reactor up to power, making a good head of steam, and then sprang all eight steam-relief valves open at the same time.

Behavior of the torus under this "worst case" condition had not been predicted by even its strongest detractors. It quenched the steam under the water in the half-filled torus as it was supposed to, but not thoroughly, and the steam started to swap sides, first showing up on the north side of the torus, and then on the south side. The torus became possessed by "condensation oscillation," and it started rocking, tearing the support irons out of the floor and terrifying the reactor operators. It finally settled down, but not before causing expensive damage.

On February 2, 1976, General Electric engineers in the Nuclear Energy Division, Gregory C. Minor, Richard B. Hubbard, and Dale G. Bridenbaugh, the "GE Three," resigned from the company and held a press conference.[261] They were dismayed by their design of the Mark I containment and wished to come clean. The structure was just too fragile to stay together and prevent what it was supposed to prevent in an unlikely but not impossible reactor breakdown, they announced. It was not a good day in San Jose, home of GE Nuclear. A Mark I Owners Group was formed by concerned power companies.

In 1980, not a moment too soon, GE issued three important modifications for the Mark I containment structures currently in use. The "ram's head" terminals on the steam pipes into the torus, which could exert a rocket-like thrust when steam rushed through them, were to be unbolted, thrown away, and replaced with spargers that would

261 The three merry whistleblowers formed a consulting company, MHB Technical Associates. They were hired as technical experts for, of all things, Jane Fonda's movie *The China Syndrome*.

distribute and release steam in small doses under the water. Deflectors were to be welded into the insides of each torus, to prevent water waves from developing in the wet well under steaming conditions, and the support irons underneath the torus were to be bolstered. All plant owners executed these modifications, including TEPCO for the five reactors having Mark I containments at Fukushima I.[262]

This was hardly the end of the matter, and over the next few years there were several studies, simulations, and scale-model tests of the mercilessly persecuted Mark I by Lawrence Livermore Laboratories, the University of California Berkeley, and the Nuclear Regulatory Commission. Of particular interest was how the dry well would perform with 787,200 gallons of water sloshing back and forth in it under earthquake conditions.[263] By this time, GE had introduced its radically improved design, the Mark III containment, and the Mark I was supported as a thoroughly vetted, field-modified, and improved legacy system.

All General Electric BWRs were equipped with several reactor safety systems, designed to prevent fuel meltdowns, radiation release to the environment, and general damage to the system in the event of the worst accident that could possibly happen. This hypothetical event, named the "design basis accident," was planned assuming that the operating staff had lost control of the power plant and that there were multiple equipment failures. Safety systems were designed to begin operating automatically, with no help from the staff. Any such system could be deactivated by an operator, but if there were no operators present and making decisions, the safety systems would sense the condition of the power plant using electronic instrumentation and

---

262 All the reactors at Fukushima I were BWR/4s with Mark I containments, except the 6th and last reactor installed. It was a BWR/5 with a Mark II containment. The Mark II was a complete redesign, having a lot of concrete backing up the steel structure. Reactor No. 6 came online on October 24, 1979. Reactors 3 and 5 were built by Toshiba under license from GE, and Reactor 4 was built by Hitachi with a similar agreement.

263 The report from the initial study, "Sloshing of Water in Torus Suppression-Pool of Boiling Water Reactors Under Earthquake Ground Motions" (LBL-7984), was released in August 1978. It arrived at no negative conclusions regarding the behavior of the wet well in an earthquake, but demonstrated a good agreement between the mathematical finite elements model and a 1/60 scale model of the torus given a shake equivalent to the El Centro Earthquake of 1940.

digital logic, and thus were capable of operating as designed without human intervention.

Unit 1 at Fukushima, the oldest BWR, was equipped with an isolation condenser. In any emergency shutdown (scram), the turbine shuts down and the steam line to the reactor is shut off immediately. This action prevents any radioactive debris from possible fuel leakage from contaminating the turbine, but it also isolates the reactor vessel from its normal cooling loop. Even in full shutdown, a reactor generates significant heat for several days, so something must take the place of the cool water that returns from the steam condenser underneath the turbine.[264] The isolation condenser was designed for this purpose, operating as an emergency alternative to the primary cooling loop without needing any external power.

The condenser is located in the reactor building, on the refueling floor, above the reactor vessel and the dry well. A pipe runs from the top of the reactor to a small condenser in a tank of water. The cool-water return pipe from the condenser connects to the bottom of the reactor. As the water in the tank absorbs heat from the reactor, it boils off and leaves the building through a pipe in the wall, blowing off into the air. There is enough water kept in the tank to cool the reactor for three days, and there is an external connection for a fire truck to refill it. Water circulates through the condenser loop by gravity. The hot water rises from the reactor top into the condenser, and the heavier cooled water is pushed by gravity into the bottom of the reactor. A remote-controlled valve in the cold leg is used to turn the system on and off. The isolation condenser is simple and seemingly foolproof.[265]

In addition, all of the reactors at Fukushima that were in service were equipped with high-pressure coolant injection systems (HPCI)

---

264 In fact, the water returning from the condenser is too cool to be put back in the reactor vessel. A typical GE BWR uses five stages of pre-heater in the return leg of the primary cooling loop. Heat for these pre-heaters comes from "extraction steam," or steam directly out of the first (high-pressure) stage of the three-stage steam turbine. The extraction steam is then fed back into the condenser and cooled with the rest of the steam from the turbine. Newer BWRs have a simplified system using electrically driven heaters.

265 The same scheme, called "thermo siphon," was used to cool the engine of the Model T Ford automobile without the use of a motor-driven water pump. Reactor No. 1 had two isolation condensers, in case one failed. Each water tank held 28,002 gallons of water and was capable of handling 116 tons of steam per hour.

for emergency use. The HPCI (pronounced "hip-see," with the accent on the first syllable) is designed to inject great quantities of water into the reactor while it is maintaining normal operating pressure in shutdown mode, assuming that there is not a major steam pipe broken. This system does not rely on external electrical power to run the high-pressure water pump. It is turned by its own, dedicated steam turbine, which is connected to the steam pipe on top of the reactor, exhausting into the wet well. It spins up 10 seconds after a scram, and delivers cooling water taken from the pool in the torus at the rate of 3,003 gallons per minute.[266] For it to work, nine valves must be open.

Units 2, 3, 4, and 5 were BWR/4s, and were nearly twice as powerful as Unit 1. An early BWR/4 was rated at 2,381 megawatts thermal, whereas Unit 1 was rated at 1,380 megawatts thermal. These reactors dispensed with the isolation condenser for emergency use, and instead used the more aggressive reactor isolation cooling system (RCIC, pronounced "rick-see"). Its operation is similar to the HPCI, using a steam-driven pump to circulate water through the reactor vessel when the normal cooling loop is shut off, and it can even compensate for coolant leaking from broken pipes. Another nine remote-control valves must be open for it to operate, and its use must be monitored and controlled, else it will overfill the reactor and send water down its own steam pipe. It takes water from a dedicated water tank, which is also used to maintain the water supply in the torus "wet well."

If the pressure in the reactor vessel reaches the danger level, 1,056.1 pounds per square inch, a steam-relief valve opens and blows down through a pipe into the torus, where the steam and gas are released underwater and are calmed down.[267] The combination of dry and wet well containment is large enough to reduce the high pressure originating in the reactor vessel, but if the pressure in the torus should go beyond the design pressure of 62.4 pounds per square inch, a valve must open or the torus will explode. Under this condition of extreme emergency, the steam and gas are sent up the ventilation stack and

266 The number given, 3,003 gpm, is only for Reactor No. 1. No. 2 and No. 3 pump at the rate of 4,249 gpm.

267 That describes the first steam-relief valve. There are eight in a BWR/3 and 16 in a BWR/4, with progressively larger blow-off pressures. The last one opens at 1,130.1 psig.

into the environment, possibly containing radioactive fission products. This is a last-ditch measure, meant to keep from causing a major break in the containment structure which would allow uncontrollable leakage of the entire reactor contents outside the reactor building. To make it up the stack, the pressurized gas and steam must break a rupture disc in the pipe, calibrated to fail only at desperately high pressure.

The engineers at GE tried to think of everything that could possibly happen, and a system or a workaround was designed for problems that seemed highly unlikely. Units 2 through 6 at Fukushima were even equipped with residual-heat removal systems using four electrically driven pumps to cool down the reactors using seawater. There was one common weakness, however, for all the safety systems: they all depended on electricity.

Each valve in the complex maze of piping required electricity to open or close it.[268] If the plant scrams in an emergency, then it stops providing electrical power to itself. In this condition, it switches to external power off the grid or to a cross-wired connection to the reactor next door. If no power is available on the power-plant site or from the area outside the plant, then each reactor has two emergency diesel-powered generators that come on automatically. Either generator is capable of handling the electrical needs of the entire plant, in case one breaks down or will not start. If the backup generators will not start, then the last resort is a room filled with lead-acid storage batteries, kept charged up at all times. They will supply direct-current power to the control room for eight hours, which is plenty of time to open the necessary valves and start the emergency cooling systems, which will then run on their own without any electricity.[269] The control-room lighting, the system-monitoring

268 The larger valves are opened and closed by air pressure supplied by an electrically driven air compressor. For air-operated valves to be manipulated in emergency conditions, compressed-air bottles are provided, but the air valve is still opened or closed remotely by an electric signal. It is possible to open up the RCIC system manually, but it is not easy. If there is already radiation leakage in the reactor building, then the time limit for people working in the area to open a valve is strictly limited by the permissible dose and the dose-rate in the building.

269 It says eight hours in the battery advertisements and in the plant specifications, but performance may vary. Running a reactor on storage batteries can be as disappointing as a laptop computer in an airport waiting area when your flight has been delayed. It will power-down due to low battery just as your iPod dies.

instruments, and the valve actuators are built to run on the DC current from the batteries under this extreme emergency. Condensate or coolant pumps will not run on the batteries.

It took over 12 years to complete the Fukushima I power plant, with the last, Unit 6, starting commercial operation in October of 1979. Although each reactor was a General Electric BWR, no two reactors were just alike. There were constant improvements implemented by GE in the 1970s, and the technical sophistication of the reactor systems increased as units were added to the plant. The TEPCO engineers made certain that the plant met the special requirements for reactors built in a heavy earthquake zone. The earth was bulldozed off the site so that the reactors could be built on solid bedrock, reducing the horizontal earth movement during an earthquake to a minimum, and a tsunami wall was built down on the beach, protecting the plant from an ocean wave as high as 18.7 feet.[270] Two breakwaters, one north and one south, spread out from the beach and onto the sea floor to prevent waves from silting up the water intakes.

Each reactor was equipped with two large water-cooled diesel backup generators, located in the lowest part of the plant, completely underground, in the basement of the turbine building facing the ocean. Electric pumps brought in seawater to cool the engines and dumped it back into the ocean. The battery rooms and the electrical switchgear spaces were also in the basement, on the inboard side of the turbine building. The control room was two stories up, connected to the turbine building. As much equipment as possible was shared between pairs of reactors, especially the vent stacks. Units 3 and 4, for example, shared one 600-foot stack, exceptionally well braced for earthquakes.

There were exceptions. Units 2 and 4 had one water-cooled diesel and one air-cooled diesel. The Unit 4 air-cooled engine was down for maintenance and was in pieces on the floor. Unit 6, the more advanced BWR/5, had two water-cooled diesels in the basement, but there was a third generator located above ground in an auxiliary building. It was air-cooled.

270 The United States is not above building a nuclear power plant on the beach in the Ring of Fire. The Diablo Canyon Nuclear Power Plant in California is built facing the Pacific Ocean, using it as the source of coolant. Fukushima has an 18.7-foot tsunami wall, but Diablo Canyon has a 32-foot tsunami wall. There is no evidence of a wave this high ever having washed over the Diablo Canyon property.

In March 2011, Units 4, 5, and 6 were down for refueling and maintenance. Unit 4 was in the middle of refueling, with the dry-well lid and the reactor vessel cover unbolted and placed aside using the overhead crane in the reactor building. The refueling floor was flooded, and all the fuel had been moved to the adjacent fuel pool for a cool-down period. The only things turned on in the Unit 4 building were the overhead lights and the coolant pumps for the open fuel pool, keeping water moving over the spent fuel as its residual heat tapered off exponentially. Units 5 and 6 had just finished refueling, and Unit 5 was undergoing a reactor-vessel leakage test. Units 1, 2, and 3 were running hot, straight, and normal, and three 275-kilovolt line-sets were humming softly.

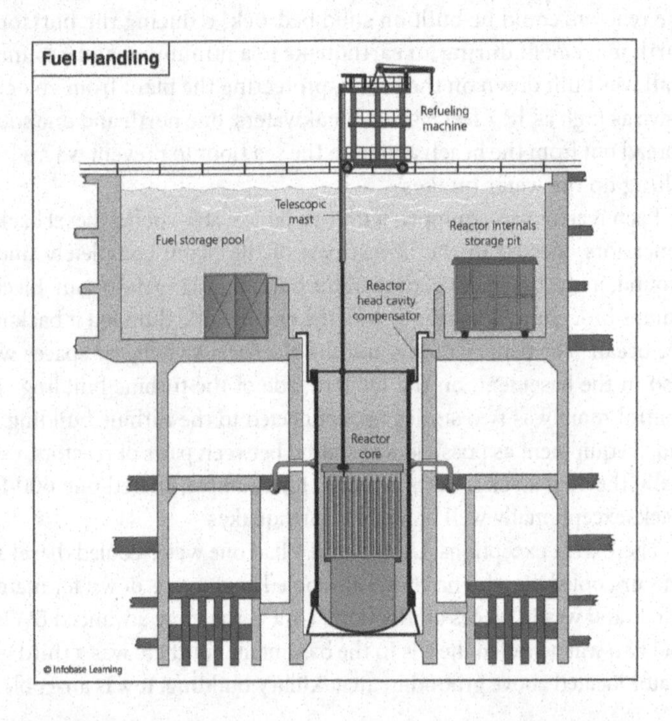

Units 5 and 6 at Fukushima Daiichi were in the middle of refueling when the Tohoku earthquake struck Japan. Both reactors were completely inert, with no fear of a meltdown or a hydrogen explosion. The fuel had been emptied from the reactor vessel and transferred to the storage pool using the refueling machine.

On March 9, 2011, 70 miles offshore, the Pacific Plate tried to slip under the Okhotsk Plate, 20 miles under the ocean floor. A magnitude 7.2 earthquake hit Japan. It caused the reactors on the northeast coast to scram due to indications from the ground-motion sensors, including Units 1, 2, and 3 at Fukushima, and it made the news, but nobody was hurt. Three more earthquakes the same day shook the ground. It was just another day in Japan, and life resumed a normal path after the bothersome disturbances. The reactors immediately restarted and resumed power production.

Earthquake prediction is a science in Japan, explored with more enthusiasm than anywhere else on Earth. Accelerometers are spread over Japan and out into the sea floor, and tsunami warning buoys are anchored offshore. These sensors can detect ground or ocean floor movement and send signals back to the Japan Meteorological Agency (JMA) at the speed of light over electrical cables. The earthquake shock travels much slower, at the speed of sound through rock (about 3.7 miles per second), and, depending on how far out the disturbance is, there can be minutes of warning issued by JMA. That is enough time to crawl under something solid or race out the front door of a building. The entire country is wired with earthquake alarms, designed to go off upon ground-movement detection from the array of accelerometers.

On Friday, March 11, at 2:46:43 P.M. Japan Standard Time, two days after the four minor earthquake shocks, Mikoto Nagai, head of the Emergency Response Team in Sendai, was at his desk on the third floor of an earthquake-proof building, sipping coffee. A lot of engineering thought had gone into how to make a building withstand ground accelerations. As Japan rebuilt after having been bombed to the topsoil during World War II, most of the new structures were constructed to sway without the foundation crumbling and the vertical support beams splintering. The early-warning earthquake alarm went off. Nagai put down his cup and looked up at the LED display bolted to the wall. It flashed 100 followed by a 4. In 100 seconds, a hit from a magnitude 4.0 earthquake was expected. The display quickly changed its mind. Make that a 6.0. No, an 8.0. Nagai stood up, and his coffee cup bounced sideways off the desk. Bookshelves collapsed, the internal wall in front of him came down, and people started screaming.

The Pacific plate had successfully relieved the east-west tension and hit Japan with its biggest earthquake ever recorded. It was 9.0 on the dimensionless Moment Magnitude scale.[271] In three minutes, the eastern coastline of Japan fell 2.6 feet, and Japan moved 8 feet closer to California.[272] The rotational axis of the Earth tilted by 10 inches. Roads were churned, high-voltage power lines were downed, and 383,429 buildings were destroyed.

The point in Japan nearest the epicenter of the earthquake was Onagawa in the Oshika District, and on a point of land jutting out into the Pacific Ocean was constructed the Onagawa Nuclear Power Plant by the Tohoku Electric Power Company, down on the beach. It consists of one BWR/4 and two BWR/5s, built by Toshiba under contract with General Electric. The last one started operation on January 30, 2002. As it was the newest reactor of the group, it has the most updated earthquake hardening techniques applied to it, and it has a substantial, 46-foot tsunami wall between it and the surf. The earthquake rolled through Onagawa, scrammed all three reactors, and subsided without doing any damage to the power plant. All the workers' homes within driving distance of the site, however, were leveled to the ground.

About 22 seconds after it hit Onagawa, the ground-shock hit Fukushima I, which was twice the distance from the epicenter. An inspector for the Nuclear Industrial and Safety Agency, Kazuma Yokata, was permanently stationed in the office building at Fukushima I, in the no-man's land between Unit 1 and Unit 5. He heard the alarm go off, but he was not overly concerned until the ceiling appeared to be coming down on him. He cringed as the L-shaped brackets holding up the bookshelves ripped out of the wall and his thick binders containing rules and regulations started flying.

There were 6,413 workers on the Fukushima I site that day. One of them, Kazuhiko Matsumoto, was in the turbine building for Unit

271 One is tempted to write "Richter scale." Somehow, it sounds better, but Richter is now considered to be an obsolete relic from 1935. It has been replaced by the Moment Magnitude scale, which expresses the energy released from an earthquake on a base-10 logarithm scale. There are correction factors added to the formula, but it is basically the same value as the old Richter scale, +/-0.6. A 9.0 event is 100 times more powerful than a 7.0 event.

272 The sinking of Japan had an effect on the rotational speed of the Earth. Just as a spinning ice skater speeds up by pulling in her arms, the Earth's rotation sped up slightly, decreasing the length of a day by 1.8 microseconds as the mass of Honshu Island drew downward, toward the center of the Earth.

6, finishing some work on air ducts. He suddenly found that it was impossible to remain standing on the sparkling clean deck, and he had to cling to a wall to keep from being dribbled on the floor like a basketball. The lights went out, and the windowless expanse of the turbine hall went black. In a few seconds the emergency lights turned on, and over the loudspeaker came a simple instruction: "Get out."

Fukushima I was built to withstand a horizontal ground acceleration due to an earthquake of 0.447g (1g = 32 feet per second per second). Unfortunately, this 9.0 earthquake came in at 0.561g.[273] The reactors, particularly the three earliest units that were running at full power, were treated roughly. Some pipe runs ripped out of wall anchors, all external power lines went down, and anything not bolted down went flying. Fortunately, almost everything in a nuclear plant *is* bolted down. All 12 available emergency generators came on after a few seconds with the control rooms running on batteries. Over the next minutes, several aftershocks hit the island, with magnitudes up to 7.2.

With full AC power from the emergency generators, the three reactors that had been running at full power experienced orderly shutdowns, with the cores being cooled by the usual means, and everything was under control. At Unit 1, the completely passive isolation condensers were doing their job, cooling down the reactor core after shutting down from running at full power. There was no need to turn on the HPCI, at least not yet.

In the opinion of the reactor operators, the isolation condenser was doing its job too well. The temperature was falling too rapidly, and, with the steam condensing in the reactor vessel, a pipe could be collapsed from the vacuum it created. Over-thinking the simple, hard-wired digital logic that had turned it on, an operator put his hand on the switch handle that would stop the isolation condenser coolant flow and turned it off. Then, the remotely controlled flow valves MO-3-A and MO-3-B closed.[274]

---

273 The Diablo Canyon Nuclear Power Plant, the American equivalent of Fukushima I built on the coast in California, is built to withstand a horizontal ground acceleration of 0.750g.

274 The closing of these valves is surprisingly complex. The operators were "cycling" Isolation Condenser A, turning it on and off, trying to slow the temperature descent, with Condenser B turned off. Both condensers, A and B, had been turned on automatically by the scram-control logic. The last operator action was to cycle Condenser A down by turning off the cold-leg return valve, which was motor-operated.

In major commercial reactor accidents, there always seems to be a single operator action that starts the downward spiral into an irrecoverable disaster. In the case of Fukushima I, closing those two valves at Unit 1 was the turning point. With that simple action, overriding the judgment of the automatic safety system, an operator doomed Fukushima I to be the only power plant in Japan that suffered irreparable damage due to the Tohoku earthquake of 2011.[275]

At 3:27 P.M., 41 minutes after the earthquake, a tsunami hit the beach at Fukushima I with a towering wave, 13 feet high. The wall built in front of the plant kept the wave from harming anything. Eight minutes later, a second and then a third wave hit. At 49 feet high, they went over the 18.7-foot wall and inundated the entire plant.[276]

The water-intake structures for all six reactors were collapsed by the wave, the water pumps were blown down, and any electrical service outside the buildings was shorted out by the salt water and then torn away. In six minutes, all the underground diesel generators were

---

275 There is a strange twist here. The operators at Unit 1 were desperate to restart the isolation condensers or to initiate the HPCI, but there was no DC current from the backup batteries. They stormed what was left of the plant parking lot, where the 6,413 workers had left their cars, and started pulling batteries out of the cars that remained on site, and, where available, they took jumper cables out of trunks. They stacked batteries on the floor in the control room, connecting ten 12-volt batteries in series using the jumper cables to make 125 volts DC for the valve-control motors. There was not enough current to keep the system alive for more than a few minutes, but for an instant the control panel lighted up and they could see that the isolation condenser valves MO-2A and MO-3A indicated closed. At 6:18 P.M. they were able to apply power to the motors and open the valves. A lookout reported steam coming from the condenser pool, indicating success. Seven minutes later, an operator reached for the switch and closed the valves, just as the automobile battery-pack died. No reason for this action has been determined. The isolation condenser valves could not be reopened.

276 The precise height of the second and third waves is not known, and it is estimated by the level of silt on the buildings at between 46 and 49 feet. The wave meter at the plant was capable of recording a wave as high as 24.6 feet, where it jammed right before it was carried away by the wave. The tsunamis that hit Fukushima I were the sum of several circular waves originating sequentially at large sea-bed movements in the quake epicenter and in the Japan Trench, running north-south between Honshu and the epicenter. The wave transmitted across the Pacific Ocean as the circular wave components increased in radius, eventually hitting the California coast. No damage was done on the eastern side of the Pacific. A tsunami of the type that hit Japan starts out with a wavelength of about 120 miles, moving at 500 miles per hour. When it approaches shallow water, the wavelength compresses to 12 miles, the wave-height expands, and the speed reduces to 50 miles per hour.

flooded, and the emergency AC power failed. One diesel-powered generator, the air-cooled unit located above ground at Unit 6, remained online, providing power for Units 5 and 6. Units 3 and 4 were now on DC power, enabling operators to read instruments in the control room and manipulate remote-control valves until the batteries lost power, and now was a good time to make sure all the valves were in an open/close condition that would do the most good, keeping the core of Unit 3, recently operating at full power, from melting. In Units 1 and 2, the battery room was flooded, and the plant was in total blackout. No valves could be turned on or off, and the status of reactor systems was not available on the control panels. They were stuck with whatever configuration was in effect when the lights went off, and that meant that Unit 1 was coming down off full power with nothing to cool its 69 tons of hot uranium oxide fuel, continuing to generate megawatts of power. The isolation condenser was shut off. In Unit 2, at least the RCIC was left running when the power failed, but without some tweaking, it too would fail eventually. TEPCO advised the Japanese government that an emergency condition existed at Fukushima I.

The tsunami rushed inland, to the ancient tsunami warning stones and beyond, carrying everything with it and drowning the Earth beneath it. Fishing boats and ocean-going ships hit the beaches and kept going. Down came houses, factories, and entire towns. Cars, trucks, and trains were moved like toys in a fire-hose spray. Power transformers blew up as electrical lines touched the ground, gas lines broke, and fires broke out, taking out any last burnable structures that had not washed away.

The wave came in, and then it went out, taking everything that would float out to sea. An estimated 18,000 human beings were washed into the Pacific Ocean. The loss of life was devastating. Two operators drowned at Fukushima Unit 4, trapped in the turbine building as the water quickly rose in the basement.

The immediate crisis at Fukushima I was a need for AC power to manage the cooldowns in Units 1, 2, and 3. Unit 1 was in total blackout with no passive systems running, and in 2 and 3 the water circulated through the torus pool was eventually going to have absorbed enough heat to start steaming. All the reactor interconnection cables, allowing the units to share 6.9 kilovolt and 480 volt power, had been lost in the tsunami. An obvious solution was to bring in portable diesel generators and hook up to whatever wiring stubs were left sticking out of the buildings, but this was not going to be simple. All roads into Fukushima I were either completely washed away, blocked by collapsed buildings, or jammed by fleeing people. Appropriate generators were available, but they were too heavy to be flown in by helicopter. They could only be transported by wide trucks on a smooth highway.

The plant wiring was also a problem. Temporary cables would have to be installed, first running from the plant parking lot to the standby liquid-control pumps for Unit 2.[277] Cables were available, but they were four inches in diameter, 656 feet long, and weighed more than a ton. Unreeling the cables and running through debris field covered with collapsed buildings and newly established lakes would have to be done without any powered equipment. No trucks, cranes, or bulldozers were available, and hidden beneath the ground clutter were manholes with the covers blown off. Everything about establishing AC power involved tremendous adversity, and it was going to take time.

In Unit 1 there was no instrument feedback revealing the state of the systems and no lighting in the control room. The operators could only look at the dead instruments using flashlights. By three hours after the earthquake, all the steam-relief valves had pried open and the water had boiled out of the reactor core. An hour and a half later, the fuel, still generating power at a fractional rate but naked of liquid coolant, started to melt away the zirconium sleeves on the fuel pins. The red-hot zirconium began to react chemically with the steam around it, oxidizing and leaving hydrogen gas in place of the steam. The zirconium core supports started to get soft and sag, and entire fuel assemblies started coming apart and tumbling down into the bottom of the reactor vessel. There were 400 fuel assemblies, and each one was 171 inches long. Compressed by the weight of the fuel, the wrecked mixture of uranium oxide, zirconium

277 These pumps (SLC) were not flooded or damaged, and if they could be powered up, it would save Reactor 2 by cooling it down with seawater. Reactor 1 was in greater need of help, but starting the SLC pumps for Reactor 2 looked like something that could be accomplished. Unfortunately, the seaward intakes for the pumps were blocked.

oxide, and melted neutron control blades increased its temperature. Fuel started to melt.

The combination of steam pressure and hydrogen gas pressure vented from the isolated reactor vessel exceeded the designed yield strength of the torus several times over. There was no electricity to open any valves, so the normal severe emergency action of venting the torus safely up the vent stack could not be initiated. General Electric's Mark I containment, made of steel one inch thick, split open, and the soluble and volatile components of fission products, set free by the absence of any zirconium cladding, were sprayed into the reactor building. Included with it was hydrogen gas, mixing with the oxygen-containing air in the large space above the refueling floor. Unit 1 was now a bomb, set to go off and heavily contaminated with fission products.

The operating staff at Unit 1 knew that after being without a cooling system for several hours, the Mark I would have to be vented up the stack, but there was no power to open the main valve, AO-72. It was an air-operated valve, but it was possible to open it by hand if they could get to it. The entire reactor building was radiation-contaminated, which was a clue that the containment structure was already broken open, but men volunteered for the hazardous job of running down pitch-dark hallways, through a maze of doorways and passages, to the valve, open it by turning on a compressed-air line, and rush back, receiving the maximum allowable dose for the entire month in a few minutes. First, a gasoline-engine air compressor would have to be located and connected to the line. Every detail took time.

The entire area around Fukushima would have to be evacuated before it was legal to vent the containment, and government permission had to be verified. The TEPCO office in Tokyo finally gave the go-ahead at 9:03 A.M. on March 12, the next day. At 2:30 P.M., after heroic effort, the torus in Unit 1 was vented up the stack shared with Unit 2, but it was too late to prevent damage to the plant.

At 3:30 P.M., the men at Fukushima I had bucked all odds and installed external AC power to the standby pumps at Unit 2. With great effort, fire hoses had been attached to the outside access points for the condensate tanks in Units 1 and 2, and fire trucks were standing by to start pumping water and relieve the obvious heat buildup inside.

The men paused a moment to rest and admire their work. Six minutes later, at 3:36 P.M., the Unit 1 reactor building exploded in a spectacular geyser of debris, sending radiation-contaminated chunks of concrete and steel beams high in the air and careering through the newly installed equipment. Five men were injured, the wiring was ripped out, the generator was damaged, and the fire hoses were torn. Heavy debris came down all around for what seemed a long time. Radioactive dust from the Unit 1 fuel floated down out of the air and began to cover the entire power plant.[278] Not only had this explosion destroyed Unit 1, but from now on all work at Fukushima I would require heavy, bulky radiation suits and respirators, and now there was a new layer of movement-restricting debris on top of the already-established debris. It was a setback.

The next day, at 2:42 A.M. on March 13, the passive high-pressure coolant injection (HPCI) system in Unit 3, running on steam made from the afterglow in the fuel, finally gave out, and by 4:00 A.M. the fuel began to degrade, eventually collapsing into the bottom of the reactor vessel and generating a great deal of hydrogen gas. A fire engine was eventually able to inject seawater into the system, effectively closing the gate after the livestock had escaped. By 8:41 A.M., the operators had managed to open the air-operated torus vent valve and relieve the pressure that was building up. It was seen as a semi-miracle. Steam was seen coming out the vent stack, and the site boundary dose rate suddenly increased to 0.882 rem per hour.

At 11:01 A.M. on March 14, the day after the Unit 3 core structure melted, the Unit 3 reactor building exploded with a fireball, taking the lead over Unit 1 for the ugliest debris field. Hydrogen gas from the core deterioration had collected in the top of the building until it reached a critical concentration, somewhere over 4% in the air, and a spark must have set it off. Two fire engines were put out of commission, 11 workers were injured, the portable generators that were now collecting in the yard were all damaged, the temporary wiring was torn out,

278 Fortunately, at 3:36 P.M. on March 12, the wind at Fukushima I was blowing out to sea. It is never good to let radioactive dust settle over the ocean, but it would have been much worse to have it settle over the inland territory to the west of Fukushima. Much nuclear mischief can be lost at sea and rendered much less harmful, because the dilution factor is so enormous as opposed to when it stays on land.

and the fire hoses were ripped apart. The new debris on the ground, everything from dust to chunks of walls, was extremely radioactive. The dose rate in the Unit 3 airlock, not even entering the reactor building, was now 30 rem per hour. The absolute emergency dose allowed one worker at the plant was 10 rem. That meant that if a worker stood in the airlock for 20 minutes, he had to be relieved and sent away, and he could no longer work on the problems at the plant. Debris on the ground after the Unit 3 explosion caused a dose rate of 1 rem per hour in the yard, and all personnel outside the control room were evacuated to the Emergency Response Center, near Unit 5.

At 12:40 P.M. the Reactor Core Isolation Cooling System (RCIC) in Unit 2 had absorbed all the shutdown heat it could stand, and the coolant-pump turbine stopped turning. It had held out for 70 hours, outperforming its design. The water in the reactor vessel boiled away, overstressing the Mark I containment structure, and at 4:30 P.M. the fuel pins started to melt, eventually falling into the bottom of the vessel and vigorously making hydrogen. Fortunately for Unit 2, the explosion of Unit 1 had blown a large hole in the side of the reactor building, so all the hydrogen leaking out of the torus was able to escape freely and not collect near the ceiling. Unit 2 never exploded, but its radioactive steam, iodine, and xenon were able to escape into the environment along with the hydrogen. Plans to vent the torus were cancelled when the pressure inside was found to be too low to open the rupture disc on the vent stack.

As the situation at Units 1, 2, and 3 continued to deteriorate, Unit 4 remained serenely innocent. All its fuel had been removed and stored in the fuel pool on the top floor in the reactor building. The cooling water surrounding the fuel was at 80.6° Fahrenheit, the tops were off the reactor vessel and the dry well portion of the containment structure, and nothing was anywhere near a crisis condition. The electrical power was gone, but Units 1, 2, and 3 were in continuous crisis, and they obviously needed more attention than Unit 4. The operating staff pitched in to help the units that were in deep trouble.

The fact that there was no power meant that the Unit 4 vent-stack damper valves had no compressed air holding them closed. They were, in fact, hanging open. Unit 4, for reasons of economy, shared a vent stack with Unit 3. There was no backflow damper installed, so when the overpressurized torus in Unit 3, heavily invested with hydrogen, was vented up the stack, using the correct procedures by the book, half of the vented gas went up the stack and half went back through the Unit 4 vent pipe. The hydrogen and radioactive steam proceeded through the relaxed valves in the Standby Gas Treatment System filters, up two stories, and out the exhaust air ducts on the fourth floor of the Unit 4 reactor building, where it collected at the ceiling and awaited an ignition spark.

At 6:14 A.M. on March 15, four days after the earthquake, the Unit 4 reactor building exploded with a mighty roar, much to the surprise of everyone working at Fukushima I. Having no theory as to what had just happened, the operators at Units 5 and 6 quickly climbed to the tops of the reactor buildings and hacked large holes in both roofs to let out the hydrogen, which did not exist, thus inflicting the only damage that the two newer reactors sustained in the Tohoku earthquake.

The cross-contamination of hydrogen and radioactive steam from Unit 3 was not figured out until August 25, and on March 15 the only plausible explanation was that the water in the fuel pool must have leaked out through a crack caused by the earthquake. The spent fuel, removed from the reactor core only days before, must have overheated and caused its zirconium cladding to generate hydrogen from the remaining steam in the pool. A helicopter flyover confirmed a great deal of radioactivity in the remains of the upper reactor building. A great deal of effort and time was spent in vain, trying to reload the fuel pool with water using helicopter drops and water cannons. It turned out that the fuel in Unit 4 was in fine condition, and the high radioactivity over the building resulted from dissolved fission products delivered to the space above the refueling floor by steam from Unit 3.

After the upper floor on the Unit 4 reactor exploded, there was basically nothing left to happen that could further degrade Fukushima I. The three reactors that were operating when the earthquake hit had melted down and were left an enormous liability for TEPCO. It would be feasible to rebuild Unit 4, because only the roof and walls covering the refueling floor had been blown up, but the radioactivity spread all over the power-plant grounds would make it an impractical work environment. Units 5 and 6, the newest reactors in the plant, could be brought back online with some rebuilding of the seawater intakes, new outside pumps, and an enhanced tsunami wall.

Unfortunately, the wind had shifted to an east-to-west direction during the disaster. Starting at the Fukushima beach, a swath of farmland running northeast about six miles wide by 25 miles long, washed of human habitation by the tsunami, may be too contaminated by fission product fallout to be repopulated immediately. It may be turned into a nature preserve. Every time it rains at Fukushima I, more radioactive dust is washed down to the shore and out into the ocean, causing issues for Japan's sizable fishing industry.

None of the spent fuel at Fukushima I, in cooling pools or dry storage, was damaged and no fission products from it leaked into the environment. All of the radioactive contamination was from damaged, hot fuel exposed to steam, which was allowed to escape from Mark I containment structures, stressed beyond the imaginations of the engineers who had designed them. No one had considered that a reactor coming down off full power could be denied electricity for more than a few minutes, given the multi-level, parallel-redundant systems built to prevent it. After Unit 1 blew up, refilling the condensate tanks from external sources and wiring up emergency generators was delayed, and the remaining reactors fell like dominoes. Without an ultimate heat sink, the core structure in a nuclear reactor that has recently generated a billion watts of power will eventually melt from the delayed energy release in the fuel. If the workers had been able to refill the condensate tanks in Units 2 and 3, there would have been a lot of steam, but the vapor would not have contained any dissolved fuel, and it would not have been radioactive. With externally provided water and electricity, Units 2 and 3 would have survived.[279]

---

279 The reactors would have survived, but the steam turbine rotors would have been ruined. A major job of the diesel backup generators is to keep the turbine turning, if even at a very slow speed, using the electrically driven "jacking gear." A large steam turbine must be kept turning at all times. If allowed to sit still for a few hours, the rotors will sag, losing balance and rubbing against the metal shell that covers the turbine. In this distorted condition, the turbine cannot be restarted. The backup batteries will not maintain this essential service to the turbine.

Nothing melted through the bottom of a steel reactor vessel at Fukushima. After accidents at Three Mile Island and Fukushima, fears of a "China syndrome" melt-through begin to seem unfounded. There is simply not enough heat generated by tons of hot fuel to make a hole in the bottom of a water-moderated reactor.

The workers at Fukushima I were dedicated to the tasks of bringing the reactors at the plant under control, with cold shutdowns a goal, and they worked without sleep, food, a way to get home, news from loved ones, or a remaining dwelling place, in a potentially dangerous radiation field. Two operators in the Unit 3 and 4 shared control room wore glasses, and their respirators would not fit correctly over their spectacles. Air containing radioactive dust leaked in to their breathing apparatus. One received a dose of 59 rem, and one received 64 rem. These were serious but not immediately fatal lung exposures. With workers retreating inside as the result of various explosions, the Emergency Response Center became heavily contaminated with radioactive dirt, and there were no controls in place to prevent radiation exposures. Workers who never worked near a reactor received substantial internal radiation doses. One woman, for example, received a dose of 1.35 rem.

As of September 1, 2013, there have been 10,095 aftershocks from the Tohoku earthquake of March 11, 2011.

With four nuclear power plants in a direct line of the tsunami waves following the Tohoku earthquake, what was it specifically about Fukushima I that caused it to be destroyed that made it different from other reactors? TEPCO, having ignored studies warning of a major earthquake and a tall tsunami, was widely blamed for the disaster. None of the preparations for a major offshore earthquake were adequate, and the probability of an earthquake disaster, and the likelihood of a corollary tsunami, was not taken seriously. The nuclear regulatory structure in Japan as well as the prime minister himself were under fire for not having insisted on more rigorous safety inspections and preparations, but what caused this highly localized breakdown when seventeen other nuclear power plants around Japan were rattled in the same earthquake? Those plants located on the east coast were even hit with the same tsunami.

Consider the fate of Fukushima II, a nuclear power plant built by TEPCO, seven miles down the coast from Fukushima I. It is a newer power plant, having four General Electric BWR/5 reactors with Mark II containment structures, built under license by Toshiba and Hitachi. The reactors came online between 1981 and 1986, producing a combined 4.4 billion watts of electrical power. All four units were operating at full power on March 11, 2011, when the Tohoku earthquake struck Japan.

Immediately during the earthquake, all four units scrammed and automatically began emergency cool-down measures. Three out of four off-site power sources went down, and the emergency diesels started. A tsunami warning was issued, indicating that a wave at least 10 feet high was on its way. All operators were called to the control rooms, and everybody else was evacuated to high ground. In 36 minutes, the waves started coming. The highest in the area of the main buildings was 49 feet, and it swamped the 17.1-foot tsunami wall, covering the entire plant in seawater.

The tsunami knocked out the majority of emergency diesel generators and above-ground seawater pumps, but Unit 3 kept both its generators and its pumps, and Unit 4 still had one generator. The last remaining high-voltage electrical line out of the plant site after the earthquake remained operable. Using this off-site power and cross-connections with the three remaining diesels, all control-room instruments and controls remained in operation. All four reactors were depressurized, and coolant injection was established using the condensate water tanks, exactly as detailed in the written emergency procedures. Unit 1 required manual actuation of some motor-driven valves, but there was no radiation leakage or loss of lighting in the reactor building, and the workers were not dosed.

Unit 3 used its still-operational seawater pumps to achieve a cold shutdown, but Units 1, 2, and 4 used a spray of water from the condensation tanks to cool off the hot water pooled in the Mark II containment wet wells. New seawater pumps and a lot of temporary electrical cable were urgently needed, and TEPCO managed to find these supplies and lower them onto the plant site using helicopters the next day. About 200 workers, unimpeded by radiation contamination on the ground, installed the new motors and 5.6 miles of new electrical cable in 36 hours. By 3:42 P.M. on March 14, three days after the earthquake, all four units were being cooled using the seawater-driven Residual Heat Removal Systems, and the reactors were in cold shutdown on March 15.

The important differences between Fukushima I and Fukushima II seem to center on the ages of the two plants and the resulting differences in design sophistication. Nuclear-plant designs, all of which are experimental, rapidly evolved in the 1970s as lessons were learned and things that worked well were kept while things that did not work well were redesigned. The Mark II containment structure was a welcomed improvement and simplification of the much-debated Mark I.

Using air-cooled instead of seawater-cooled emergency diesel generators, located out of the basements and above ground, was also important. Being washed over by a wave was different from being soaked in a permanent pool of salt water, particularly if the engines used electrically driven pumps to circulate water out of the ocean, where the intakes were destroyed by the tsunami, and the electrical distribution boxes were left under water.

Fukushima II was also simply luckier than Fukushima I, which had its entire electrical switching yard wiped out, including the inter-unit electrical connections. Unit 1 at Fukushima I, having the smallest of the six reactors, had a lesser heat load to manage using the same Mark I containment structure that Units 1 and 3 had. Its set of two isolation condensers, unique to it at the plant, was ingenious and robust, and there was no reason why it could not have saved the reactor from destruction and brought it down to a cold shutdown without the convenience of emergency electrical power, if only it had not been turned off. The fate of Fukushima I, the safety reputation of American-designed light-water reactors, the remains of the nuclear power industry, and the background radiation in the Sendai Province could all have been different.

Take a step back and contemplate the sampling of nuclear wreckage that has been laid out before you in the ten chapters of this too-brief narrative. You can see patterns developed in this matrix of events. There are hot spots, imprints, and repetitions. The markings are all over the developed world, left there by one very large experimental program that was trying to improve the lot of mankind, and not to destroy or degrade it. The boldness of this long-term program can raise an eyebrow or two, but from an engineering standpoint it was new, exciting, unexplored territory. In all, it killed fewer people than the coal industry, it caused less unhealthy pollution than the asbestos industry, and it cannot be blamed for global warming.[280] In the last chapter, we will discuss what it all means and reach for what conclusions there are.

280 Statistically, for every person who dies as a result of the generation of electricity by nuclear means, 4,000 people die as a result of fine-particle pollution caused by burning coal. Between 1972 and 2002, 730,000 lawsuits were filed because of asbestos dust inhalation, costing the asbestos industry $70 billion. The deaths caused by asbestos are in the half-million-people range. Global warming is believed to be caused by a buildup of carbon dioxide in the atmosphere, caused by our digging or drilling up and burning carbon that was sequestered underground. This carbon was geologically kept out of the total carbon inventory on the surface, but the need for power in the industrialized world has made it necessary to release energy any way we can. Most of the extra carbon dioxide comes from burning coal, gasoline, and natural gas. Nuclear power generation does not oxidize anything, but derives energy from nuclear fission. Some have pointed out that building a nuclear plant requires a lot of fuel to pour concrete and workers drive to the plant in cars, but these are ridiculously weak arguments against non-polluting nuclear power.

Chapter 11

CAUGHT IN THE RICKOVER TRAP

"If a man fires at the past from a pistol, the future will fire at him from a cannon."

—Abutalib

It was summer, 1985. We had successfully conceived, designed, birthed, programmed, built, worried over, installed, documented, and exhaustively tested the Safety Parameter Display System for the two General Electric BWR/4 reactor units at Georgia Power's E. I. Hatch Nuclear Power Plant down in Appling County, Georgia. It was time to celebrate! Instead, the three of us, I, Jeff Hopper, and Mark Pellegrini, retired to our rat-hole motel in Hazlehurst, over in the adjacent Jeff Davis County, to catch up on our sleep.

There is seldom a great deal of civilization near a nuclear plant, and it was a half-hour to Hazlehurst, which was the closest place with a motel. I found my room, collapsed in bed, and extinguished the light. As soon as the springs stopped groaning, I noticed a periodic chirping noise. Eek. . . eek. . . eek. . . eek. Was it a cricket, or was it a dry bearing in the air conditioner blower? My attention was riveted to the sound, and I could not help but try to analyze it. Having just come down off an extended computer system verification exercise, I found that I had to submit it to a test. I turned on the light. The chirp stopped instantly. Yep. It was a cricket. Problem solved. Lights out. I drifted off to sleep and suffered through an incubus of an enormous squirrel-cage fan with a dry bearing, moving closer and closer.

These computer systems were built to satisfy new control-room instrumentation requirements set forth by the Nuclear Regulatory Commission in NUREG-0696, "Functional Criteria for Emergency Response Facilities," written in response to inadequacies discovered in the TMI-2 disaster of 1979. Our sparkling new systems, bolted strongly to the control-room floor, each collected 128 analog and 512 digital data in real time from each reactor system and displayed them on demand using color video monitors. The equipment had been shake-tested on an earthquake simulator in California, run continuously for years in our lab back in Atlanta, and endured handling by inexpensive student labor from Georgia Tech. It was as solid and glitch-free as a dedicated team of engineers could possibly make it. I was overflowing with confidence that it would perform perfectly, as I would remind anyone standing near it.

The next day began an adventure that went down in the history of nuclear power as the "Rat Cable Problem."

We wanted to give the systems ample time to run before we returned to examine the overnight performance logs, so we lingered over breakfast and retired to our rooms to relax. I'm not sure what the other guys did, but I watched cartoons on the television. That afternoon, we drove to the plant. Pellegrini fretted over the party that the control room operators must be planning for us.

The parking lot at the plant was blazing hot, with the power of the sun reflected back at us, making the air seem to sizzle and boil as we walked the distance. There was some race of enormous beetles living on the plant site—they were huge black things with horns. Pellegrini feared that they were mutants caused by irradiated bug-DNA. I scoffed, but they *were* big, and they were marching eight deep in a line across the parking lot, toward the river, moving perpendicular to our trajectory as we made a straight line for the guard shack.

They were a determined bunch of insects, keeping a disciplined column at a constant speed on the egg-frying concrete. We were equally determined. Our tracks were about to cross, and I was not about to hesitate as we intersected the beetle path. My eyes locked on an individual. He was not going to hesitate either. Estimating his speed and mine, it looked as if he would be walking either over the top of my right shoe or under it. Crunch. The big insect's life ended suddenly, right under my foot. It was a bad omen. I wondered silently as we completed the hot trek. Why do silly omens seem to gain importance as we draw nearer to an active fission process?

We reached the electronics bay adjacent to the control room. Hopper turned on the maintenance terminal and started reviewing the system's performance over the last 18 hours. It was bad. The error log was filled with digital data reception failures, and the size of the log was growing as we watched. It was not every reception. It was just every once in a while, a parity check would fail.[281] It could go a minute or an hour between errors, or a burst of errors could occur, as the digital data set was polled once per second, and, even more unnervingly, the failures appeared to be completely random and did not follow any particular pattern or appear to be connected to any one particular system failure.

The digital data were collected using a wired loop running all over the plant and passed on to the computer systems by a Cutler Hammer Directrol Multiplexer Communications Station, bolted to the back wall of the electronics bay. A fat, multi-conductor cable (no. 5769) ran between the Directrol and the ROLM 2150 I/O expansion box. Both the Directrol and the expansion box had been exercised continuously back at our shop in Atlanta for years without a single error. Had something broken at the plant? We decided to let it keep running to see if

281 Digital data consisted of the open/closed status of valves and the on/off status of electrical motors and solenoids, collected at nodes all over the plant. Each datum was a 16-bit digital "word" with an extra parity bit. If the binary bits in the 16-bit word added to an even number, the parity bit was set to a one. If the addition was an odd number, the parity bit was zero. Upon reception back at the computer, the binary bits were added up, and the odd/even character of the resulting number was compared to the parity bit. If the results did not agree, then the datum had been corrupted somewhere, having either dropped or gained an odd number of spurious bits. This condition was logged as an error, and a repeat transmission was requested.

the system self-regulated, and also so we could collect more data and try and figure out what the problem was.

The errors would not stop coming. The next day, the error log was in crisis mode, and it was eating up all the reserve space on the disc. Another week of this, and the working space on the disc would be gone. The operating system would lock up. The log was now larger than the 30-minute sliding record of everything in the plant that was held on the disc as a "black box" recording.[282] Why? It had to be the only one difference between the development setup in our shop and the operational installation in the power plant.

Back at Georgia Tech, we had built a special room in the high-bay of the Electronics Research Building, complete with enhanced air conditioning and a raised computer-room floor. The room was a good replica of the Plant Hatch electronics bay, except for one detail. There was an open pipe through the wall, leading to the outside. It had been added to the building so that power lines from a gasoline-powered 400-hertz generator could be connected. The pipe was not in use, except as an access walkway for vermin. Somehow, a brown rat, *rattus norvegicus*, had found our lab, and he visited often. By the time we discovered him, he had chewed up the insulating sheath on our very expensive cable no. 5769, connecting the ROLM I/O box to the Directrol. Why he so enjoyed the taste of plastic insulation on that particular cable, I could not comprehend. It had zero food value, but he made a meal of it.[283]

Damage to the cable was strictly cosmetic, and we wrapped some electrical tape over the wound. It still worked perfectly, but we were too embarrassed to deliver it to the plant in that condition, so we called ROLM Military Computers in Cupertino, California, and ordered a

---

282 We started the project specifying a 32-megabyte hard disc. Remember, this was the early 1980s, and 32 megabytes was considered to be a large data-storage capacity. It was housed in a box that fit in a 19-inch rack. Casey Lang, head of the software development group, came to me requesting a 50-megabyte upgrade to the specification. That was huge! It was hard to see how we could ever fill a monstrous 50-megabyte bucket, but I changed the order. It is amazing how data capacities have drifted up in the decades since then.

283 Having had 30 years to think about it, I now believe that the rat's interest in the plastic cable insulation was caused by salt left on the surface by people having handled the cable. Human sweat contains a lot of salt, and when the water evaporates, the salt is left on the surface. Both ends of the cable had been handled many times, both at the factory in Cupertino and at the application point in Atlanta.

new unit. They sent it promptly, and this was an improved example of the cable. The Army had apparently given them grief about the weight of their parallel-digital cables, so ROLM had built new ones using a smaller gauge of wires. It was slender and svelte, not as clumsy and bulky-looking as the original cable, and each individual wire was tested for continuity and lack of cross-talk. Once the new cable passed all the tests, we installed it in to the power plant.

However, not everything was perfect. The problem we soon found with the new cable was that, thanks to the slimmer design, there was a reduced cross-sectional area of the wires used. A digital computer signal is sent over an electrical circuit, and as such it needs a transmission wire and a return wire, or ground. In a wired digital circuit of many parallel paths, such as a data-bus or an interconnection cable as we were using, a common ground connection is used by all signals. In our original cables, the wire gauge was overkill, with more copper than was necessary for the low-current digital signals, and a single ground wire was big enough to carry all the signals firing at once. With the wire size reduced to a gauge that was still good for the signal current, the smaller ground wire was incapable of returning more than a few signals at the same time. If, by pure chance, most of the sent bits at a particular instance happened to be ones, then the signal voltage would drop below the logic threshold, and at the receiving end they would be interpreted as zeros. The reduced-diameter cable would test perfect if every wire and return was actuated individually, but under certain, seemingly rare conditions of data structure, it could fail.[284]

We sent home an emergency request for the rat-bitten cable to be returned to us, re-installed it, and experienced no further data drop-outs. The electricians at the E. I. Hatch Nuclear Power Plant found our predicament amusing. The celebration party never happened, but the laughter we generated still rings in my ears.

The lesson of the Rat Cable Problem gave me a sense of sympathy for the engineers at General Electric as they watched their dreams of robust reactor systems, designed to work under the worst conditions

284 We were not the only ones bitten by the ground-return problem. The S-100 "Altair" data bus (IEEE696-1983), introduced in 1974, was an extremely popular design feature used in many micro-computer applications in the late seventies through the eighties, but it suffered from too few grounds.

imaginable, crumble to pieces at Fukushima I when the unimaginable happened. Trying to build something that will work perfectly for all time is a noble goal, but it is simply impossible. The issue that causes failure can be as grand as an earthquake and tsunami or as mundane as a brown rat.

Admiral Rickover's nuclear submarine power plant was such a design. It was radical in every detail, and a great number of innovations were necessary to make it work. It started with the idea of using water as both the coolant and the neutron moderator. That would work, but it would mean that the fuel had to be enriched with an unnatural concentration of uranium-235. Some material would have to be devised that would hold the fuel in place with water running among small, cylindrical rods of uranium, and it could not parasitically absorb neutrons. Every neutron was precious. It also had to be able to withstand high temperature and be strong enough to hold the reactor core together without dominating the space. Zirconium fit the list of requirements, but there were no zirconium mines, refineries, or fabrication techniques. Rickover had to invent it all from scratch. He came up with the idea of control rods made of hafnium, which was another material that was not available at the hardware store. His exotic machine was entirely successful, driving his submarines, catapulting the United States Navy further into world domination, and it did not harm a single sailor. Rickover's system test program was, without question, as rigorous and complete as could be accomplished.

The civilian nuclear-power industry, also starting from scratch in a very small, experimental step-off, was not the sort of enterprise that could develop new metals and radical designs of things that had never been built before. The utilities that were bold enough to try nuclear power were pleased when the naval reactor technology was declassified and turned into a stationary power plant at Shippingport, Pennsylvania, in 1957. It was a small plant, generating only 60 megawatts of electricity, but it never had a problem in 25 years of service. Seeing this as a good sign, the United States and eventually the world eventually stopped experimenting with different reactor concepts and settled on Rickover's submarine unit as a standard for how nuclear power should be applied to the need for reliable, clean power.

Was this a good idea, or did the world's utilities fall into a trap? The other radical idea for nuclear power, such as the liquid-metal-cooled fast breeder, had also been tried many times with consistently unfavorable results, just as Rickover had predicted long ago, when the Navy wanted to try it in submarines. The liquid-metal technology, as he pointed out, was expensive, prone to disasters, and extremely difficult to repair when something broke. Having a coolant that would catch fire when exposed to air did not seem right. Rickover was correct on that observation. Why would he be wrong about the pressurized-water reactor?

There have been trillions of problem-free watt-hours generated by scaled-up Rickover plants, but there may be a problem area that was not evident when submarine reactors were tiny, 12-megawatt machines, but that was revealed when the Rickover model was enlarged multiple times over for industrial use. The reactor core, the uranium fuel pellets lined up in zirconium tubes and neatly separated from each other, is terribly sensitive for such an otherwise robust machine. Let the coolant come off the fuel for a few minutes, even with the reactor shut completely down, and the entire, multi-billion-dollar machine is in irreversible jeopardy. The high-temperature zirconium alloys in an overheated reactor core oxidize, losing their metallic strength, generating explosive hydrogen gas, and contributing to high-pressure conditions in the isolated reactor vessel. The delicately structured core collapses, and the soluble fission products are able to mix with the escaping remnants of the coolant. It has happened as recently as 2011. To start the destruction sequence requires a lot of bad luck and human intervention. It is part of the nuclear power plant that does not easily forgive errors or dampen out mistakes. This part of the reactor design, the orderly matrix of thin tubes filled with fuel, is a weakness in an otherwise robust system.

There have been many engineering fixes and modifications to correct these problems, but ironically, these fixes can then present new issues, as they are complex add-ons, cluttering up an otherwise simple design with a maze of pipes and hundreds of additional valves, tanks, electrical cables, pumps, turbines, filters, re-combiners, and compressed-air tubing. Most of the plumbing in a nuclear plant has nothing to do with generating electricity. It is part of the fix that keeps the reactor core from melting down under unusual circumstances.

These complex light-water-reactor designs, the boiling-water reactor and the pressurized-water reactor, have been pursued with such enthusiasm over the past sixty years, one could assume that there is no other reasonable way to build a civilian power reactor. Alternate designs, such as the graphite, as was used in Windscale, and the liquid-metal-cooled reactor at Fermi 1, have proven impractical and have fallen away.[285] If only there were a proven reactor design that was in no danger of melting the fuel, collapsing the core structure, and generating hydrogen, it would solve many problems that bedevil the current crop of world-wide power reactors.[286]

The concept of this radical design, a reactor that has no metal core structure and no meltable fuel, dates back to World War II, when anything conceived and built at Los Alamos, New Mexico, was top secret and available only to those with a need to know. By the 1950s, the idea was further developed and incorporated into the Aircraft Nuclear Propulsion project, a program that was so far out on the technical limb, anything developed for it was easily dismissed when the project, again secret, folded up in 1961. In the 1960s the usable concepts developed in the defunct atomic airplane quest were built into an experimental power reactor in Oak Ridge. It was the answer to all problems that had to be addressed in the light-water-reactor designs, including the availability of uranium and the expense of enriching it, but it was too late. Westinghouse, General Electric, Babcock and Wilcox, Combustion Engineering, and a host of copying manufacturers in Europe had already thrown their resources into light-water-reactor development. We had fallen too far down into the Rickover trap to escape,

285 Another alternate design, the Canadian CANDU reactor, continues on. Although its coolant/moderator, heavy water, is extremely expensive, the concept of a power reactor that runs on natural uranium with constant refueling has been attractive to countries such as India and China. A byproduct of CANDU power generation is bomb-grade plutonium-239.

286 The nuclear-power industry is acutely aware of these weak points and has not been sitting idle. The newer Generation 3 reactors, such as the Westinghouse AP1000, address the issues pointed out here with several innovations. The emphasis is on passive systems, not requiring electricity, to keep water covering the fuel and preventing core damage. Two of these reactors are currently being built at the Vogel Nuclear Power Plant near Augusta, Georgia. The nuclear division of Westinghouse is now a Japanese-owned company.

and in 1969 the Oak Ridge experimental power reactor was quietly shut down and dismantled. Ten years later, a brand-new B&W power plant at Three Mile Island, Pennsylvania, was a total loss and a clean-up liability because its reactor core had overheated. The bottom-line expense of nuclear power was rightfully called into question.

Reactor designs with liquefied core configurations were started at Los Alamos in 1943. The first was LOPO (LOw POwer), consisting of a stainless steel sphere filled with 14% enriched uranyl sulfate dissolved in water and surrounded by a reflector made of beryllium. The fuel consisted of the world's entire stock of enriched uranium at the time. The purpose of the reactor was to measure characteristics of uranium fission, but it was also the first reactor using a single fluid for fuel, neutron moderator, and coolant. It became known as Water Boiler, not because the mixture got hot enough to boil, but because the water broke down into hydrogen and oxygen under the heavy gamma-ray bombardment during fission, and gas would bubble to the surface. It was thus the first observation of radiolysis. Subsequently, any reactor using water as a moderator was equipped with a recombiner, a catalyst screen that encouraged the hydrogen and oxygen to re-form into water.

LOPO begat HYPO, and HYPO begat SUPO, research reactors with increasing mechanical sophistication and power. By 1953, the scientists and engineers working at the Los Alamos Lab began to think about a civilian power source based on nuclear fission. In the early fifties, there was barely enough known reserve of uranium to make bombs for the United States, much less to make electrical power for the world for centuries, but there would be enough plutonium produced artificially to at least power the Western Hemisphere far into the future. Almost 100% of the world's stockpile of plutonium happened to be in Los Alamos, New Mexico, and there was no reason not to use some of it to build a prototype power reactor.

The first plunge into the Liquid Metal Fuel Reactors (LMFR) was LAMPRE, the Los Alamos Molten Plutonium Research Reactor.[287]

287 A bridge between LOPO and LAMPRE was the Los Alamos Power Reactor Experiment (LAPRE) in 1955. Its liquid fuel was uranium oxide dissolved in phosphoric acid. The fuel was so corrosive, the entire inside of the primary loop, including the reactor vessel, had to be pure gold. The experimental budget of the Los Alamos Scientific Laboratory in the 1950s could cause heart palpitations.

It was the first reactor ever built that used molten metal, a eutectic alloy of plutonium and iron, as the fuel. There was no fear of the core melting down, because the core *was* melted. The reactor would run at 1,200° Fahrenheit, a temperature that was impossible for any reactor with solid, structured fuel, but it would make very efficient steam for running a turbo-generator. The plutonium-239 fissioned efficiently using fast neutrons, so there was no need for a moderator, and there were plans to add a uranium-238 breeding blanket to the reactor so that it would produce extra plutonium as well as power. Any problem of having the reactor melt was taken care of by making it out of tantalum-tungsten alloy, which would melt somewhere above 6,000° Fahrenheit. It would be possible to slowly draw off the molten plutonium fuel through a pipe at the bottom of the reactor core, filter out fission products, add plutonium to replace that which had fissioned, and pump the fuel back into the reactor.[288]

By the time LAMPRE first achieved hot criticality on March 27, 1961, the concern for uranium scarcity was over, and Rickover's PWR had become the darling of the civilian nuclear-power industry. LAMPRE-2 was planned for, but funding for exotic reactor projects was tight, and the ambitious follow-on project was dropped. Rickover was pleased. He despised the frontier experiments at Los Alamos, Oak Ridge, and Idaho as a silly and impractical waste of federal money.

The next big step was the Direct Contact Reactor (DCR) at Oak Ridge, where the Aircraft Reactor Experiment (ARE) was underway. The ARE was a hyper-exotic setup, meant to become a jet engine in a strategic bomber, using a molten fuel made of uranium fluoride, sodium fluoride, and zirconium fluoride, moderated with beryllium oxide, with liquid sodium as a coolant. The metal structure was made

---

288 That was the plan, but there was a lot to learn about molten fuel, and in reality LAMPRE-1 started out smaller and less complicated than was laid out in the proposal. The power was dialed back from 60 megawatts to 1 megawatt, and the core structure was reduced to tantalum tubes containing the molten fuel. Unforeseen problems turned up, as are expected in a leading-edge experimental program, but much was learned in two fuel loadings and three years of experience. The project was terminated in 1964, before the third fuel loading could be tested, and LAMPRE-2 was changed to the Fast Reactor Core Test Facility (FRCTF). The FRCTF project was abandoned, 70% completed, as was the Molten Plutonium Burnup Experiment (MPBE). The light-water reactors from Westinghouse and GE were succeeding beyond the Atomic Energy Commission's wildest dreams, and everything else fell by the wayside.

of Inconel 600 alloy, and the thing ran for 1,000 hours at a temperature of 1,580° Fahrenheit. In the history of the art, nuclear reactors did not get much fancier than the ARE.

The goal of the DCR was a high-efficiency, fast-neutron power reactor using molten plutonium fuel. The fuel was to be a plutonium-cerium-cobalt alloy, and in this innovation the fuel would also be the primary coolant, pumped around in a loop. Using the principle of "critical shape" that was employed in all the fuel-processing plants, the plutonium would be in a critical configuration, generating power by nuclear fission, only when it happened to be in the reactor core, which was a sphere. In the spherical reactor tank, the surface-area-to-volume ratio was at a minimum, and a large percentage of fission neutrons were able to propagate fission. Going around in the coolant pipes, the plutonium was thinned out, and most of any neutrons were lost out the walls of the pipes. There was no need for a moderator, because plutonium-239 was the fuel, and there was no core structure. This was a major idea.

Instead of there being a heat exchanger to pass off the high temperature created by fission, the fuel mixture was mixed with liquid sodium in the loop outside the core by a jet pump. The heat transfer from the direct contact of the hot fuel and the coolant was 10 to 100 times more efficient than using a metallic heat exchanger. The hot sodium then transferred the heat to water in an external steam generator, so that a standard turbo-generator could be used to make electricity.

A gravity-drop separator then took apart the sodium/plutonium mixture, putting the sodium back into the continuous jet pump loop and the cooled plutonium, still melted, back into the reactor. Another great feature of this plan was that the fission products would stick to the sodium and be scrubbed out of the fuel loop. A blanket of depleted uranium around the core would breed new plutonium fuel. With radioactive products continuously processed out of the fuel, there would be no buildup of radioactivity to escape, and there was no chance of the molten metal being able to blow off like steam and spread anything over the environment. The primary loop could, in fact, run at atmospheric pressure. Each DCR in a power plant would be small, only four feet wide, and generating about 227 megawatts of heat. A power plant would consist of 20 reactors, each encased in a 60-foot-long tube, sunken into the ground. It was almost too good to be true. The Atomic Energy Commission applied for the patent in 1961.

By August 1960, a mockup of the DCR had been built at the Los Alamos Science Laboratory and the new subsystems for the reactor were being tested. In May of 1962, a functioning DCR, code-named the Pint Bottle Experiment (PBX), was ready to be built. The budget for experimental reactors, however, was collapsing as the research projects were growing numerous. The nuclear-reactor market was expanding all by itself, without any DCRs, PBXs, or LAMPREs to help it along. The need for anything better than a Westinghouse PWR or a General Electric BWR was vanishing. Not only was there enough uranium to run the world into the far future, but the cost of it was dropping fast. PBX died on the drawing board. A new report predicted that by the year 2000, half of the electricity used in the United States would be generated by nuclear fission.

While the Los Alamos lab was suffering from a mismatch of funds-to-ambitions, the Oak Ridge National Laboratory was working on a very interesting mutation of the aircraft reactor. The nuclear bomber had been euthanized as soon as John F. Kennedy was sworn in as president, but lessons learned while operating the ARE led to a new, radical design for a civilian power reactor, the Molten Salt Reactor Experiment (MSRE). It was possibly the most important advance in nuclear-reactor design in the 20th century.

At the end of the nuclear airplane era, the ARE was dismantled, and Oak Ridge secured funding to construct the MSRE in the same building. One fluid, a mixture of salts composed of fluorine compounds of fuel plus neutron moderator, was pumped into the round reactor chamber, where the fuel would fission, and out into a salt-to-sodium heat exchanger. The cooled but molten salt would then re-introduce into the reactor in a loop configuration.[289] Operating temperature was 1,300° Fahrenheit. Steam was then made in a liquid-sodium-to-water heat exchanger. Neither the primary fuel loop nor

289 In the Oak Ridge MSRE, criticality was established in the reactor vessel by sending the fuel through a perforated moderating core made of pyrolytic graphite. A cleaner design would make use of the good moderating qualities of the lithium in the salt and do away with any core structure, but this was the first attempt to build a molten salt reactor, and all advantages could not be accomplished in the initial experiment.

the secondary sodium loop inflicted any pressure on the reactor structure.[290]

The innovation of this reactor was the fuel. Instead of using uranium or plutonium, it used thorium-232. It turns out that thorium is four or five times more common than uranium, and there is a large reserve concentration of it in the United States. Mined uranium is over 99% uranium-238, and less than 1% is the usable isotope, uranium-235. All of the thorium available is thorium-232. There is no unusable thorium in nature. No isotope separation or enrichment is necessary.

Thorium-232 is not a reactor fuel. It will not fission, but upon capture of a neutron, it develops into uranium-233, which is as fissile as uranium-235. The conversion of thorium into uranium can take place in the reactor core, using surplus neutrons produced in the fission process.

When uranium-233 fissions, it produces 10 times less radioactive fission product than does uranium-235. Moreover, the cumulative fission product that it does produce has a half-life 100 times shorter than that produced by uranium-235. The danger of the radioactive waste is gone after 300 years, whereas the waste from uranium-235 fission remains dangerous for 30,000 years.

There was no fuel cladding, no zirconium egg-crates holding the fuel in a rigid matrix, no steam in the reactor vessel to float away with fission products into the atmosphere, and, of course, there was no danger of the fuel melting. In the power-plant embodiment of the MSRE, the fuel would be continuously reprocessed by running the primary loop through a chemical scrubber to extract fission products and inject fresh thorium. An atomic bomb cannot be made from thorium-232, because it does not fission, and it becomes uranium-233 only in the reactor core under neutron bombardment.[291]

---

290 The plumbing and reactor vessel did require a special, high-performance nickel alloy to withstand the molten salt. Haynes International developed Hastelloy-N for the MSRE project, and it has found use in other nuclear power applications worldwide.

291 The problem with using U-233 in a bomb is that the bridge nuclide between Th-232 and U-233 is protactinium-233. The Pa-233 is very active, with a half-life of 27 days, and it beta-minus decays into U-233. Unfortunately, its decay also involves a 317 kev gamma ray, and while the half-life is days, all of the protactinium never really goes away. The energetic gamma activity makes it very dangerous to work with, and a lot of shielding is necessary. Inside the reactor primary loop, there is nothing to worry about. The fuel is processed by machinery, with no human interaction; but to make a bomb, a lot of fabrication and machining is necessary. Plutonium, for all its faults, is easier to work with than U-233.

The advantages of this reactor design seem overwhelming, particularly after the turn of the century, when we have seen an entire power plant go down because of melted core structures and broken steam systems spreading fission products. These weaknesses that destroyed light-water reactors would not exist in the molten-salt reactor.

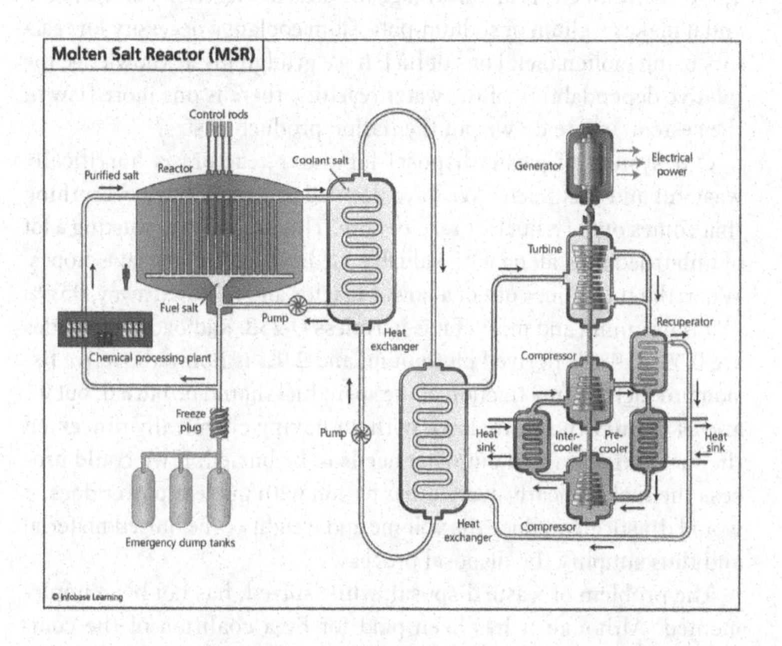

The Molten Salt Reactor is an old, forgotten design from the Age Of Wild Experimentation, when radical ideas of fission power were explored with physical embodiments. Given the problems we have experienced with the standard water-cooled reactors in the past 35 years, the MSR begins to look better and better. With the fuel dissolved in the primary coolant, there is no worry of melting.

The MSRE ran for four years. The program was shut down in 1970, and by 1976 any trace of the reactor was gone. There was no need for an improved way of making power by fission, and all the eggs by now were in one basket. An entire industry, from fuel-pellet manufacture to steam-generator fabrication, had been built around the water-cooled reactor designs used in Rickover's nuclear navy, and there would be no turning back. We effectively had to dance with the reactor we came with.

It could have been worse. The water reactors were well designed, and they have given us 40 years of reliable electrical power in the United States. A good thing that you can say about water is that it is not sodium, which is an isolated drawback to any molten-fuel reactor concept. The high operating temperature of sodium, which is over 1,000° Fahrenheit, is an advantage for heat-to-electricity conversion, and it makes sodium or sodium-potassium coolant a necessity for reactors using molten metal or salt fuel. If we grudgingly acknowledge the relative dependability of the water reactors, there is one more flaw in the system: where do we put the fission product wastes?

Our system of waste disposal for water reactors is horrifically wasteful and inefficient. We have decided to simply bury everything that comes out of a nuclear reactor core. This amounts to wasting a lot of unburned fuel, along with valuable medical and industrial isotopes. When the fuel comes out of a power reactor and is stored away, 95.6% of it is uranium, and most of it is harmless U-238. Radioactive nuclides are 0.5%, 0.9% is derived plutonium, and 2.9% is non-radioactive fission products. A tiny fraction of the spent fuel should be buried, but we are set to bury the entire load, without having chemically processed the fuel and separated out what needs to be buried. If we could process the fuel, as nearly every other nation with nuclear power does, it would drastically reduce the volume and weight of the buried material and thus simplify the disposal process.

The problem of waste disposal, while solved, has not been implemented. Although it has been paid for by a coalition of the commercial nuclear power utilities in the United States, the spent fuel repository built under Yucca Mountain in Nevada is currently having trouble accepting fuel deliveries. The state of Nevada has changed its welcoming position to the facility after spending $12 billion of the power companies' money to study the site and dig the tunnels. Although a federal law designating the Yucca Mountain facility as the nation's nuclear-waste repository is still in effect, usage of the facility is being blocked. As this controversy continues, nuclear waste builds up in dry storage casks at every light-water reactor in the United States.

The option of processing the waste down into extremely small parcels and easing the burden of burying it was once a goal in the

United States. Learning from all previous attempts to process spent reactor fuel as a commercial venture, the Allied Corporation, the Gulf Oil Company, and Royal Dutch Shell combined resources and began construction of a sleek, very sophisticated chemical plant in South Carolina, named the Barnwell Nuclear Fuels Plant. It was fully automated using computer controls and gravity-driven processes through seamless stainless steel pipes and tanks. In 1977 it was nearly completed, and with the permission of the Nuclear Regulatory Commission, it ran a test load of spent fuel through the plant. It performed perfectly, and a license for its operation was assumed to be on the way.

On April 7, 1977, United States President James Earl Carter announced at a special press conference, "We will defer indefinitely the commercial reprocessing and recycling of plutonium produced in the U.S. nuclear power programs. The plant at Barnwell, South Carolina, will receive neither federal encouragement nor funding for its completion as a reprocessing facility."

This announcement hit the nuclear industry like a two-by-four to the back of the head. There was no way not to process plutonium out of nuclear fuel. It is a byproduct of all commercial power generation using uranium as fuel. The operating license for the plant was denied, on presidential order.[292] It turned out that the reason for this strange action, stopping an industrial plant from operating after a quarter of a billion dollars had been spent building it, was President Carter's fear of nuclear-weapons proliferation via easy access to plutonium. He was not necessarily afraid that the United States would build weapons using plutonium, as we were running two large, federally owned plants to turn out bomb-grade plutonium by the ton. He was afraid of smaller countries using their own fuel reprocessing for this purpose, and he wanted to symbolically show them that if we did not do it, then they should follow our example and not touch their nuclear waste.

The rest of the world went on about their business, and there is no record of anyone giving a thought to the President's symbolic gesture.

---

292 Not exactly a presidential order. Barnwell and the Clinch River Breeder Reactor were shut down by presidential veto of S. 1811, the ERDA Authorization Act of 1978, preventing the legislative authorization necessary for constructing a breeder reactor and a reprocessing facility.

The Barnwell plant suffered drastically from the lack of an operating license, and although the next president, Ronald Reagan, lifted the ban on commercial fuel reprocessing, the concept of the federal government being able to shut it down at its discretion discouraged any further idea of operating it as a business, and its major customer, the Clinch River Breeder Reactor, had also been cancelled by President Carter. If there could be no civilian fuel reprocessing, then there was no sense in building a civilian breeder reactor. A fast breeder reactor runs on plutonium, and it must be extracted from the breeding blanket. The Barnwell plant was decommissioned in 2000. It had been a fine, constructive idea, a large investment, and 300 workers' jobs, all gone down the drain.

The premise of the Carter administration's decision to shut down civilian fuel reprocessing was incorrect, and it shows that a little knowledge is dangerous. Yes, spent uranium fuel from a commercial power reactor does contain plutonium, but it is not "bomb-grade" plutonium. In a plutonium-production reactor, specifically built to make plutonium for nuclear weapons, the fuel is natural uranium, containing very little fissile uranium-235. It burns up quickly, in weeks of running at full power, and it is changed for fresh fuel on a regular basis. The plutonium is separated chemically from the spent fuel. The plutonium is primarily the nuclide Pu-239, with a small amount of Pu-240. Pu-240 is made when a neutron is captured by Pu-239, presumably after it has been made by neutron capture by U-238. U-238 becomes U-239, and it does two quick beta-minus decays into Pu-239, using neptunium-239 as the bridge nuclide.

It is important to minimize the Pu-240 contamination, which is why the fuel is not allowed to linger in the neutron-rich environment of an operating reactor core. Pu-240 fissions spontaneously, not waiting for a neutron trigger, and in a bomb it would cause the device to engage in runaway fission and melt before it ever had a chance to explode. The presence of a Pu-240 contaminant, which cannot be separated from the desired Pu-239, is what forced the original design of the complex, difficult implosion method of setting off an atomic bomb in World War II. Minimizing the Pu-240 content is the reason for quick turnaround in the refueling of a plutonium production reactor.

A civilian power reactor does not turn around fuel quickly. It is usually changed out on a three-year schedule. Refueling requires the reactor to be taken offline, cooled down, and dismantled. You are not making electricity and the money from selling it when the reactor is down, so the time between refuelings is drawn out as long as is practical. Because it stays in the neutron environment for so long, the plutonium-239 is heavily invested with plutonium-240.[293] Its only application is for reactor fuel.[294] Nobody has ever built a nuclear explosive device using plutonium taken from a civilian power reactor.[295]

So, the current status of commercial nuclear power in the United States sums up bleakly as this:

1. America's 100 operating nuclear power reactors are bloated examples of Rickover's celebrated submarine power plants, increased in size to the point where the core structures are the weak point.

2. There is not a single reactor fuel reprocessing plant in the United States, making us unique in the nuclear-powered

---

293 "Bomb-grade" plutonium is defined as 92% Pu-239. Commercial reactor-derived plutonium is 60% Pu-239. The United States, just to show off, once tested a plutonium-based bomb made of 85% Pu-239 having an explosive yield of less than 20 kilotons.

294 Uranium fuel mixed with recovered plutonium is called "MOX." MOX fuel was being used in reactor units 3, 5, and 6 at Fukushima I when it was hit by the Tohoku earthquake and tsunami, and this caused a great concern of several milligrams of plutonium dust possibly escaping into the atmosphere. (With 550 plutonium-based bombs having been tested above ground or underwater, there were probably already several tons of plutonium in the atmosphere.) MOX is commonly used in European power reactors.

295 The President's heart was in the right place, but his nuclear waste isotope was wrong. What he should have been worried about was not plutonium, but neptunium. Neptunium-237 shows up in spent fuel, and it can be extracted using published chemical processes. No tedious isotope extraction is necessary, because it is all Np-137, which is just as fissile as Pu-239, U-235, or U-233. Np-137 makes a very low neutron background, so it can be used in a simple assembly bomb, just like U-235. There is evidence of an Np-137 atomic bomb test in the U.S., but it is, of course, classified SECRET. The neptunium bomb is something to worry about.

world, causing our reactor waste to be mostly inert filler, and discarding unused fuel.

3. That does not matter, because we presently have no place to bury the waste, even if to do so would be grotesquely inefficient. The waste is being stored in dry casks on the property of every nuclear power plant in the United States, waiting to be hauled away.

But all is not lost, and nuclear engineers and scientists, or what is left of them, are not sitting idle. There are currently in design or test at least five new power reactors, and they are all small, modular units, as was carefully planned for the Direct Connection Reactor back in 1960. It was a brilliant concept, to install from two to twenty tiny reactors at a power plant instead of four huge ones. Small reactors have small problems, small explosions, small coolant drips, and small investments. An entire reactor can be built in a production-line factory, loaded onto a truck, and taken to a pre-made hole in the ground. The difference in efficiency of building a small reactor out of standardized parts in a factory instead of welding together a unique mountain of plumbing in the field is mind-boggling.

This important concept, of minimally sized simple power units had been exploited by the U.S. Army in its Engineer Reactors Group beginning in 1954.[296] Its reactors provided reliable power for Army installations from the Panama Canal Zone to McMurdo Station in Antarctica. After a stunning list of accomplishments, including a nuclear power plant that could fit on the back of a truck, the program was laid to rest in 1977, due to budget cuts.

The companies that are planning to make modular reactors available in the competitive market are all private companies, and not any government or military organization. Three of them are in the United States. The NuScale Power Company in Corvallis, Oregon, is working on

296 Right in the middle of the development program, the Army's SL-1 reactor exploded in Idaho. The Army took this accident to mean that their reactor was *too* simple, and they dialed back the requirement of using as few moving parts as possible.

a 45-megawatt power reactor enclosed in a steel tube. Gen4 Energy, Inc. in Santa Fe, New Mexico, has designed a 25-megawatt modular reactor. Generation mPower LLC in Bedford County, Virginia, is planning to put six very small reactors in the ground in Tennessee where the Clinch River Breeder Reactor was supposed to have been.

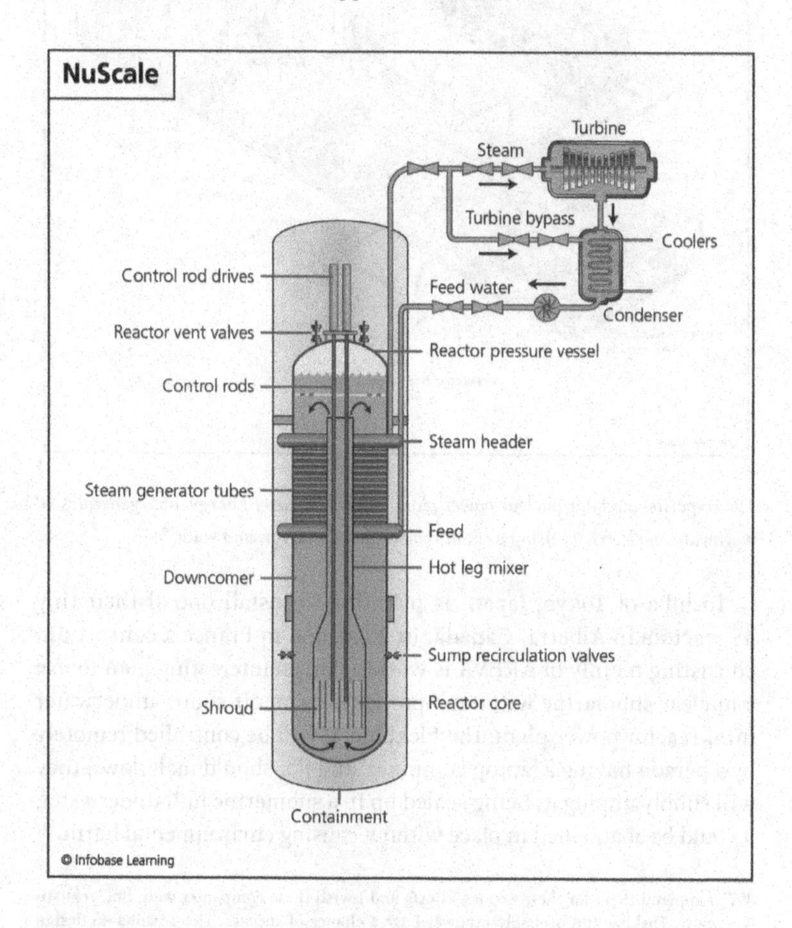

The NuScale power plant is simplified down to the point where the steam generator is built into the reactor vessel, and the power level of a single unit is such that it cannot build up enough delayed fission to melt the mechanism. The modular reactor may be an idea whose time has come.

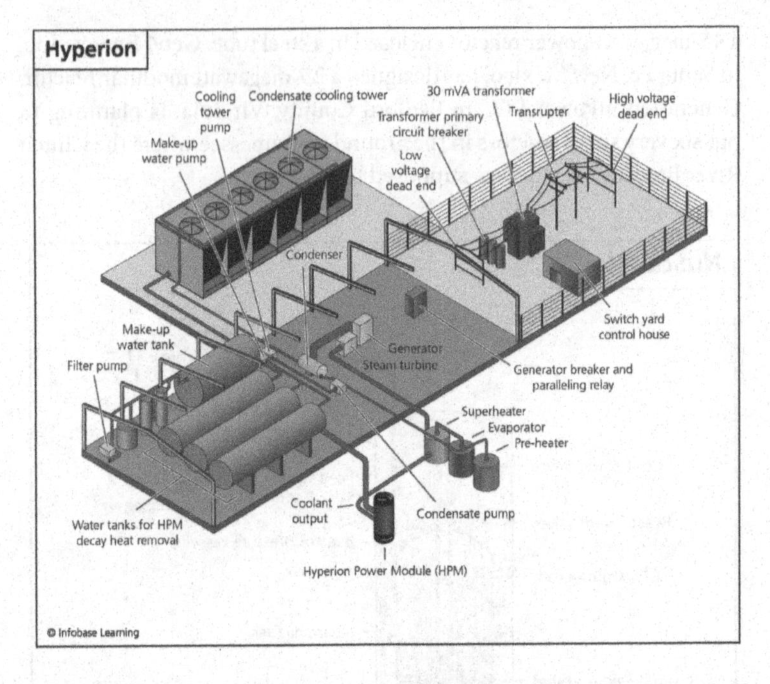

The Hyperion modular nuclear power plant, offered by Gen4 Energy, Inc., generates 70 megawatts of electricity using a small reactor in an underground vault.

Toshiba of Tokyo, Japan, is planning to install one of their tiny 4S reactors in Alberta, Canada, in 2020, and in France a consortium consisting mainly of AREVA is working on an interesting plan to use a nuclear submarine without a propeller as an off-shore, underwater mini-reactor power plant, the Flexblue. It will be controlled remotely by a person having a laptop computer, and if it should melt down they will simply unplug it. Being sealed up in a submarine hull underwater, it could be abandoned in place without causing environmental harm.[297]

297 I applaud them for their various efforts and I wish these companies well, but, realistically, Toshiba and probably mPower have a chance of success. The Toshiba 4S design raised some licensing concern at the USNRC. It uses one central control rod, the part of the oversimplified SL-1 reactor that caused its destruction in a fatal steam explosion. We had pledged to never do that again. Toshiba will prove that its 4S reactor will not suffer from the single-control problem, but the other small companies are somewhat underfunded, and all are competing for what may be an initially limited market.

An even better development is the Generation IV International Forum, a coalition of nine countries, including the United States, which was brought together by the Department of Energy in January 2000. The purpose of this group of scientists and engineers is to identify realistic targets for research and development of a new generation of nuclear power plants, using all that we learned in 50 years of experimentation and experience. The goal is to develop and build these new power plants by the year 2030, without discouraging the building of Generation III reactors, such as the Westinghouse AP1000s now under construction in Georgia.

The list of exotic reactors being studied by the Forum includes sodium-cooled, gas-cooled, and lead-cooled fast reactors, supercritical water reactors, and very high temperature reactors, but right at the top of the pile is the molten-salt thorium-fueled reactor. The old, nearly forgotten concept of constantly melted fuel may find a new, productive life in the 21st century. It and the other revived reactor designs could help save us from packing more carbon dioxide into the atmosphere than nature can handle. The dangers of atomic bomb fabrication, flying nuclear weapons around in airplanes, Soviet engineering and bureaucracy, and ingesting radium will be in history books, along with the curious recreation of crashing train locomotives into each other.

As the nuclear engineering community lifts its graying head and looks to the future, remember one thing. If the person sitting next to you seems concerned with the radioactive fish from Japan, the air over the Tokyo Olympics heavy with fallout, or the contaminated junk that washes ashore in Oregon, then caution him or her not to eat a banana. It is crawling with potassium-40, a naturally occurring radioactive nuclide that spits out an impressive 1.46 MeV gamma ray. Neither radiation dose, from eating a banana or a bluefin tuna contaminated with cesium-137 (0.662 MeV gamma), is considered to be the slightest bit dangerous. In fact, tuna fish have been contaminated with radioactive cesium for the past 60 years or so, ever since the oceanic nuclear weapons tests from long ago, and it is used as a radioactive tag to trace migratory routes. The destruction of the Fukushima I nuclear plant may have added to the countdown period when all the detectable cesium-137 will have decayed away, but the danger remains indetectably slight.

The real danger is that any engineering discipline can fall into its own Rickover Trap. We do not, for example, necessarily burn gasoline at the rate of 134 billion gallons per year in the United States because it is the best way to power an automobile: we do so because we have been doing it a long time, and the infrastructure is in place. As is the case of pressurized water reactors, it has worked well for us for a long time, but there could be a better way to do it.

The dangers of continuing to expand nuclear power will always be there, and there could be another unexpected reactor meltdown tomorrow, but the spectacular events that make a compelling narrative may be behind us by now. We have learned from each incident. As long as nuclear engineering can strive for new innovations and learn from its history of accidents and mistakes, the benefits that nuclear power can yield for our economy, society, and yes, environment, will come.

BIBLIOGRAPHY

Author's Note

1. Cruz, Euler and Cesário, Rafael. *Accident at Russia's Biggest Hydroelectric: Sayano-Shushenskaya—2009 August 17.* Brazil: Cruz and Cesário, 2009.
2. Sayano-Shushenskaya. http://www.youtube.com/watch?v=yfZoq68x7lY

Introduction: Bill Crush and the Hazards of Steam Under Pressure

3. Reisdorff, James J. *The Man Who Wrecked 146 Locomotives: The Story of "Head-On Joe" Connolly.* David City, NE: South Platte Press, 2009.

Chapter 1: We Discover Fire

4. Harve, David I. *Deadly Sunshine: The History and Fatal Legacy of Radium.* Gloucestershire, UK: Tempus Publishing Limited, 2005.
5. Hill, Colin K. *The Low-Dose Phenomenon: How Bystander Effects, Genomic Instability, and Adaptive Responses Could Transform Cancer-Risk Models.* Thousand Oaks, CA: SAGE Publications, 2012.
6. Kaperson, Roger E. *The Social Amplification of Risk and Low-Level Radiation.* Thousand Oaks, CA: SAGE Publications, 2012.
7. Mahaffey, J. A. *The History of Nuclear Power.* New York: Facts On File, 2011.
8. Mullner, Ross. *Deadly Glow: The Radium Dial Worker Tragedy.* Washington, DC: American Public Health Association, 1999.
9. Schoffelmayer, V. H. *Radium Mine in the Ozarks*, Chicago, IL: Technical World, Vol. 18, 1913.
10. Slovic, Paul. *The Perception Gap: Radiation and Risk.* Thousand Oaks, CA: SAGE Publications, 2012.

Chapter 2: World War II, and Danger Beyond Comprehension

11. Bernstein, Jeremy. *Plutonium: A History of the World's Most Dangerous Element.* Ithaca, NY: Cornell University Press, 2007.
12. Coster-Mullen, John. *Atom Bombs: The Top Secret Inside Story of Little Boy and Fat Man.* 2008.
13. Glasstone, Samuel. *The Effects of Nuclear Weapons.* Washington, DC: United States Atomic Energy Commission, 1962.
14. Goudsmit, Samuel A. *Alsos.* Woodbury, NY: AIP Press, 1996.
15. Groves, General Leslie M. *Now it Can Be Told: The Story of the Manhattan Project.* New York: Harper, 1962.
16. Kelly, Cynthia G. *The Manhattan Project: The Birth of the Atomic Bomb in the Words of its Creators, Eyewitnesses, and Historians.* New York: Black Dog and Leventhal Publishers, Inc., 2007.
17. Morgan, Karl Z. and Peterson, Ken M. *The Angry Genie: One Man's Walk through the Nuclear Age.* Norman, OK: University of Oklahoma Press, 1998.
18. Powers, Thomas. *Heisenberg's War: The Secret History of the German Bomb.* Cambridge, MA: Da Capo Press, 1993.
19. Richardson, David. *Lessons from Hiroshima and Nagasaki: The Most Exposed and Most Vulnerable.* Thousand Oaks, CA: SAGE Publications, 2012.
20. Robinson, George O. Jr. *The Oak Ridge Story: The Saga of a People Who Share in History.* Kingsport, TN: Southern Publishers, Inc., 1950.
21. Russ, Harlow W. *Project Alberta: The Preparation of Atomic bombs for Use in World War II.* Los Alamos, NM: Exceptional Books, 1984.
22. Serber, Robert. *The Los Alamos Primer: The First Lectures on How to Build an Atomic Bomb.* Berkeley, CA: University of California Press, 1992.
23. Sparks, Ralph C. *Twilight Time: A Soldier's Role in the Manhattan Project at Los Alamos.* Los Alamos, NM: Los Alamos Historical Society, 2000.
24. Zoellner, Tom. *Uranium: War, Energy, and the Rock That Shaped the World.* New York: Penguin Group, 2009.

Chapter 3: A Bit of Trouble in the Great White North

25. Allen, Thomas B. and Palmar, Norman. *Rickover: Father of the Nuclear Navy.* Washington, DC: Potomac Books, Inc., 2007.
26. Carter, Jimmy. *Why Not the Best? The First Fifty Years.* Fayetteville, AK: University of Arkansas Press, 1996.
27. Godbold, E. Stanly, Jr. *Jimmy & Rosalynn Carter: The Georgia Years, 1924-1974.* New York: Oxford University Press, 2010.
28. McFarlane, Harold, et al. *Controlled Nuclear Chain Reaction: The First 50 Years.* La Grange Park, IL: American Nuclear Society, 1992.
29. Rockwell, Theodore. *The Rickover Effect: How One Man Made a Difference.* Lincoln, NE: iUniverse, Inc., 2002.

Chapter 4: Birthing Pains in Idaho

30. Cordes, G. L.; Bunch, D. F.; Fielding, J. R.; and Warkentin. *Radiological Aspects of the SNAPTRAN-2 Destructive Test.* Idaho Falls, ID: NRTS, 1967.
31. Marcus, Gail H. *Nuclear Firsts: Milestones on the Road to Nuclear Power Development.* La Grange Park, IL: American Nuclear Society, 2010.

32. McKeown, William. *Idaho Falls: The Untold Story of America's First Nuclear Accident.* Toronto, Canada: ECW Press, 2003.
33. Penner, S. S. *Advanced Propulsion Techniques.* New York: Pergamon Press, 1961.
34. Stacy, Susan M. *Proving the Principle.* Idaho Falls, ID: Idaho Operations Office of the Department of Energy, 2000.
35. Tucker, Todd. *Atomic America: How a Deadly Explosion and a Feared Admiral Changed the Course of Nuclear History.* New York: Free Press, 2009.
36. BORAX-1a. http://www.youtube.com/watch?v=qIl97ByeU_M
37. BORAX-1b. http://www.youtube.com/watch?v=yUhVGH-WHKk&feature=relmfu
38. EBR-I. http://www.youtube.com/watch?v=KncekYvqyWs&feature=related
39. nuclear bomber part 1. http://www.youtube.com/watch?v=S86zRMehs-A&feature=relmfu
40. nuclear bomber part 2. http://www.youtube.com/watch?v=m614VDfJ9oU&feature=relmfu
41. nuclear bomber part 3. http://www.youtube.com/watch?v=rsCw0s0BJKY&feature=related
42. nuclear bomber part 4. http://www.youtube.com/watch?v=SZ-kkogj9Ro&feature=relmfu
43. nuclear bomber part 5. http://www.youtube.com/watch?v=_cnFtYuqXX8&feature=relmfu
44. SL-1 accident. http://www.youtube.com/watch?v=Q0zT9ARfsT4&feature=related
45. SL-1 part 1. http://www.youtube.com/watch?v=gMNqPUT-yP0&feature=related
46. SL-1 part 2. http://www.youtube.com/watch?v=PCfnurdwPwc&feature=relmfu
47. SL-1 part 3. http://www.youtube.com/watch?v=XfZu_s8Cl74&feature=relmfu
48. SL-1 part 4. http://www.youtube.com/watch?v=KLWVGEFhxgc&feature=relmfu
49. SL-1 part 5. http://www.youtube.com/watch?v=1NZfkmPNiuk&feature=relmfu
50. SL-1 part 6. http://www.youtube.com/watch?v=ajQV7bx8pWk&feature=relmfu
51. SL-1 part 7. http://www.youtube.com/watch?v=_pAJAIXMeCI&feature=relmfu
52. SPERT 1. http://www.youtube.com/watch?v=0FIhafVX_6I&feature=related
53. SPERT 2. http://www.youtube.com/watch?v=aqg7FEirbK4&feature=related

Chapter 5: Making Everything Else Seem Insignificant in the UK

54. Arnold, Lorna. *Windscale 1957: Anatomy of a Nuclear Accident.* New York: Palgrave Macmillan, 2007.
55. Goodchild, Peter. *Edward Teller: The Real Dr. Strangelove.* Cambridge, MA: Harvard University Press, 2004.
56. Jensen, S. E. and Nonbol, E. *Description of the Magnox Type of Gas Cooled Reactor (MAGNOX).* Roskilde, Denmark: Nordic Nuclear Safety Research, NKS-2, 1999.
57. Windscale. http://www.youtube.com/watch?v=ElotW9oKv1s&feature=relmfu

Chapter 6: In Nuclear Research, Even the Goof-ups are Fascinating

58. Ashley, R. L., et al. *SRE Fuel Element Damage: An Interim Report.* Canoga Park, CA: Atomics International, NAA-SR-4488, 1959.
59. Ashley, R. L., et al. *SRE Fuel Element Damage: Final Report.* Canoga Park, CA: Atomics International, NAA-SR-4488, 1961.
60. Babcock, W. *State of Technology Study—Pumps: Experience with High Temperature Sodium Pumps in Nuclear Reactor Service and their Application to FFTF.* Richland, WA: Battelle Memorial Institute, BNWL-1049, 1969.
61. Calahan, Jim. *Sodium Fast Reactor: Safety #2.* Chicago, IL: Argonne National Laboratory, 2008.
62. Cochran, Thomas B., et al. *Fast Breeder Reactor Programs: History and Status.* Princeton, NJ: International Panel on Fissile Materials, 2010.
63. Fillmore, F. L. *Analysis of SRE Power Excursion of July 13, 1959.* Canoga Park, CA: Atomics International, NAA-SR-5898, 1959.
64. Fuller, John G. *We Almost Lost Detroit.* New York: Ballantine Books, 1975.
65. Maffei, H. P.; Funk, C. W.; and Ballif, J. L. *Sodium Removal Disassembly and Examination of the Fermi Secondary Sodium Pump.* Amford, WA: Hanford Engineering Development Laboratory, HEDL-TC-133, 1974.
66. Santa Susana 1. http://www.youtube.com/watch?v=jAHmaEs5cYU&feature =related
67. Santa Susana 2. http://www.youtube.com/watch?v=jMadIhVBRpU

Chapter 7: The Atomic Man and Lessons in Fuel Processing

68. blanked out. *Compliance Investigation Report, United Nuclear Corporation Scrap Recovery Facility, Wood River Junction, Rhode Island, License No. SNM-777: Type "A" Case—Criticality Incident.* Santa Fe, NM: Division of Compliance Region I, 1964.
69. NA. *Lessons Learned from the JCO Nuclear Criticality Accident in Japan in 1999.* Vienna, Austria: IAEA, 2000.
70. Ackland, Len. *Making a Real Killing: Rocky Flats and the Nuclear West.* Albuquerque, NM: University of New Mexico Press, 1999.
71. Breitenstein, Bryce D. *1976 Hanford Americium Exposure Incident: Medical Management and Chelation Therapy.* Oxford, UK: Pergamon Press Ltd., Health Physics, Vol. 45, No. 4, Oct 1983.
72. Brown, W. R. *1976 Hanford Americium Exposure Incident: Psychological Aspects.* Oxford, UK: Pergamon Press Ltd., Health Physics, Vol. 45, No. 4, Oct 1983.
73. McLaughlin, Thomas P., Monahan, Shean P., Pruvost, Norman L., Frolov, Vladimir V., Ryanzanov, Boris G., and Sviridov, Victor I. *LA-13638: A Review of Criticality Accidents.* Los Alamos, NM: Los Alamos National Laboratory, 2000.
74. McMurray, B. J. *1976 Hanford Americium Exposure Incident: Accident Description.* Oxford, UK: Pergamon Press Ltd., Health Physics, Vol. 45, No. 4, Oct 1983.
75. Medvedev, Zhores A. *Nuclear Disaster in the Urals.* New York: W. W. Norton & Company, 1979.
76. Soran, Diane M. and Stillman, Danny B. *An Analysis of the Alleged Kyshtym Disaster.* Los Alamos, NM: Los Alamos National Laboratory, LA-9217-MS, 1982.

77. Thompson, Roy C. *1976 Hanford Americium Exposure Incident: Overview and Perspective.* Oxford, UK: Pergamon Press Ltd., Health Physics, Vol. 45, No. 4, Oct 1983.

Chapter 8: The Military Almost Never Lost a Nuclear Weapon

78. blanked out. *Project Crested Ice: The Thule Nuclear Accident (U), Volume I, SAC Historical Study #113.* Headquarters Strategic Air Command: History & Research Division,1979.
79. Dobson, Joel. *The Goldsboro Broken Arrow: The Story of the 1961 B-52 Crash, the Men, the Bombs, the Aftermath.* Lexington, KY: Lulu.com, 2011.
80. Hunziker, Maj Gen Richard O., et al. *Project Crested Ice.* Kirtland AFB, NM: USAF Nuclear Safety, 1970.
81. Maggelet, Michael H. and Oskins, James C. *Broken Arrow: The Declassified History of U. S. Nuclear Weapons Accidents.* Lexington, KY: Lulu.com, 2007.
82. Maggelet, Michael H. and Oskins, James C. *Broken Arrow Volume II: A Disclosure of Significant U. S., Soviet, and British Nuclear Weapon Incidents and Accidents, 1945-2008.* Lexington, KY: Lulu.com, 2010.

Chapter 9: The China Syndrome Plays in Harrisburg and Pripyat

83. Gray, Mike and Rosen, Ira. *The Warning: Accident at Three Mile Island.* New York: W. W. Norton & Company, 1982.
84. Medvedev, Grigori. *The Truth About Chernobyl: An Exciting Minute-by-Minute Account by a Leading Soviet Nuclear Physicist of the World's Largest Nuclear Disaster and Coverup.* New York: Basic Books, 1989.
85. Murphy, G. A. and Cletcher, J. W. II. *Operating Experience Review of Failures of Power Operated Relief Valves and Block Valves in Nuclear Power Plants.* Oak Ridge, TN: ORNL, NUREG/CR-4692, 1987.
86. Paatero, Jussi, et al. *Chernobyl: Observations in Finland and Sweden.* Research Triangle Park, NC: RTI Press Book, 2011.
87. Pryor, Andrew J. *The Browns Ferry Nuclear Plant Fire.* Boston, MA: Society of Fire Protection Engineers, Technology Report 77-2, 1977.
88. Walker, J. Samuel. *Three Mile Island: a Nuclear Crisis in Historical Perspective.* Berkeley, CA: University of California Press, 2004.

Chapter 10: Tragedy at Fukushima Daiichi

89. NA. *Lessons Learned from the Nuclear Accident at the Fukushima Daiichi Nuclear Power Station: Revision 0.* Atlanta, GA: INPO, 2012.
90. NA. *Special Report on the Nuclear Accident at the Fukushima Daiichi Nuclear Power Station: Revision 0.* Atlanta, GA: INPO, INPO 11-005, 2011.
91. NA. *Reactor Concepts Manual: Boiling Water Reactor (BWR) Systems.* Chattanooga, TN: USNRC Technical Training Center.
92. NA. *BWR/6: General Description of a Boiling Water Reactor.* San Jose, CA: General Electric Nuclear Energy.
93. Aslam, M.; Godden, W. G.; and Scalise, D. T. *Sloshing of Water in Torus Pressure-Suppression Pool of Boiling Water Reactors Under Earthquake Ground Motions.* Berkeley, CA: Lawrence Berkeley Laboratory, LBL-7984, 1978.

94. Ford, Daniel. *The Cult of the Atom: The Secret Papers of the Atomic Energy Commission.* New York: Simon & Schuster, 1984.
95. Grimes, C. I. *Safety Evaluation Report: Mark I Containment Long-Term Program, Resolution of Generic Technical Activity A-7.* Washington, DC: USNRC, 1980.
96. Holman, G. S., McCauley, E. W., and Lu, S. C. *Three-Dimensional Linear Analysis of Fluid-Structure Interaction Effects in the Mark I BWR Pressure Suppression Torus.* Lawrence Livermore Laboratory, UCRL-82217, 1980.
97. Martin, R. W. and McCauley, E. W. *The Effects of Torus Wall Flexibility on Forces in the Mark I Boiling Water Reactor Pressure Suppression System—Part I.* Livermore, CA: Lawrence Livermore Laboratory, UCRL-52506, 1978.
98. Meng, Steve. *The Nuclear Accident at the Fukushima Daiichi Nuclear Power Station.* Atlanta, GA: INPO, 2012.
99. Mohrbach, Dr.-Ing. Ludger. *Earthquake and Tsunami in Japan on March 11, 2011 and Consequences for Fukushima and Other Nuclear Power Plants.* Essen, Germany: VGB PowerTech, 2011.
100. Nelson, Larry. *Boiling Water Reactor Basics.* San Ramon, CA: GE Global Research, 2008.
101. Stevens, Gary L. *NUREG-0651—Torus (Suppression Chamber) Portion of the Mark I Containment.* NRC/RES/DE/CIB, 2011.
102. West, J. W., et al. *EBWR: The Experimental Boiling Water Reactor.* Chicago: Argonne National Laboratory, 1957.

Chapter 11: Caught in the Rickover Trap

103. Bunker, Merle E. *Early Reactors.* Los Alamos, NM: Los Alamos Science Laboratory, 1983.
104. Harper, Johnny R. and Garde, Raymond. *Decommissioning the Los Alamos Molten Plutonium Reactor Experiment (LAMPRE I).* Los Alamos, NM: Los Alamos National Laboratory, LA-9052-MS, 1981.
105. Harvey, David W., et al. *Historical Context of the Omega Reactor Facility, Technical Area 2: Historic Building Report No. 234.* Los Alamos, NM: Los Alamos National Laboratory, 2004.
106. Kiehn, R.M. *LAMPRE: A Molten Plutonium Fueled Reactor Concept.* Los Alamos, NM: Los Alamos Science Laboratory, LA-2112, 1957.
107. Mahaffey, James A. *E. I. Hatch Nuclear Power Plant Emergency Response Data Systems Maintenance Manual, Units 1 and 2.* Atlanta, GA: GTRI, 1986.
108. Mahaffey, James A. *The Future of Nuclear Power.* New York: Facts On File, 2012.

INDEX

ILLUSTRATION CREDITS

Illustration credits are listed in the order the images appear in the art inserts

Tho-Radia ad (Courtesy of Wikimedia Commons/Rama)
Undark ad (Courtesy of Wikimedia Commons)
Daghlian mock-up (Courtesy of the Los Alamos National Laboratory Archives)
Slotin mock-up (Courtesy of the Los Alamos National Laboratory Archives)
Slotin setup table (Courtesy of the Los Alamos National Laboratory Archives)
The Castle Bravo bomb (Courtesy of the Los Alamos National Laboratory Archives)
NRX reactor, Chalk River, 1955, after rebuild (Courtesy of Library and Archives Canada)
NRX reactor, ports to the core (Courtesy of Library and Archives Canada)
NRU reactor, Chalk River, under construction (Courtesy of Library and Archives Canada)
BORAX-I (Courtesy of Idaho National Laboratory)
SL-1 accident (Courtesy of Idaho National Laboratory)
SL-1 poster (Courtesy of Idaho National Laboratory)
Reactor refueling face, Oak Ridge (Photo courtesy of the author)
Windscale, tight shot from the rear (Copyright © Nuclear Decommissioning Authority)
Windscale, wider shot (Copyright © Nuclear Decommissioning Authority)
Windscale Unit 1, aerial shot (Copyright © Daily Mail/Rex/Alamy)
Santa Susana Field Laboratory (Courtesy of Wikimedia Commons)
Damaged fuel rod, SRE (Courtesy of Wikimedia Commons)
Americium extraction hood, Hanford (Courtesy of AP Photo)
Room 180, Building 771 (Courtesy of Wikimedia Commons)
HEPA filter room, Building 771 (Courtesy of Wikimedia Commons)
Workers evaluating radiation at JCO, 1999 (Courtesy of AP Photo/JCO Co.)
H-bomb being unloaded from B-52 (Courtesy of Wikimedia Commons)
President Carter at TMI, April 1, 1979 (Copyright © ZUMA Press, Inc./Alamy)
Chernobyl-4, aerial shot after smoke has cleared (Copyright © Igor Kostin/Sygma/Corbis)
Chernobyl-4 burning (Courtesy of Sovfoto/UIG via Getty Images)
The author running the SPDS at Hatch (Photo courtesy of the author)
The next three photos are all of the inside of a BWR/4 (Photos courtesy of the author)
Diagram of Mark I containment (Courtesy of United States Nuclear Regulatory Commission)
The Fukushima plant, March 24, 2011 (Courtesy of Air Photo Service/EPA/Landov)

All interior images first appeared in *Nuclear Power* by James Mahaffey, published by Facts on File, and have been modified by originating artist Bobbi McCutcheon, for this book.

ACKNOWLEDGMENTS

WRITING *ATOMIC ACCIDENTS* WAS A complicated project with several opportunities to blunder per page, and I am grateful to everyone who helped me through it, particularly my three long-term friends with PhDs in physics, Don Harmer, Doug Wrege, and Monte Davis. Monte's wife, Nancy, a natural-born editor without mercy, alone found over 1,000 misspellings, wrong words, and typos, and she re-inserted most of the hyphens she had taken out of *Atomic Awakening*. Deepest thanks to Chris Rowe, retired from the Santa Susana Field Laboratory, and Kamara Sams, in Environmental Communications at the Boeing Company, for setting me straight about the Sodium Reactor Experiment and SSFL. I appreciate Jarmok Kivinen for his inside look at Finland during the Chernobyl-4 disaster. I offer a shout-out to all who participate in the American Nuclear Society Social Media for providing me with more material on Fukushima than I could possibly use, and to the entire crew at the Georgia Tech Research Institute who participated in project A-3026 thirty years ago. Sincere thanks to my superb editor at Pegasus Books, Jessica Case, to Suzie Tibor, photograph researcher, Bobbi McCutcheon, line-drawing artist, Phil Gaskill, eagle-eyed proofreader, to Maria Fernandez for the beautiful layout, and to the world's best literary agent, Jodie Rhodes. Writing this book would not have been possible without my wife and my muse, Carolyn Mahaffey, and the quiet encouragement from Turtle and Lance.